AF690904

Radar Countermeasures for Unmanned Aerial Vehicles

Related titles on radar:

Advances in Bistatic Radar Willis and Griffiths
Airborne Early Warning System Concepts, 3rd Edition Long
Bistatic Radar, 2nd Edition Willis
Design of Multi-Frequency CW Radars Jankiraman
Digital Techniques for Wideband Receivers, 2nd Edition Tsui
Electronic Warfare Pocket Guide Adamy
Foliage Penetration Radar: Detection and characterisation of objects under trees Davis
Fundamentals of Ground Radar for Air Traffic Control Engineers and Technicians Bouwman
Fundamentals of Systems Engineering and Defense Systems Applications Jeffrey
Introduction to Electronic Warfare Modeling and Simulation Adamy
Introduction to Electronic Defense Systems Neri
Introduction to Sensors for Ranging and Imaging Brooker
Microwave Passive Direction Finding Lipsky
Microwave Receivers with Electronic Warfare Applications Tsui
Phased-Array Radar Design: Application of radar fundamentals Jeffrey
Pocket Radar Guide: Key facts, equations, and data Curry
Principles of Modern Radar, Volume 1: Basic principles Richards, Scheer and Holm
Principles of Modern Radar, Volume 2: Advanced techniques Melvin and Scheer
Principles of Modern Radar, Volume 3: Applications Scheer and Melvin
Principles of Waveform Diversity and Design Wicks *et al.*
Pulse Doppler Radar Alabaster
Radar Cross Section Measurements Knott
Radar Cross Section, 2nd Edition Knott *et al.*
Radar Design Principles: Signal processing and the environment, 2nd Edition Nathanson *et al.*
Radar Detection DiFranco and Rubin
Radar Essentials: A concise handbook for radar design and performance Curry
Radar Foundations for Imaging and Advanced Concepts Sullivan
Radar Principles for the Non-Specialist, 3rd Edition Toomay and Hannan
Test and Evaluation of Aircraft Avionics and Weapons Systems McShea
Understanding Radar Systems Kingsley and Quegan
Understanding Synthetic Aperture Radar Images Oliver and Quegan
Radar and Electronic Warfare Principles for the Non-specialist, 4th Edition Hannen
Inverse Synthetic Aperture Radar Imaging: Principles, algorithms and applications Chen and Martorella
Stimson's Introduction to Airborne Radar, 3rd Edition Baker, Griffiths and Adamy
Test and Evaluation of Avionics and Weapon Systems, 2nd Edition McShea
Angle-of-Arrival Estimation Using Radar Interferometry: Methods and applications Holder
Biologically-Inspired Radar and Sonar: Lessons from Nature Balleri, Griffiths and Baker
The Impact of Cognition on Radar Technology Farina, De Maio and Haykin
Novel Radar Techniques and Applications, Volume 1: Real Aperture Array Radar, Imaging Radar, and Passive and Multistatic Radar Klemm, Nickel, Gierull, Lombardo, Griffiths and Koch
Novel Radar Techniques and Applications, Volume 2: Waveform Diversity and Cognitive Radar, and Target Tracking and Data Fusion Klemm, Nickel, Gierull, Lombardo, Griffiths and Koch
Radar and Communication Spectrum Sharing Blunt and Perrins
Systems Engineering for Ethical Autonomous Systems Gillespie
Shadowing Function from Randomly Rough Surfaces: Derivation and applications Bourlier and Li
Photo for Radar Networks and Electronic Warfare Systems Bogoni, Laghezza and Ghelfi
Multidimensional Radar Imaging Martorella
Radar Waveform Design Based on Optimization Theory Cui, De Maio, Farina and Li
Micro-Doppler Radar and its Applications Fioranelli, Griffiths, Ritchie and Balleri
Maritime Surveillance with Synthetic Aperture Radar Di Martino and Antonio Iodice

Electronic Scanned Array Design Williams
Advanced Sparsity-Driven Models and Methods for Radar Applications Li
Deep Neural Network Design for Radar Applications Gurbuz
New Methodologies for Understanding Radar Data Mishra and Brüggenwirth
Radar countermeasures for Unmanned Aerial Vehicles Clemente, Fioranelli, Colone and Li
Holographic Staring Radar Oswald and Baker

Radar Countermeasures for Unmanned Aerial Vehicles

Edited by
Carmine Clemente, Francesco Fioranelli,
Fabiola Colone and Gang Li

The Institution of Engineering and Technology

Published by SciTech Publishing, an imprint of The Institution of Engineering and Technology, London, United Kingdom

The Institution of Engineering and Technology is registered as a Charity in England & Wales (no. 211014) and Scotland (no. SC038698).

© The Institution of Engineering and Technology 2021

First published 2021

This publication is copyright under the Berne Convention and the Universal Copyright Convention. All rights reserved. Apart from any fair dealing for the purposes of research or private study, or criticism or review, as permitted under the Copyright, Designs and Patents Act 1988, this publication may be reproduced, stored or transmitted, in any form or by any means, only with the prior permission in writing of the publishers, or in the case of reprographic reproduction in accordance with the terms of licences issued by the Copyright Licensing Agency. Enquiries concerning reproduction outside those terms should be sent to the publisher at the undermentioned address:

The Institution of Engineering and Technology
Michael Faraday House
Six Hills Way, Stevenage
Herts, SG1 2AY, United Kingdom

www.theiet.org

While the authors and publisher believe that the information and guidance given in this work are correct, all parties must rely upon their own skill and judgement when making use of them. Neither the authors nor publisher assumes any liability to anyone for any loss or damage caused by any error or omission in the work, whether such an error or omission is the result of negligence or any other cause. Any and all such liability is disclaimed.

The moral rights of the authors to be identified as authors of this work have been asserted by them in accordance with the Copyright, Designs and Patents Act 1988.

British Library Cataloguing in Publication Data
A catalogue record for this product is available from the British Library

ISBN 978-1-83953-190-3 (hardback)
ISBN 978-1-83953-191-0 (PDF)

Typeset in India by MPS Limited

Contents

About the editors

Carmine Clemente is a Reader at the Department of Electronic and Electrical Engineering, University of Strathclyde, the United Kingdom. His research is focussed on sensor signal processing for defence and civilian applications, with a specific interest on radar signal processing. He directs the Sensor Signal Processing & Security Labs at the University of Strathclyde; he is a senior member of IEEE and serves as associate editor for the *IEEE Transactions on Aerospace and Electronic Systems* and *Frontiers in Signal Processing*. Dr Clemente has authored over 140 peer-reviewed publications and book chapters.

Francesco Fioranelli is an assistant professor in the Microwave Sensing Signals and Systems (MS3) section within the Department of Microelectronics at the Delft University of Technology (TU Delft). His research interests and current projects include distributed radar and automatic target classification working at the interface of machine learning and radar signal processing in automotive and health-care applications, among others. Dr Fioranelli is a senior member of the IEEE and author of over 100 peer-reviewed publications and book chapters.

Fabiola Colone is an associate professor at the Department of Information Engineering, Electronics and Telecommunications (DIET) of Sapienza University of Rome, Italy. Dr Colone's research activity is devoted to radar systems and signal processing and, on these topics, she has authored over 150 peer-reviewed publications and book chapters. Dr Colone has been co-recipient of the 2018 Premium Award for Best Paper in IET Radar, Sonar & Navigation. Since 2017 she is a member of the Board of Governors of the IEEE Aerospace and Electronic System Society (AESS) in which she has served as vice president for Member Services. She is a senior member of IEEE from 2017 and a member of the IEEE AESS Radar System Panel from 2019. Dr Colone is an associate editor for the *IEEE Transactions on Signal Processing*, the *Int. Journal of Electronics and Communications* (Elsevier) and *Frontiers in Signal Processing.*

Gang Li is a professor in the Department of Electronic Engineering, Tsinghua University, China. His research interests include radar signal processing, remote sensing, distributed signal processing, sparse signal processing and information fusion. He has authored over 160 journal and conference papers. He is the author of

Advanced Sparsity-Driven Models and Methods for Radar Applications (London, UK: IET – SciTech Publishing, 2020). He is an associate editor for the IEEE *Transactions on Signal Processing* and a member of the Sensor Array and Multichannel (SAM) Technical Committee of the IEEE Signal Processing Society. He is the recipient of the National Science Fund for Distinguished Young Scholars of China and the Royal Society-Newton Advanced Fellowship.

List of reviewers

Name	Affiliation
Luca Pallotta	University of Roma Tre
Pia Addabbo	University "Giustino Fortunato"
Michail Antoniou	University of Birmingham
Stephen Harman	Aveillant Ltd
Alessio Balleri	Cranfield University
Tatiana Martelli	Sapienza University of Rome
Adriano Rosario Persico	Aresys Srl
Christos Ilioudis	University of Strathclyde
Julien Le Kernec	University of Glasgow
Duncan Robertson	University of St Andrews
John Soraghan	University of Strathclyde

Introduction

Christopher J. Baker[1]

Over the last 10 years or so, the numbers of unmanned aerial vehicles (UAVs) or 'drones' have changed from being just a few specialist systems, used for scientific data gathering and military purposes, to them proliferating in huge numbers. They are used across a broad range of different leisure, commercial and military activities. UAVs can be used for the movement of items in factories for manufacturing, passenger and freight transportation, which can take various roles in the agriculture and forestry industries (dispensing seeds, watering and monitoring crops), remote sensing for the oil and gas industries, traffic flow monitoring, support of emergency services, hobbies, security, military and many other applications. This has come about due to the development of low-cost, high-performance stable platforms, employing equally low-cost communication and navigation systems supplemented by simple to use software and interfaces. As a consequence, there is a need to be able to monitor the rapidly changing use of airspace, especially at low and normally neglected altitudes to ensure that UAVs do not compromise safety or are used for malicious purposes. Radar is the only sensor that is able to perform this function on a 24-h, all-weather and wide-area basis. Hence, this book, concerned with radar surveillance of UAVs, has been compiled using contributions from the leading experts around the world to create a single body of knowledge on this important, yet still emerging, topic.

In this opening chapter a brief history of the development of UAVs, putting today's explosive growth in numbers into the wider context, is outlined. It is not the intention to provide a comprehensive account of the development of all UAVs as this would require a separate and somewhat sizeable text. Here, some of the key milestones are noted as a means of the charting development that has led to the UAVs of today and why there is a need for specialist radar surveillance systems. Challenges facing successful radar surveillance both for countering UAVs and enabling their legitimate use are introduced and discussed. Further, the consequences associated with the way in which the occupancy of low-altitude airspace is changing are considered, especially in relation to how this may open up opportunities for a wider exploitation of radar surveillance. Lastly, some speculation as to future challenges for the radar surveillance of UAVs are presented and examined.

[1]School of Engineering, University of Birmingham, Birmingham, United Kingdom

I.1 UAV development

The origins of unmanned flight can be traced back to around 400 BC with the 'Chinese flying top' during the Warring States period. The Chinese flying top, essentially, was a stick with a propeller on its top as illustrated in Figure I.1.

The propeller was spun round by hand and then released to make it fly. This would also make the first flying vehicle, both unmanned and a helicopter. Much later Leonardo da Vinci produced sketches of an 'aerial screw' in the late 1400s – see Figure I.2. This appears to take inspiration from the Archimedes screw, first espoused around 200 BC. Again, it has more in common with helicopters than conventional aircraft. da Vinci never made his sketch into a real flying machine and if he had, it, almost certainly, did not have the means to sustain flight other than fleetingly.

Much later, in the eighteenth century the Russian polymath, Mikhail V. Lomonosov, produced and demonstrated to the Russian Academy of Sciences,

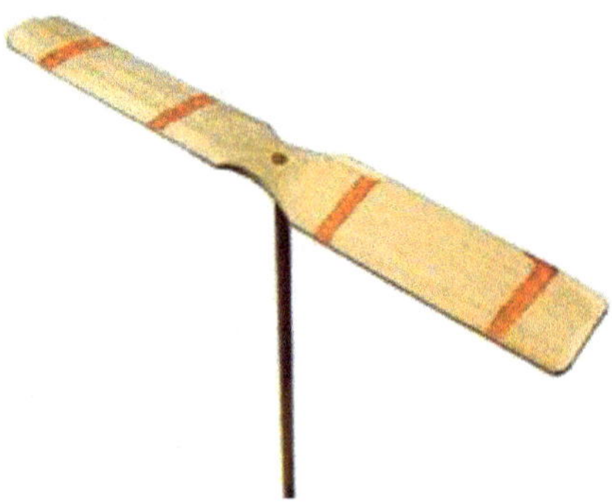

Figure I.1 The Chinese top

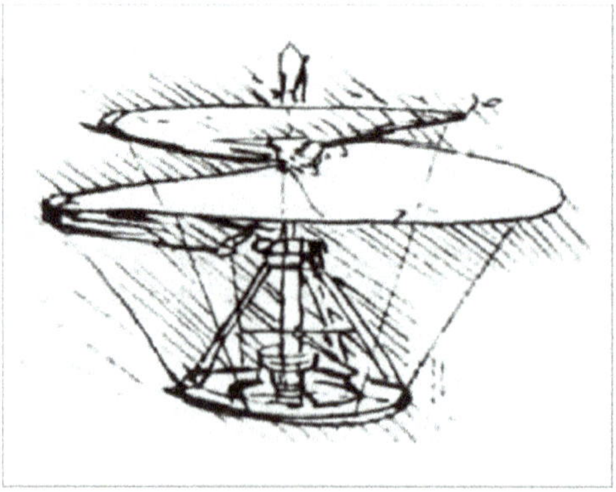

Figure I.2 The Leonardo da Vinci flying screw

what is thought to be the first self-propelled version of a flying screw, powered by a spring, Figure I.3.

In 1784, two Frenchmen, Launoy and Bienvenu, devised an ingenious flying machine consisting of two propellers made of birds' feathers fixed to the tips of a shaft, as shown in Figure I.4. Two strings were twisted around the shaft, tensioning

Figure I.3 The Lomonosov flying screw

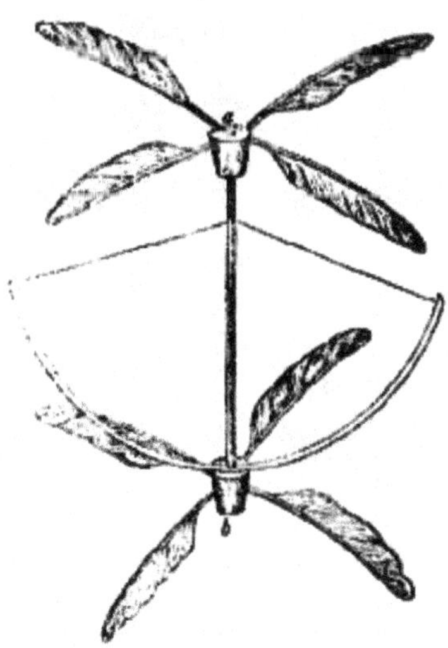

Figure I.4 The Launoy and Bienvenu flying helicopter

a spring. As the assemblage was straightened out, the spring caused the propellers to rotate for a few seconds, sufficient to send the device spinning a few metres. Launoy and Bienvenu presented their invention to the French Academy of Sciences.

In the nineteenth century, in the United Kingdom, Sir George Cayley, an English engineer, was a key figure in the development of flight, setting out much of the basis of modern flight. In 1799, he outlined the concept that established the configuration of an aeroplane as a fixed-wing flying machine with separate systems for lift, propulsion and control. His initial sketches, such as that shown in Figure I.5, led to models, powered by elastic, that were able to demonstrate unaided flight. These early examples of flying machines are closer to the helicopter and the small unmanned quad and hexacopter aircraft of today.

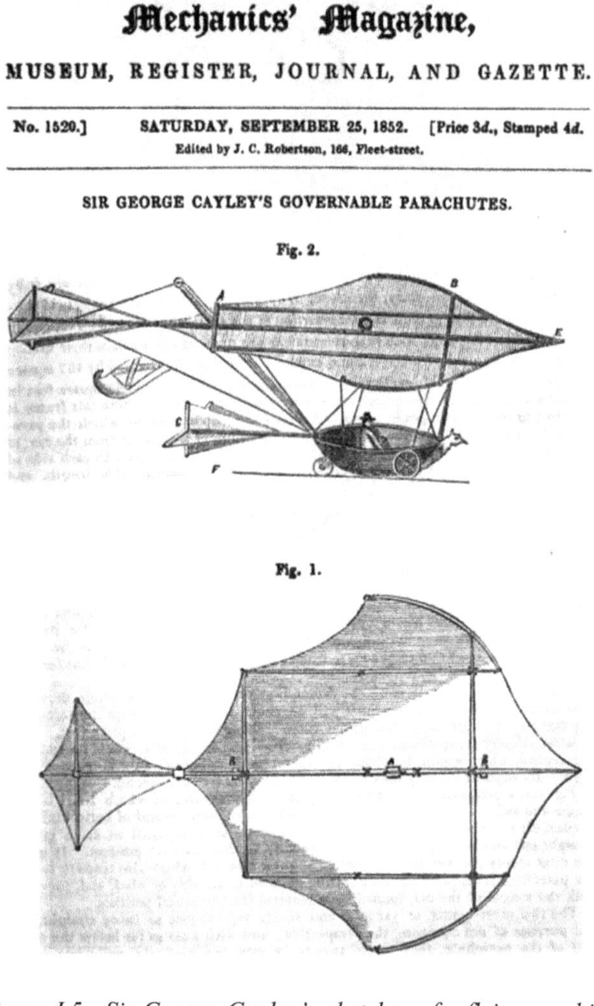
Mechanics' Magazine,

MUSEUM, REGISTER, JOURNAL, AND GAZETTE.

No. 1520.] SATURDAY, SEPTEMBER 25, 1852. [Price 3d., Stamped 4d.

Edited by J. C. Robertson, 166, Fleet-street.

SIR GEORGE CAYLEY'S GOVERNABLE PARACHUTES.

Fig. 2.

Fig. 1.

Figure I.5 Sir George Cayley's sketches of a flying machine

In 1903, the Wright brothers made the first sustained, controlled, powered and heavier-than-air manned flight on 17 December 1903. The first flight by Orville Wright of 37 m lasted for 12 s, and in the fourth flight (of the same day) Wilbur Wright flew 260 m for a period of 59 s. Figure I.6 shows the Wright brothers aircraft, the Wright flyer, on 17 December 1903.

Their success spurred the growth that has led to the modern aviation industry. Quickly after, there appeared remotely piloted aircraft, principally developed as a means of delivering bombs in the First World War. These flying machines split into two categories: one being the precursors to missiles, i.e., non-recoverable UAVs and the other, aircraft that were designed to return to base, i.e., UAVs. The latter were motivated by a desire to avoid the loss of an expensive aircraft and by the lack of suitably trained and available pilots. An early example of the missile category was the 'Kettering Bug', first flown in 1917 and shown in Figure I.7. It was constructed by Charles Kettering, who used a fuselage made by Orville Wright and a control and navigation system designed by Elmer Sperry. Following take-off, the aircraft was guided towards its target. After a predetermined elapsed time, the wings were folded down, the engine was switched off, and the aircraft fell from the sky in the form of a bomb. The Kettering Bug was never used as the war ended before it could be put into service.

It was after the First World War that renewed interest in pilotless aircraft, motivated by the need to train gunners. The 'Queen Bee' was the first example, developed in the United Kingdom in 1935, that employed a remote-control system in a De Havilland DH 82 Tiger Moth. The gunners had to 'offset-aim' so that the Queen Bee could return and be reused. It is a curiosity that we have the term,

Figure I.6 The Wright flyer

Figure I.7 The Kettering Bug

Figure I.8 The Queen Bee

'drone', in common use today, given that the sole purpose of a drone, a male honey bee, is to impregnate the Queen Bee! Perhaps, it was the continuous buzzing of the aircraft that led to the adoption of 'drone' and well after the heyday of the pilotless Queen Bee? A number of theories for the term 'drone' coming into common use have been postulated but supporting evidence seems to be lacking (Figure I.8).

Soon after, there followed many examples of similar remotely piloted or radio-controlled aircraft that were primarily used as targets for training purposes. Equally, there were parallel but separate developments leading to the modern missile.

In 1960, the advent of the 'Reconnaissance UAV' in the form of the BQM-34 Firebee was seen, as shown in Figure I.9. This UAV could be launched by catapult

Figure I.9 The BQM-34 Firebee

Figure I.10 The AeroVironment Pointer

or from the ground or a ship but was most commonly launched from a DC-130 aircraft. In the Vietnam War, the Firebee carried out over 3,400 sorties helping to locate air defence batteries. In the 1970s, Israel was a leader of UAV technology, and two notable systems were Mastiff and Scout. These were used to gather information as to the location of radio emitters and their parameters. In the 1980s, the US furthered the development of UAVs in support of various conflicts. This led to the Pioneer, Predator and Global Hawk systems, which also eventually evolved armed combat versions such as the MQ-9 Reaper.

In the latter part of these conflicts, there was a need for cheaper systems that could be more easily launched and were simple to use. These systems were powered by batteries and had camera payloads, the attributes of the modern commercially available UAV that can now be bought in stores anywhere. Examples include the AeroVironment Pointer, Figure I.10, which operated at altitudes of up to 150 m and had a range of around 8 km.

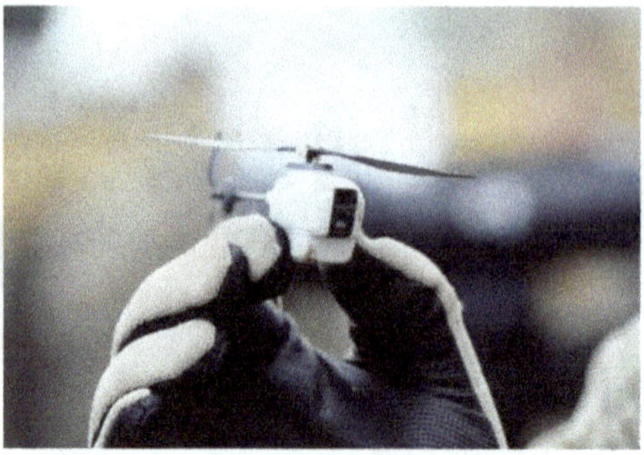

Figure I.11 The Black Hornet Nano

More recently, the Black Hornet Nano, developed by Prox Dynamics AS of Norway, took the genre to new extremes – see Figure I.11. This is micro-unmanned aircraft with dimensions of, approximately, 10 by 2.5 cm, weighs a little over 16 g, incorporates a camera and can fit in the palm of a hand.

Thus, a number of terms have been almost interchangeably introduced and used for this wide variety of unmanned aircraft. The principle ones are drone, remotely piloted vehicle, Unmanned Air System (UAS), Uninhabited Air System (also UAS) and UAV. Here, we will primarily use UAV.

Starting with the Chinese flying top and arriving at the black hornet nano, it is easy to see that

1. the capability and aerodynamic performance of UAVs has greatly improved;
2. the ability to squeeze increasingly complex and capable payloads onto smaller and smaller platforms has dramatically increased;
3. there is a tremendous variety of UAV types, with huge differences in size, weight, flight ability and application.

All these serve to pose significant challenges in ensuring reliable surveillance and safe use of airspace. Most of the early developments were pioneered for military applications. More recently, with costs reducing, UAVs started to be developed for scientific observation. However, the most dramatic developments have been over the last 10–15 years, and there is now the ability to create UAVs with remarkable performance in small packages. This has been brought about by the advent of high-speed digital processing, miniature high capability communications and IT coupled with sophisticated software for command and control, all at low cost. As a result, it is now possible to produce sophisticated small UAVs easily affordable by large tracts of the general public. Consequently, this has led to a massive level of proliferation that dwarfs anything seen before with UAVs now selling in their tens of millions all over the world. Around 20 different manufacturers hold most of the market with well-known names such as Parrott, DJI and Yuneec being especially prominent. Figure I.12 shows the DJI Phantom 4 PRO and Mavic quadcopter UAVs. The DJI phantom is physically small with an overall span

Figure I.12 The DJI Phantom 4 PRO and Mavic UAVs

of around 35 cm and a weight in the region of 1.3 kg. Although these UAVs are physically very small, they are able to carry a good quality camera and can support small additional payloads. Most of the UAV is made of plastic, with the exception of the battery and parts of the motor assemblies.

It is these quadcopters, hexacopters and fixed-wing UAVs, produced in great numbers, that present a particular challenge for radar surveillance as they change the way in which low-altitude airspace is occupied.

I.2 Radar surveillance of UAVs

Essentially, small UAVs are flying batteries with plastic appendages and hence they present a very small radar cross section (RCS). The RCSs of these small UAVs are in the order of −20 to −30 dBsm, so tiny that they make stealthy aircraft appear positively enormous!! To detect such very small RCS targets, it is necessary to design a radar system with very high sensitivity. However, as the RCS of small birds can be of the same order as that of the UAVs, they will also be detected creating a confusing picture on a radar operator display. Information on the RCS of single birds is fairly scant but generally they are the same order as UAVs and there is a slight increase in RCS in moving from a transmission frequency of 1 GHz, upwards. Higher frequency radar systems are more sensitive to smaller RCS birds and are even able to detect swarms of insects. In fact, all moving targets with a similarly sized RCS, falling within the radar illumination, will be detected, including ground targets such as cars, people, animals and even items such as air-conditioning fans. This is further complicated by the need to detect UAVs to a given maximum range, such that the heightened sensitivity at shorter ranges means that smaller and smaller RCS targets may be detected at shorter and shorter ranges.

A second characteristic of small UAVs is that they tend to fly at low altitudes, typically at 100 m or below. Further, most UAVs are relatively slow-moving with maximum speeds of around 20 m/s and often lower if there is a prevailing head-wind. As radar targets, UAVs are often said to be low, slow and small.

As a consequence of these three defining target attributes, radar systems have a number of challenges that directly arise, which are as follows:

1. Detection of very small RCS targets against what can be a strong clutter background.
2. Accurate extraction of location data, especially elevation in the presence of multipath.
3. Reliability of tracking.
4. Because so many different types of target are detected, there is a vital need for reliable classification.
5. UAVs can be launched from anywhere and hence, they can 'pop-up' at short ranges, placing a premium on short reaction times.

Each of the challenges mentioned earlier will be considered in a little more detail, although, as will become clear, it is their combination that, ultimately, presents key new difficulties in designing radar systems that are able to provide reliable UAV surveillance.

The first requirement to be met in the design of any radar system is the assurance of an adequate signal to interference plus noise ratio. The interference may be in the form of clutter or other received electrical signals, either accidentally or deliberately. For larger RCS targets such as aircraft, this is likely to mean that the radar is insensitive to smaller air objects such as single or small numbers of birds and UAVs and it is typical that performance is noise rather than clutter limited at maximum range. However, as the target RCS becomes smaller and smaller, first, there has to be some compensating factor to maintain an adequate signal-to-noise ratio (SNR). The main options are increasing transmitter power, pulse length, antenna gains, PRF or a combination of these. Of course, there are all sorts of limits that might apply, particularly in relation to size, weight, power and cost, and ultimately the maximum detection range will be limited. The choice of frequency is a prime factor in any radar design. The optimum frequency for a UAV surveillance system is far from clear. Indeed, the literature reports systems working for a lower frequency of around 100 MHz to an upper frequency of 94 GHz, i.e., pretty well the whole of the spectrum used for radar!! The choice of antenna, especially in relation to gain and beam widths, will have a bearing not only on the SNR but also will be a factor in determining likely levels of clutter and multipath. Low-frequency systems are more likely to have wider beams that will intercept the ground introducing significant clutter. High-frequency systems may be able to avoid this but will have the usual problems at long detection ranges and in foul weather. It is also, generally, cheaper and easier to generate high powers at lower frequencies but this is traded for larger and potentially more cumbersome hardware. Whatever the radar design, the very small RCS of the targets means that the maximum detection ranges will be relatively short and even tens of km may not prove feasible.

Given that clutter is likely to be strong, detection in the Doppler domain has many attractions and is the method of choice reported in almost all most publications. A further motivation for this is the need for classification and the obvious exploitation of propeller sidebands, as a differentiator between UAVs and birds. However, the sidebands are of a much lower RCS than the main UAV body, then the criteria for adequate SNR become even more demanding. This might be as much as a further 20–30 dBs, making the radar sensitivity requirements even harder to achieve at long range. The relatively slow speeds and hence even slower radial velocities, as seen by a radar system, means that attention must be paid to the minimum detectable velocity so that the UAV is not filtered out for long periods making the maintenance of continuous tracking much more difficult.

Tracking is important as the different forms of trajectory type between birds and UAVs are also a valuable discriminator. Thus, even for a radar that has adequate SNR, the resulting probability of detection will be reduced when the UAV (or any target) is embedded in stationary clutter. Of course, as with all radar systems, clutter has internal motion due to wind, which broadens the clutter spectrum. This is a function of the local environment and prevailing environment and environmental conditions and can reduce the probability of detection still further.

A high Doppler resolution can provide a means of detecting targets as their radial velocity reduces but this has to be coupled with sophisticated signal processing to minimise the minimum detectable velocity. Equally, if the Doppler resolution is too high, the target Doppler signature can exhibit migration and hence spread energy into multiple resolution cells resulting in no improvement in detection performance. Overall, there has to be a careful balance between range, azimuth, elevation and Doppler resolutions that allows targets to be sufficiently separable in four-dimensional space, whilst being able to accumulate sufficient echoes the characteristics of which enable UAVs to be reliably detected and differentiated from birds and other small moving targets.

Accurate UAV position information is necessary for many applications that fall into counter-UAV or Air Traffic Management (ATM) categories. ATM regulations, as they develop, may well specify minimum separations between UAVs. Counter UAV technology may require the radar to perform broad area surveillance such that it hands over coordinates of a UAV location to an optical sensor. This will need to be accomplished with radar accuracies that enable the optical sensor to lock on to the UAV for seamless tracking and confirmation of target identification. The precise application will determine the precise requirements for positional accuracy. It may also relax the required levels of classification performance unless a true all-weather capability is needed. Whilst accuracy is largely determined by beam widths and SNR, it is likely that there will be a ground bounce component in received echoes. This is because UAVs (and birds) typically fly at altitudes of around 100 m and can sustain flight at altitudes of just a few metres. A ground bounce term can severely disrupt accurate measurement of altitude. Altitude is an obvious discriminant helping to differentiate between targets on the ground and in the air. Indeed, if this can be done with high confidence, it can remove a large number of false targets and greatly improve the operational capability of the radar system. Note, however, it is not always simple to measure target altitude with complete confidence due to the local land relief and to trees, buildings and other tall objects. Equally, very low-altitude targets that are following a system of roads are likely to be vehicles and hence could be filtered out on this basis. However, this could be a problem to the UAV that wishes to be concealed and hence follows a road at low altitude!

Ground bounce rarely takes the form of the classic 'Lloyds mirror' cases often used in textbooks. It is more likely to exhibit a combination of coherent and incoherent forward scattering from the ground, over an extended patch, that will not just affect altitude estimation but will also add to target echo fluctuations, hence impacting detection performance. Indeed, it is more typical that detection environments will contain objects from which multiple bounces can occur between the object and a UAV causing additional and sometimes quite wildly varying target fluctuations – a further complications for detection and classification.

Thus, a UAV surveillance radar system with high-detection sensitivity may detect many candidate targets of which, perhaps, only one is a UAV. The detection of multiple targets may also make the target association component of tracking potentially much more troublesome due to the sheer numbers of detections. Overall, there will be a need to minimise track initiation times, track confirmation times and to have a reliable means of track elimination. Given the slow speed of UAVs and the need to revisit for scanning radar systems, the total time to track confirmation can easily become the order of many seconds and will be a factor in determining latencies and system response times and, ultimately, system utility.

Depending on the application there will almost certainly be a need to know what the target is. For example, if the radar is part of a wider counter dUAV system that has to cue other sensors, the control of false target reports to operationally acceptable levels becomes crucial. If there are too many false target reports, the operational effectiveness of the system will be compromised, eventually to the point where user confidence is lost. Equally, ATM systems will only want to provide reporting on UAV movements. Ornithologists will have an interest in categorising different species of birds and evaluating their numbers and behaviours. The design of the radar and signal processing will necessarily influence the resulting performance.

The characteristics of low, slow and small targets need to be considered collectively when the goal is to detect and locate a UAV as a function of time. As already stated, the high sensitivity required to detect small UAVs creates a need for discrimination so that targets can be differentiated from one another to some required level. Categories of discrimination may include UAV and non-UAV or even into much finer categories such as UAVs, birds, vehicles, people and animals. There may also be great utility in determining the type of UAV or the type of birds. In short, effective radar surveillance of UAVs mandates a target classification capability. For many applications, this will need to be at a very high classification performance level as the false reporting of targets may result in low and possibly unacceptably low operational effectiveness. Radar target classification, with a few notable exceptions, has been a notoriously challenging area of radar research and development.

The need to be able to classify UAVs from other targets was recognised early by the research community. Many of the first publications, specifically examining radar surveillance of UAVs, measure and use the micro-Doppler responses of UAV propeller blades as a basis for classification. As already mentioned, the reliable detection of UAV propeller blades may require a further 20–30 dBs of additional radar sensitivity, illustrating a fundamental link between detection and classification that should be considered in the design of a UAV surveillance radar system.

The central importance of classification and the potential to achieve high performance levels is key to successful system design. Figures I.13 and I.14 illustrate spectrograms for a time series of Doppler profiles for a bird and a DJI Inspire UAV and represent the essential information that the radar signal processing has to exploit to perform classification.

A casual glance at the two figures is sufficient to reveal very significant and obvious differences between the responses for a bird and a UAV. The multiple

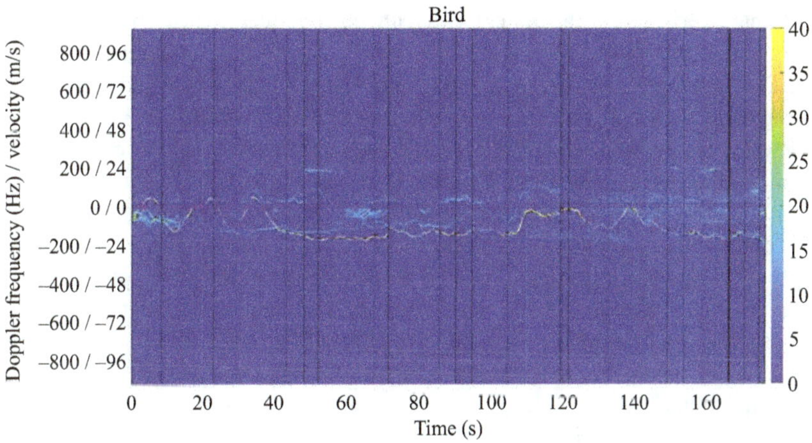

Figure I.13 Time series of Doppler profiles showing the echo history for a bird

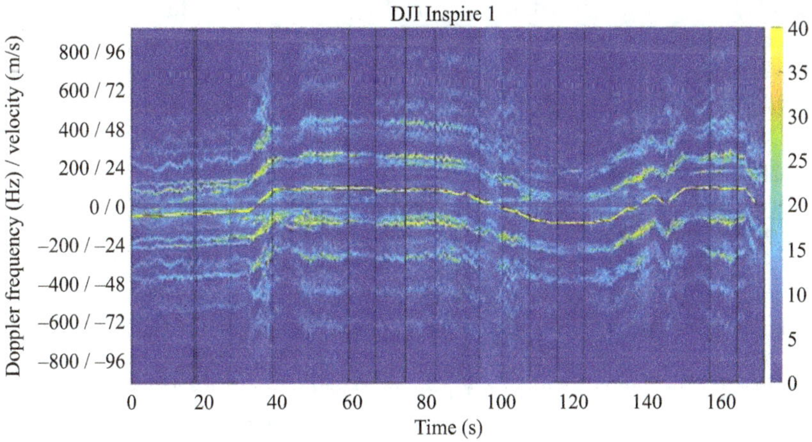

Figure I.14 Time series of Doppler profiles showing the echo history for a DJI Inspire 1

echoes from the propeller blades clearly stand out and can be used for classification. Indeed, it is hardly surprising that so much research has been devoted to classification approaches that exploit the returns from UAV propeller blades to provide differentiation from birds. However, the lower level of echo strength for the propeller blades can mean that the maximum detection ranges, based on the body of a UAV, may be less than the maximum classification range. As a consequence, this needs to be factored into the way in which the information can be exploited in a counter UAV or UAV surveillance system which may have to wait until a UAV is sufficiently close to correctly classify with high confidence.

It also suggests that alternative options for classification ought to be explored. At longer ranges, where, perhaps, propeller blades cannot be detected reliably, other characteristics of the trajectories can be exploited. These may be simple differentiators that filter out features such as overly strong echo strengths (indicative of large aircraft), velocities outside the range of a small UAV and zero altitude moving targets. Equally, they might be subtler characteristics, such as length of track (birds often have short flight durations as they forage for food) or differences in the 'stability' of the trajectories. As Figure I.13 indicates, birds typically have a more erratic trajectory than the smoother path flown by UAVs. Further, the non-rigid bodies of birds can exhibit a wider Doppler profile, especially when the bird is flapping its wings, although the widening of the propeller blade sidebands is also noted on occasion in Figure I.13. Suitable exploitation of such features can extend the maximum classification range and hence improve the effectiveness of the radar sensor.

As a direct consequence of the information contained in the echoes from birds, UAVs and other moving objects, there is clearly optimism for effective classification. Many techniques have been examined from conventional neural networks to convolutional neural networks, machine learning and statistical decision-making. The results presented in the research literature and in this book show great promise. However, there remain challenges, many of which derive from the 'open loop' nature of the problem. Indeed, it is a reminder that all radar performance is, in fact, a function of the scene being interrogated, although often presented as a statistical average. This is even more apparent when attempting classification. A region where there are typically large numbers of birds, such as nesting sites, will inevitably lead to more false reports than a region where birds are less commonly found. The types of birds vary from region to region and this variation may be extremely pronounced in different parts of the world. It is well known that birds are much more active in the skies at dawn and dusk and hence there can be a timing factor governing performance. All these mean that averages have to be treated with great care and that significant deviations from the average can be experienced.

The subtlety of echo characteristics might also lead to speculation that different types of birds can be differentiated from one another. The physical flight profiles of larger birds are quite different to those of smaller birds. The mating strategies of birds also differ between species and, potentially, such behaviours could be measured and analysed as a means of extending radar ornithological research. The fact that this is speculative also highlights how much we do not know about detecting and extracting reliable information from such targets. For example, the seemingly simple question, as to what frequency to use, is a case in point. There is no clear answer to this at the moment. No doubt, part of the answer will be dictated by the application, including economic factors, and part will be informed as further research improves the understanding of these phenomena.

I.3 Scope of this book

This book provides a timely and up-to-date account of the current state of the art. In Chapter 1, an overview is given of methods aimed at countering UASs. This chapter examines the range of UAVs to be countered and makes the important point that a UAS implies much more than just the flying component. There is the data link, control, human in the loop and other ancillary equipment and support elements that also have to be factored in when considering counter technologies. The chapter goes on to identify a range of surveillance sensors that might be employed as a cooperative set of interlinked systems and how they may be deployed within operational scenarios. As a result, their command and control in the form of an OODA loop has as much bearing on mission success as do the sensors themselves.

Chapter 2 examines the relationships between user requirements, affordability and system design. In particular, it highlights the inherent complexity of approach demanded by the wide variety of UAV types and possible roles they may have exacerbated by the pace at which UAVs themselves are so rapidly evolving. It then examines the role and design of radar surveillance sensors and concludes that (a) designs need to be specific to the demands of the characteristics of UAVs mentioned earlier in this introduction and (b) may require a large range of radar designs to cope with the larger and increasing range of UAV types.

In Chapters 3–8, a number of different types of radars for UAV surveillance are introduced and examined. First, in Chapter 3, the advantages of high spatial and Doppler resolutions using the mm-wave band are considered for UAV detection and classification, especially the discrimination of birds and UAVs. Most usefully the authors present a comprehensive survey of published work on mm-wave radars applied to UAV detection and classification and conclude with a review of classification methods appropriate to the application of UAV surveillance. Overall, it is shown how the mm-wave band is well suited to UAV detection and classification, provided the requirements can be satisfied with the relatively short ranges achievable.

The radar concept employed for UAV surveillance described in Chapter 4 is the one based on interferometry. The authors show how this is an approach that has much merit, particularly as micro-UAVs often perform tight manoeuvres that leave them embedded in clutter. In such circumstances, interferometry enables fine measurement in angle, such that, over time, radial range, radial range-rate, angle and angle rate can all be measured with high accuracy. After CFAR detection, this information processed using an advanced tracking algorithm to output target reports. The concept has been implemented and shown to be an effective approach both via simulation and also a hardware realisation within a laboratory environment.

The approach considered in Chapter 5 is somewhat different, exploiting radar illuminators of opportunity in a passive bistatic configuration. Specifically, the radar used is of a staring type providing continuous illumination of the area of interest. This chapter introduces the passive bistatic radar concepts and explains how this works in the case of a radar illuminator which has the advantage of an appropriately designed waveform. It goes on to describe the design and construction of an experimental implementation. The illuminator operates in L-band and targets are shown to be detected out to ranges of a few hundred metres. In this way, the viability of the approach is demonstrated, whilst acknowledging that further work is needed to move towards operationally deployable systems. The passive radar approach is also considered within Chapter 6 but this time the illuminators of opportunities are DVB-T signals. The authors begin with a system-level computation of coverage and show through detailed calculations that detection ranges of up to 8 km are possible. Then they go on to describe the detailed design of a multichannel experimental receive system and layers of processing that alleviate characteristics typical of this type of radars such as direct signal breakthrough. The inherent poor angular resolutions at DVB-T illumination frequencies are partially overcome with an antenna array and a multichannel digital receiver. The experimental results are impressive and show detections out to 5 km. It is interesting to speculate on the numbers and strength of echoes that might be received from birds at these frequencies and how the RCS of birds compares to that of the UAVs.

Chapter 7 extends the DVB-T passive radar concept to include the exploitation of multiple frequency bands. The premise being that frequency diversity offers improvements in coverage and surveillance performance, especially against UAV targets of very small radar cross section. The selected bands (DVB-T, DVB-S, WiFi, DAB, GNSS and GSM) are all at the lower end of the electro-magnetic spectrum, but as the authors point out, this offers scope for an integrated single receiver hardware architecture. Through experimentation, the authors examine the different roles that the different frequency bands might play and conclude that the more powerful DVB-T signals provide the primary surveillance function with weaker WiFi and DVB-S signals enabling higher location accuracies and therefore enhancement of tracking performance together with target characterisation, albeit at shorter ranges.

The passive radar theme continues in Chapter 8 but here the illuminators of opportunity, in the form of GNSS signals, are located in space. The resulting geometry causes some marked difference in coverage and also facilitates detection by both traditional backscattering and forward scattering. Low transmit powers means that realistic ranges are in the tens-of-metres category but coverage is universal and near continuous. The forward scatter mode of detection is shown to be more of a secondary, reinforcement, form of detection being too sporadic to be relied on for primary detection. Computations show how accurately target detections can be located in range and angle. Finally, the computations are validated experimentally and demonstrate good agreement thus establishing the complementary viability of a space-based approach.

Having examined a number of different radar surveillance sensor concepts that yield detections of UAVs, we come back to the problem of differentiating these from detections of birds that have similar echo strengths. This is the theme of the next few chapters. Chapter 9 examines the characteristics of echo signatures resulting from scattering from UAVs and birds. The chapter usefully includes a comprehensive review of the existing literature and a taxonomy of UAV types that acts as a reminder of the vast arrange of air vehicle designs that are available. The authors present a series of simulation and measurement results that highlight the variability of signatures as a function of both radar parameters and local environment. Just as importantly, they also show that the principle difference between echoes from UAVs and birds lies in the form of their micro-Doppler characteristics and that this provides an obvious basis for their differentiation.

Chapter 10 continues the theme of evaluating UAV responses to extract more information, this time to differentiate between different types of UAV. The basis of this approach is to recognise, for example, that quadcopters, with four sets of rotor blades and hexacopters, with six sets of rotor blades, are likely to generate radar micro-Doppler features that differ from one another. This generates differences that can be observable in spectrograms from which classification features can be derived. Using a laboratory level experiment with a radar operating at a transmission frequency of 25 GHz, they show that high levels of classification performance can be achieved under a variety of combinations of different UAV types. They also examine performance as a function of an SNR. This indicates that classification levels start to deteriorate as the SNR falls below 10 dBs. It is intriguing to speculate on the main micro-Doppler contributors to classification as well as the effects of key radar parameters such as the transmission frequency.

Chapter 11 deals with the specifics of classification between UAVs that may be carrying different payloads. The premise is that different weight UAVs will require different levels of power and this will affect their micro-Doppler signature as seen by radar. Equally, the differently weighted UAVs will also have different trajectory behaviours that can also be exploited by a discriminator. They use a multistatic radar to make measurements at a transmission frequency of 2.4 GHz. Spectrograms show a marked difference both monostatically and bistatically between the echoes from a UAV with and without a payload with the propeller sidebands showing less variability under loading. This allows suitable discriminating features to be extracted and applied to a classifier with excellent performance and hints at the potential for radar surveillance to provide much more than just UAV detection and tracking.

Chapter 12 describes methods of good practice and approaches to the design and development of counter UAV systems. The authors briefly describe a novel staring radar concept and show how it can be used to detect and discriminate drones against a severe background of other targets such as birds. Specifically, there is a focus on the important topic of collecting truth data from both control and from opportune targets. After all, classification inherently requires accurate truth data to validate and understand achieved performance. Examples are provided that use real-world measurements to illustrate how labelled training data can be generated and then used to demonstrate target classification performance when using a supervised-learning-based approach.

Together, the chapters that comprise this book collectively provide a detailed overview of the state of the art of approaches to the surveillance of drones using radars. Notably, a wide range of radar types are under development and examination, raising the intriguing question as to which may eventually be routinely adopted for operational use. Equally, the classification challenge is shown to be one that has yet to be fully met and there is much scope for future improvements. Certainly, the higher the drone classification accuracy couple with the greater time between false reports will inevitably increase the range of application and rate of adoption of radar. In this way, radar is set to play a key role in surveying and understanding the occupancy of low-level airspace, just as it does in managing and controlling the use of higher level airspace as used by manned aircraft.

Bibliography

[1] Johnson W., "*Helicopter Theory*", Dover Publications, New York, NY, USA, 1980 ISBN: 0-486-68230-07.

[2] Nonami K., "*Prospect and recent research and development for civil use autonomous unmanned aircraft as UAV and MAV*", Journal of System Design and Dynamics, 1(2), pp. 120–128, 2007.

[3] Ninghao Y., Tuian L., Beibei Z. and Nan L., "*A review: UAV-based remote sensing*", IoP Conference Series: Materials and Science Engineering, 490, pp. 1–6, 2019.

[4] Sullivan J.M., "*Evolution or revolution? The rise of UAVs*", IEEE Technology and Society Magazine, pp. 43–49, 2006.

[5] Prisacariu V., "*The history and the evolution of UAVs from the beginning till the 70s*", Journal of Defence Resources Management, 1, pp. 181–189, 2017.

[6] Luo Y. and Fan H., "*The development and application of China military UAV*", International Conference on Advanced Electrical, Mechatronics and Computer Engineering, pp. 145–148, 2019.

[7] Palik M. and Nagy M., "*Brief history of UAV development*", Repüléstudományi Közlemények, 31(1), pp. 155–166, 2019.

[8] Cai G., Chen B.M. and Lee T.H., "*An overview on development of miniature unmanned rotorcraft systems*", Frontiers of Electrical and Electronic Engineering in China, 5, pp. 1–14, 2010.

[9] Keane J.F. and Carr S.S., "*A brief history of early unmanned aircraft*", John Hopkins APL Technical Digest, 32(3), pp. 558–571, 2013.

[10] Newcome L.R., "*Unmanned Aviation: A Brief History of Unmanned Aviation Vehicles*", American Institute for Aeronautics and Astronautics, 2004 ISBN: 978-1-56347-644-0.

[11] Fahlstrom P.G. and Gleason T.J., "*Introduction to UAV Systems*", John Wiley and Sons, West Sussex, UK, 2012 ISBN: 978-1-119-97866-4.

[12] de Wit J.J.M, Harmanny R.I.A. and Premel-Cabic G., "*Micro-Doppler analysis of small UAVs*", Radar Conference (EuRAD), 2012 European, pp. 210–213, Oct 2012.

[13] Molchanov P., Harmanny R.I.A., de Wit J.J.M., Egiazarian K. and Astola J., "*Classification of small UAVs and birds by micro-Doppler signatures*", EuRad Conference, Nuremberg, Germany, pp. 172–175, Oct 2013.

[14] Clemente C., Balleri A., Woodbridge K. and Soraghan J., "*Developments in target micro-Doppler signatures analysis: radar imaging, ultrasound and through-the-wall radar*", EURASIP Journal on Advances in Signal Processing, 2013 (1), p. 47, 2013.

[15] Harman, S.A., "*Comparison of staring radars with scanning radars for UAV detection*", EuRad 2015, France, pp. 185–188, Sep 2015.

[16] Jahangir, M., "*Target centric wide-area 3-D surveillance using a non-scanning multibeam receiver array*", IEEE RadarCon 2015, Arlington, VA, USA, May 2015.

[17] Ritchie M., Fioranelli F. and Griffths H., "*Micro-drone RCS analysis*", Proc. of IEEE Radar Conference, Johannesburg, South Africa, pp. 452–456, Oct 2015.

[18] Ritchie M., Fioranelli F., Griffths H. and Torvik, B., "*Monostatic and bistatic radar measurements of birds and micro-drone*", IEEE Radar 2016, Philadelphia, PA, US, pp. 1–6, May 2016.

[19] Jahangir, M. and Baker, C., "*Persistence surveillance of difficult to detect micro-drones with L-band holographic radar*", CIE 2016 Int. Radar Conf., Guangzhou, China, Oct 2016.

[20] Poitevin P., Pelletier M. and Lamontagne P., "*Challenges in detecting UAS with radar*", 2017 International Carnahan Conference on Security Technology (ICCST), pp. 1–6, Oct 2017.

[21] Vaugh C.R., "*Birds and insects as radar targets: a review*", Proceedings of the IEEE, 73(2), pp. 205–227, 1985.

[22] Torvik B., Olsen K.E. and Griffiths, H.D., "*X-band measurements of radar signatures of large sea birds*", IEEE International Conference on Radar, pp. 1–6, Oct 2014.

[23] Rahman S. and Robertson D.A., "*In flight RCS measurements of drones and birds at K-band and W-band*", IET Radar, Sonar and Navigation, pp. 300–309, 2019.

Chapter 1
Counter UAS systems overview

Daniela Pistoia[1]

1.1 Introduction

As many recent examples show, small unmanned aerial systems (UASs or drones, in the following) have become a real threat to both civil and military targets. While such advanced technology means a huge opportunity for the military and industry, its alternative, sinister use for criminal and terrorist purposes is no longer a fictitious risk, requiring a huge effort in terms of counteractions.

New developments such as advanced passive and active counter UAS multispectral technologies have been demonstrated to be the most feasible to deal with the counter-UAS challenge. Multiple domains (electromagnetic (EM), acoustic, electro-optic), multiple sensors (active radar, passive EM interceptors, acoustic sensors, infrared (IR) cameras), and multiple jamming/deception systems, integrated via a dedicated command and control (C2) capability, are key elements within this approach.

Small UAS flying near critical infrastructure or sensitive areas (e.g., government buildings, high-profile event locations, prisons, military compounds, big events) are considered as unconventional threats. Detection, identification, neutralization, and (when possible or allowed) destruction of such UAS have therefore become a priority for military forces and homeland security organizations. So far, traditional countermeasures have demonstrated their weakness in this regard. Unconventional threats require advanced solutions, and many industrial and government initiatives are rising to meet this new threat.

To neutralize such kind of threat and to perform a soft kill action, jammer and deception (namely, electronic warfare or EW) are today the most feasible approaches and many examples of in-service products exist, with several degrees of sophistication.

When considering nondestructive techniques, low-power laser technologies and low-power EM pulsed sources are emerging with the aim to negate the threat of the use of electro-optic sensors or the use of the EM spectrum (EMS).

Exploiting cyber capabilities is an important vector to counter the mini drone threat, though respective solutions are still immature and require superior knowledge, skills, and tools in the cyber domain.

[1]Chief Scientist, ELETTRONICA Group, Roma, Italy

Technologies for the destruction of drones (hard kill) include guns and high-power laser and EM-pulsed sources. Blue suicide drones, guided against the threat, are also an example of hard kill action. However, the destruction of drones is not always a good option for several reasons. First, cost-per-shot is very high and requires moderate-to-high level skills of the human in the loop as well as the type of users (military or not). Second, the rules of engagement and local regulations often forbid the use of hard kill due to the presence of civilian population, especially in urban environment or during crowded events. Third, installation and deployment constraints required by hard kill subsystems could not be met in any situation. Finally, the destruction of the drones means the destruction of evidence: reverse engineering processes are often put in place to investigate the origin of the platforms or of its components (including the payload), the modifications that have been set, and the human component behind. As explained in the following paragraphs, we have to consider that while the aircraft itself is not manned, the system is manned (at least prior to the mission to program the flight paths and execution): UASs require a certain amount of human involvement to prepare and execute the mission.

However, despite all the progresses already achieved in countering drones, the game is far from the end.

Specific challenges are continuously arising. Thanks to the evolution of materials and the availability of commercial off-the-shelf (COTS) technologies, including additive manufacturing, long endurance batteries, and commercial navigation aids, drones are gaining lower speed, lower altitude, very small RF, thermal and acoustic signatures, high maneuverability, posing new challenges in terms of detection and tracking.

Furthermore, threats are getting resilient to first-generation counter-drone systems that rely on "traditional" detection and neutralization, for example, the possibility of flying in swarm, possibly boosted by autonomous guidance, using "per-opportunity" data-link and communication means (i.e., 5G).

User requirements are evolving accordingly, posing challenges in terms of variety of applications and installation constraints: from individual soldiers to vehicles and command posts to protect from single individuals to larger critical infrastructures, including urban areas. Therefore, inherent modularity, scalability, and interoperability have to be addressed by design.

Finally, the threat is evolving so fast that procurement rules from both military and law enforcement agencies are drastically changing. Time-to-market is going to be substituted by time-to-field or time-to-operation, traditional tenders are complemented by field trials and live contests, and availability is much more than reliability and, in some cases, "something is better than nothing."

The goal of the following sections is to introduce the reader to the fluid environment of drone and counter-drone solutions, starting from today and flying deeper into tomorrow.

1.2 Too small and simple to be a threat?

UAS can be classified as summarized in Table 1.1, generally adopted by North Atlantic Organization (NATO).*

*NATO UAS Classification Guide. September 2009 JCGUAV meeting.

Table 1.1 Unmanned aircraft classification guide

Class	Category	Normal employment	Normal operating altitude	Normal mission radius	Primary supported commander	Example platform
Class III (>600 kg)	Strike/combat	Strategic/national	Up to 65,000 ft ASL	Unlimited (BLOS)	Theatre	Reaper
	HALE	Strategic/national	Up to 65,000 ft ASL	Unlimited (BLOS)	Theatre	Global Hawk
	MALE	Operational/theatre	Up to 45,000 ft MSL	Unlimited (BLOS)	JTF	Heron
Class II (150-600 kg)	Tactical	Tactical formation	Up to 18,000 ft AGL	200 km (LOS)	Division, Brigade	Watchkeeper
Class I (<150kg)	Small (>15 kg)	Tactical unit	Up to 5,000 ft AGL	50 km (LOS)	Battalion, Regiment	Scan Eagle
	Mini (<15 kg)	Tactical sub-unit (manual or hand launch)	Up to 3,000 ft AGL	Up to 25 km (LOS)	Company, Platoon, Squad	Skylark
	Micro (<2 kg)	Tactical sub-unit (manual or hand launch)	Up to 200 ft AGL	Up to 5 km (LOS)	Platoon, Squad	Black Widow

Source: NATO ATP-3.3.8.1, ed. B, Ver. 1.

Notes: HALE: high altitude long endurance; MALE: medium altitude long endurance; ASL: above sea level; BLOS: beyond line of sight; AGL: above ground level; LOS: line of sight; JTF: joint task force.

NATO UAS categories are based on unmanned aerial vehicle's (UAV) maximum gross take-off weight and typical operating altitude. Categories first focus on classes of weight, which are further divided based on the operational altitude.

Class I, which is the subject of the discussion in this chapter, refers to platform less than 150 kg (further divided down based on altitude). Under 150 kg, the NATO Certification Standards do not apply.

The maximum operating altitude of several current UAVs is shown in Figure 1.1, along with the boundaries for different classes of controlled airspace, referred to as the maximum take-off weight (MTOW). Micro, mini, and tactical UAS show clear breakpoints in mass and maximum altitude. Micro and mini UAVs are likely to be operated close to the ground and are generally not capable of reaching above 1,500 m. Tactical UAVs occupy a much broader mass range, from 15 to 600 kg, and are primarily distinguished by their ability to operate higher, approaching the boundary of airspace at 5,000+ m.

The maximum endurance of several UAVs is shown in Figure 1.2: micro UAVs typically have endurances measured in minutes, as they typically do not carry enough power to remain aloft for long time. As the mass of UAVs increases, the current maximum endurance capability also increases exponentially. Mini UAVs can typically be operated for several hours, tactical on the range of 5–10 h, and medium and high altitude from 10 h to days.

Small, mini, and micro UAVs or drones are rapidly becoming "tools of the trade" in many industries and can be categorized into segments of the market as in Table 1.2: government (including military), enterprise (corporations/businesses), and consumers (personal/hobbyist).

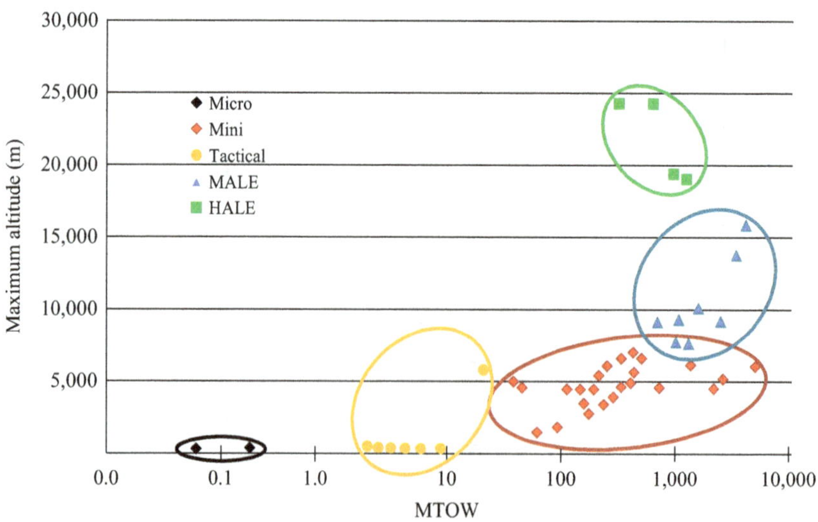

Figure 1.1 Unmanned platform's maximum altitude w.r.t. MTOW

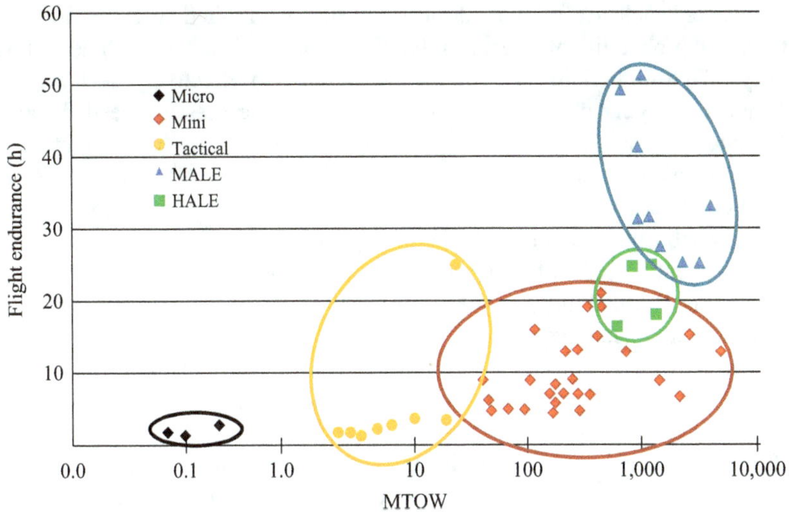

Figure 1.2 Maximum endurance of current UAVs with respect to MTOW

Table 1.2 Types of mini–micro UAS application

Segment	UAS application
Government agencies	Homeland security, weather, search and rescue (first responder), military air/sea/land applications (weaponized, surveillance, and patrolling)
Businesses/ enterprises	Agriculture, photography, video production (movie, TV, documentary), infrastructure, and building inspections (pipelines, cell towers, railroads, waterways, docks, and locks)
Personal/hobbies	Racing, personal photography, blogging, podcasts, video-blogging, experimental purposes, sports affiliated, and videoing events (including live streaming)

Since low-altitude drones fly only hundreds of feet above ground, they mostly operate outside the traditional radar coverage used to track commercial aircraft. Additionally, military air defense radar systems are usually not designed to detect aircraft with such a small radar cross-section (RCS). In other words, there is an airspace segment still out of control for both civil authorities and military air power.

According to the Centre for the Study of the Drone at Bard College "*there are more drones, made in more countries, and flown by more groups, than in any other previous conflict*." Among the 38 different types of UAS, the institute counted at least eight recreational hobby drones and possibly six unidentified homemade models, asserting "*the conflict marks the first time that hobby drones have been modified with explosives and turned into flying improvised explosive devices*."[†]

[†]2016 Field Guide "Drones Operating in Syria and Iraq" published by the Centre for the Study of the Drone at Bard College (Annandale-On-Hudson, New York).

However, such use of mini drones may not be limited to zones of war or conflict. The probability of threat proliferation to domestic areas combined with the inability of traditional airspace control and defense to effectively deal with such small and low-flying objects underlines even more the pressing need for appropriate counter-UAS technology. Many companies worldwide have therefore created intense research and development programs to provide effective solutions.

The combination of versatility, easiness to use, good endurance, low visibility, market availability, and relatively large payload makes these platforms a very dangerous threat if used with malicious intents and requires a dedicated, specialized, and "fine-tuned" capability in order to provide effective detection–classification–identification and neutralization.

When we talk about a UAS, we have not just to identify the flying platform but also to address the integration of several subsystems and people that take part to perform a single mission.

In particular, Figure 1.3 identifies the main components involved in an action performed by a UAS. It has to be considered that while the aircraft itself is not manned, the system is manned. UASs require a certain amount of human involvement to prepare and execute the mission, and this is mandatory to be remembered in putting in place any strategy of counteraction.

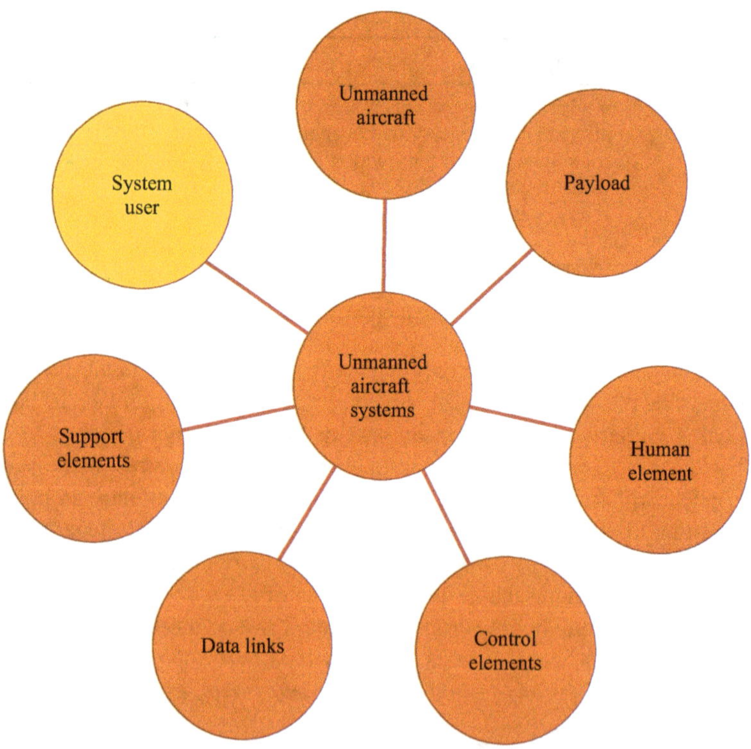

Figure 1.3 Unmanned aircraft system components

1.3 Why an integrated system of sensors and shooters is a must?

Defending against small drone threats is a complex issue, since it is not only about eliminating the drone to prevent it from completing its mission. A successful defense must ensure the immediate detection, classification and identification of the object prior to neutralize it in a secure framework for the safety of the people on the scene, as well as minimizing collateral damages.

Given the physical and kinematic characteristics of the drone and the typical modes of use, a multiple sensor approach is necessary to improve the detection capability. Several options, each with their own strengths and weaknesses, are currently taken into account.

Radar. Detection, tracking, and range estimation of aerial targets are the main tasks performed by the radar. However, mini drones are hard to detect and identify due to very low radar signatures and related difficulties to exploit the Doppler effect.

A large variety of commercial drones do not allow assessing a reference value for the RCS of such targets. In any case, as depicted in Figure 1.4, measurements of RCS on significant samples reveal that RCS is substantially independent of frequency (in the range of 1–40 GHz) and polarization and reaches a minimum value in the order of 0.01 m^2. This is due to the low reflectivity of the materials of the body. The batteries give the main contribution.

Furthermore, it is a tremendous challenge to distinguish the target from other objects, particularly in an urban environment with a high probability of false alarms. The challenges increase when trying to use a passive (bistatic) radar.

On-field experiences confirm that target discrimination task is a system issue and not a sensor issue. While mechanical rotating parts (blades) generate micro-Doppler effects that can be used for this purpose, this is not always a reliable or even available information and often it must be correlated to other measurements such as estimated trajectory, detection of radio signals, and visual confirmation (by human or by automatic algorithms or both).

Anyway, an active radar remains the prior sensor in all the solutions on the market, due to its unique capability to estimate the distance of the threat and consequently its time-of-arrival. Estimated time-of-arrival is the key element in the decision-making process of the reaction chain.

Therefore, the best choice of the radar sensor in a counter-drone solution is a critical issue. The main characteristics that are currently driving this choice (and will drive much more in the future) are as follows:

- Native 3D radar (azimuth, elevation, and range), based on an electronically scanning antenna, is able to generate one or multiple beams in azimuth and elevation. This is justified by the operational requirements to cue efficiently the camera (automatically or operator-assisted), to have a prompt confirmation of the nature of the threat, and to steer the shooters (soft/hard) in both azimuth and elevation to maximize effectiveness.

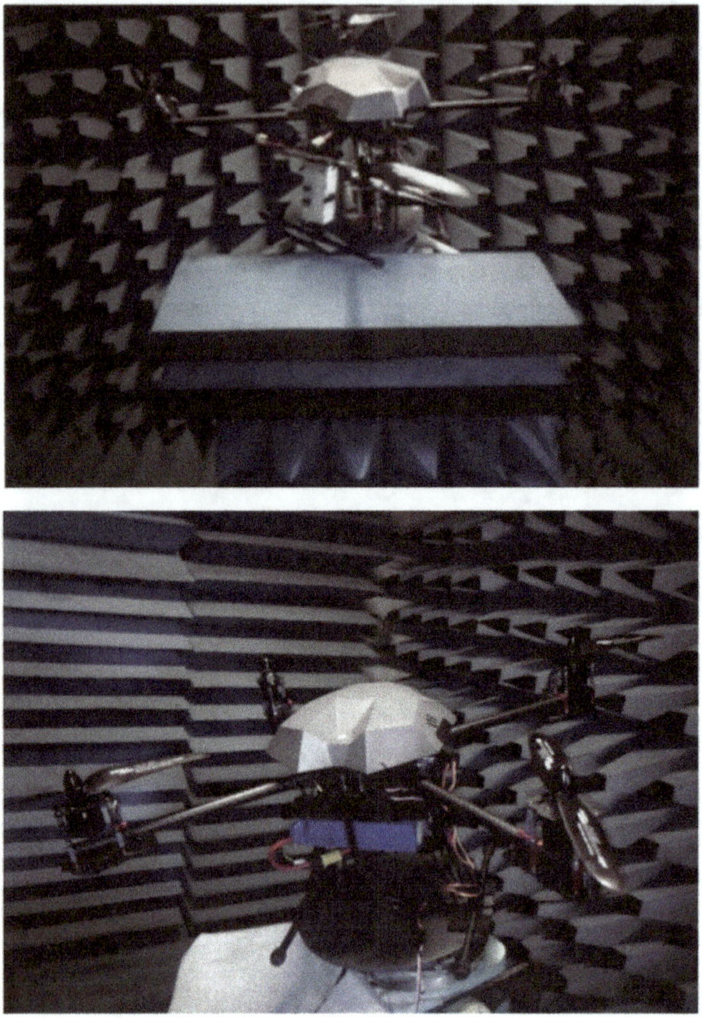

Figure 1.4 Measurement of RCS of drones at ELETTRONICA facilities (courtesy of ELETTRONICA SpA)

- Simultaneous tracking of multiple threats. This is needed to counter attacks of coordinated drones or cooperating swarms.
- Easy re-programmability via software (SW), which is a key characteristic to guarantee a real growth capability against new or emerging threats.
- AESA architecture (no rotating nor mechanical parts), which ensures the faster reaction time to alarms, the activation of the countermeasure chain, and high reliability.
- Designed to be moved or transported and deployed rapidly, even by a single soldier or operator, which requires the radar to be "tactical" by design.

- High-technology readiness level, available in operation in a short time to face the so-called Urgent Requirement often expressed by users.
- Fast delivery time, which means capability of high production rate.

The use of a 2D radar instead of a 3D might be justified by several factors such as simpler expected scenarios, reduced automation in cueing the camera, or the countermeasures, no countermeasure integrated, cost. National and international regulations are also driving factors: it is interesting to underline that 3D radars are usually classified as "military goods" and require military personnel to operate.

Camera. Together with EM sensors, thermal cameras are essential devices, usable under low-visibility conditions and at night. The main issue is that the range of threat acquisitions is less than the radar detection range. This requires the camera to be cued (in both azimuth and elevation) by the radar tracks and the radar to continuously track the target.

The challenge for IR sensors is represented by the need to reveal drones even in the presence of strong lighting conditions. Highly sophisticated cameras available on the market show some abilities to locate thermal hotspots generated by motors. Such hotspots, located in fixed positions in relation to the structure of the drone, also contribute to automatic object identification by making use of IR image reference libraries. Examples of such thermal hotspots are shown in Figure 1.5.

Usually cameras are human-operated and identification confirmation is human-assisted.

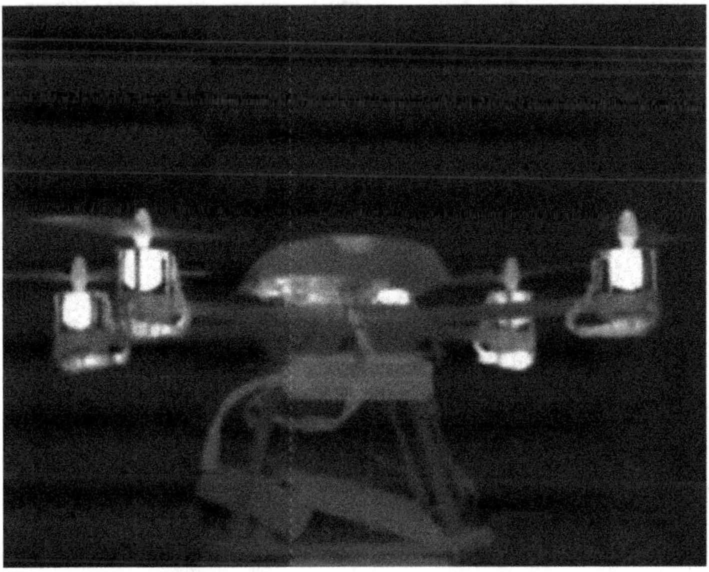

Figure 1.5 Measurement of IR signature of drones at ELETTRONICA facilities (courtesy of ELETTRONICA SpA)

EM sensor. Electronic defense systems are commonly integrated in the most effective counter-drone solutions. These devices quickly detect the presence of radio signals used to send the commands from the pilot to the drone (uplink) and the data and images from the drone to the command post (downlink). For commercial drones, the radio signals are transmitted on well-known and standardized frequencies, relatively easy to be intercepted with electronic surveillance in automatic mode, even though complex wave modulations are often superimposed by the carrier signal.

Passive interceptors of this type have the capability to automatically analyze the radio wave and allow the identification of the threat based on the transmitted waveform (noncooperative target recognition). This is a very interesting complement to the radar detection and could be also used for false alarm reduction and target discrimination from birds or even friend drones. Furthermore, passive geo-location techniques can be put in place to locate both the drone and the control station.

However, this apparent over-complexity in the counter-drone systems is even essential in terms of threat identification: the availability of preloaded data allows the system to automatically identify friend-vs.-foe drones and finally correctly generate smart jamming waveforms dedicated to the specific drone.

Acoustic. During flight, drones generate noise both in the audible frequencies and in the ultrasounds. Acoustic sensors reveal the presence of mini drones as well as help to classify the target based on noise characteristics specific to the drone model.

However, the operational range of acoustic sensors is limited to a few hundred meters. At higher distances, drones are lost in background noise. On the other hand, pulse radars have a blind spot at shorter distances. This means that the acoustic sensor, made up of an array of microphones, is the ideal complement to radar systems to cover both long and short ranges. Being relatively cheap, acoustic sensors are efficient tools for continuous surveillance of particularly sensitive areas.

Examples of acoustic signature in a typical operational background are shown in Figure 1.6.

Every sensor technology has pros and cons, touching a variety of factors: operational performance, installation constraints, life cycle costs, modularity and scalability, military or nonmilitary use, maturity.

Table 1.3 shows a qualitative comparison among the eligible sensor technologies.

The conclusion is that there is no “silver bullet” to detect–classify–discriminate–identify the threat and that a plethora of sensors, correctly and functionally integrated, are needed to enhance the probability of detection and the correct trigger of the reaction chain. As we mentioned earlier, human-in-the-loop is also a key factor, at least for today’s policies and tactics, and the complexity of the technologies put in place together with the unpredictability of the threat requires a high level of skills.

Detection and identification are essential, but they are only the preliminary steps in solving the problem of removing the rogue drone from the scene and/or its neutralization. “Hard Kill” or physical destruction options are limited to combat zones or an open field, where the consequences of falling wreckage, ordnance, or other harmful items are generally irrelevant. In an urban scenario, a different approach aiming at a “Soft Kill” philosophy is preferable. The following described options have currently been proven as feasible and effective.

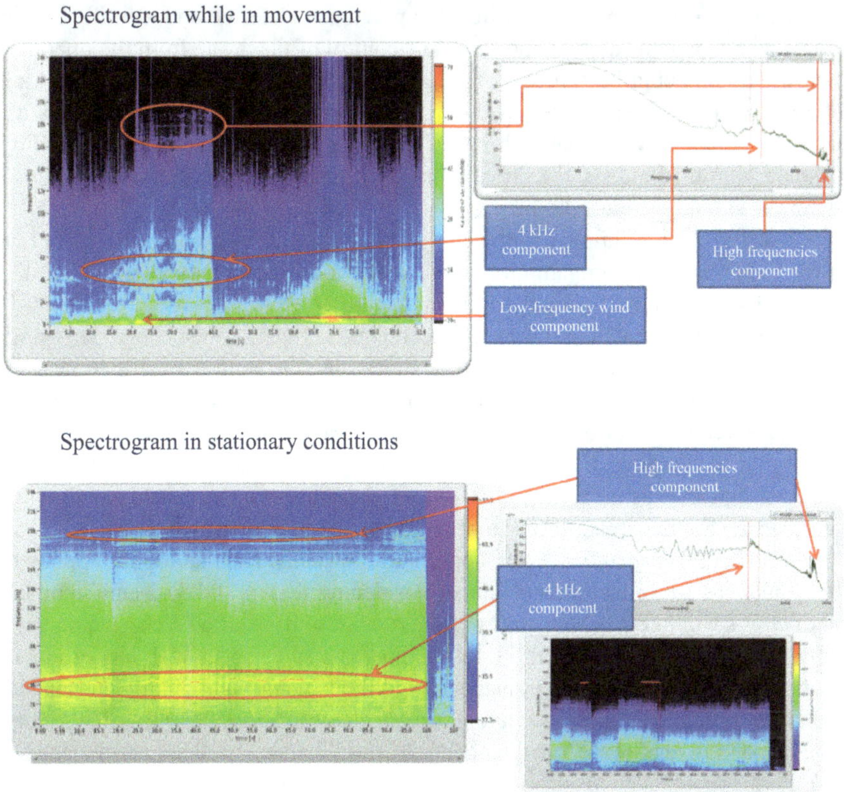

Figure 1.6 Examples of acoustic signature of drones in typical background (courtesy of ELETTRONICA SpA)

Table 1.3 Counter-drone eligible sensors—"What is best" analysis

	Active radar	Passive radar	Acoustic sensor	Infrared sensor	EM sensor (RF, Comm)
Capability of detection	GOOD	MEDIUM	POOR	MEDIUM	GOOD
Capability of identification	POOR	POOR	MEDIUM	GOOD	MEDIUM
Resistance to interference	GOOD	POOR	POOR	MEDIUM	MEDIUM
Applicability in urban environment	MEDIUM	MEDIUM	POOR	MEDIUM	MEDIUM
Installation constraints	MEDIUM	POOR	GOOD	GOOD	GOOD
Life cycle cost/complexity	POOR	POOR	GOOD	MEDIUM	GOOD
Technology maturity	GOOD	POOR	MEDIUM	GOOD	GOOD

Jamming. A first option is to affect the previously detected and identified radio signals, which would subtract and sever the control of the drone from the operator. Then the drone could be forced to land in a safe area or to crash without risking collateral damage. The simplest technique is to generate jamming signals against the control link, delivering enough power to negate the use of the EMS. According to the programmed modes, the drone then automatically enters into fail-safe mode causing it to land or return home.

This "brute force" approach however requires generating a huge amount of EM power and broad-spectrum jamming of the whole area, which may also result in the undesired suppression of friendly communications.

A more sophisticated and selective technique is the so-called Smart Jamming, which consists of jamming the control signal only in some specific timeslots, according to the specific protocol used by the radio remote control. Smart Jamming techniques require a priori knowledge of the waveform and its coding, which requires the availability of a library of control signal protocols. This is obtained by extensive and expensive a priori activity of reverse engineering and intelligence that gives additional strength to the blue forces in countering the threat.

GPS deception/spoofing. One should argue that this technique only applies to UAV large enough to be utilizing Global Positioning System (GPS). Many mini and micro UAVs will likely be too small and short range to have GPS. Well, size may not really be limiting the use of GPS. Look at modern smart phones with lots of inbuilt functions and features taking advantages from the positioning and timing information. Miniaturization of a GPS receiver and global availability of positioning systems today are integrated in many UAVs available on the open market (5–10 k€, not really toys, etc.). Furthermore, even toys can be modified on purpose to support this capability.

Negating the use of GPS, or more generally, the use of Global Navigation Satellite System (GNSS), is part of the discipline identified by NATO as "Navigation Warfare" and has been largely demonstrated to be effective against many navigation receivers.

In 2012, a research team led by Professor Tod Humphrey of the University of Texas at Austin successfully demonstrated for the first time that UAV GPS signals could be commandeered by an outside source.[‡]

The most effective albeit complex technique is GPS spoofing, provided the targeted UAS is using satellite navigation. Based on military capabilities designed to deceive adversary precision-guided munitions, the technique consists of first seducing the UAV's GPS receiver to recalculate its position, and second, deviating its path in accordance with preplanned countermeasures. To this end, a spoofing device transmits imitated satellite signals while deceiving the target with formally correct but false position data. This requires knowing the exact position and speed of the drone, which can be provided by a radar sensor. Precise scheduling of each spoofing phase is also needed to reduce the effectiveness of counter-countermeasures of certain smart, GPS-based guidance systems.

[‡]https://www.ted.com/speakers/todd_humphreys.

Direct energy weapons. In addition to these soft-kill techniques, weapons are being developed that produce a high-power microwave EM pulse that is highly effective against electronic equipment. With a specifically shaped antenna, or an emitter, the energy can be focused to produce effects within a confined area and limited range. Under certain circumstances such weapons could ideally complement the other techniques to neutralize small drones.

Under this category, it should be mentioned also the use of high-power laser. The technology, even if still not fully mature, is demonstrated to be highly effective, generating the destruction of the target. The major issue is linked to the safety of the humans operating the laser gun.

Blue drone. The blocking of the enemy drone, that is, the neutralization of the threat, can also take place physically thanks to the use of friendly drones (blue force drones→blue drones) also known as drone hunters.

The family of drone hunters is young but is populating very rapidly. Inside the family we can list the following:

Drone catcher. This is a kind of drone (often a multicopter) equipped with a net fired by a gun.

After detection, the drone catcher is rapidly deployed to intercept the moving target, under the control of a remote operator.

As soon as it gets close to the victim drone, using multispectral onboard sensors, it locks the net gun onto the target then safely captures it by firing the net, tows it away to a safe location, and releases it there.

Collision drone. This is a drone normally remote operated. After detection, the drone is rapidly deployed to intercept its prey, locks on it, and disables it by collision.

Weaponized drone. This is a kind of drone equipped with explosives. After detection, it is rapidly deployed to intercept the moving target, under control of a remote operator. As soon as it gets close to the victim drone, it fires the explosives to damage the target. This application is particularly useful to counter-drone swarm attacks (one-to-many).

All the abovementioned solutions can be upgraded to a higher automation level making a wide use of onboard artificial intelligence (AI) and advanced machine vision processing making them patrolling a given airspace.

Within the assigned area, they can autonomously detect, intercept, and neutralize any unwanted objects.

The new generation hunter drones can be operated as a standalone unit or integrated inside a higher level counter-drone infrastructure.

1.4 Operational use of counter-drone solutions

From the operational point of view, a counter-drone solution should be able to

- detect, classify, track, identify, and counter UASs (single and/or multi-UAVs) in defense scenarios using multiple technologies;
- operate from detection to counter UASs also in civil scenarios;
- deliver and update real-time operational pictures and alerts;

- integrate into a multilayer C2 system with cross-security-domain approach. Two main operational modes must be available:
  - Soft
  - Hard (integrating Soft Kill capabilities and supporting functions);
- provide a range of selective mitigation alternatives with the ability to evaluate mission success probabilities and resulting drawback;
- require limited logistic support for deployment and maintenance;
- require minimum operator effort for decision-making;
- provide dynamic scalability of sensors and effectors;
- facilitate the incorporation of counter UASs in security and defense systems for fixed and deployed asset;
- be able to operate in a wide range of climate (e.g., arctic and subtropical) and weather (e.g., rain, fog, and snow) conditions.

Therefore, a counter-drone solution (an example is given in Figure 1.7) is not just the sum of sensors and shooters, but a comprehensive and continuously evolving multidomain area protection capability aimed to detect, discriminate, identify, and neutralize drone threats.

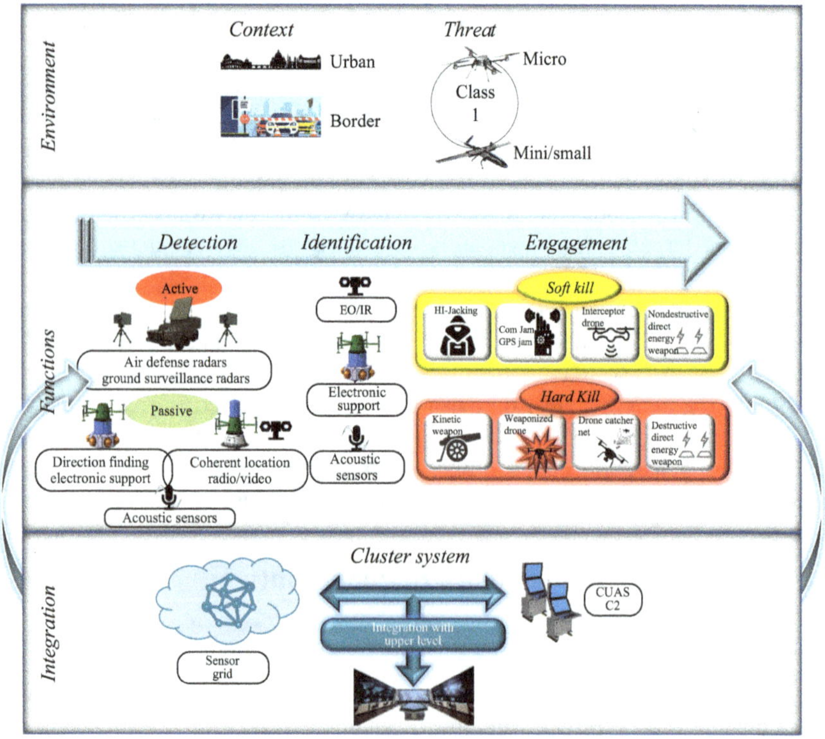

Figure 1.7 C-UAS possible solution

Sensors and countermeasures need to be functionally integrated and effectively coordinated, so they interface via a mobile ad hoc network with a C2 station, typically with a man-in-the-middle, with an intuitive and easy-to-use interface.

As none of the sensors nor the shooters previously described can be considered a "silver bullet," the core of the solution is the C2.

It is important to remember the correct definition of C2, as recommended by US doctrine.[§]

> "Refers as command and control (C2) system the facilities, equipment, communications, procedures, and personnel essential for a commander to plan, direct, and control operations of assigned and attached forces pursuant to the missions assigned."

The operational point of view is much more complicated, and at least for today and for the next decade, the decision-making process is supported by the machine but essentially conducted by the human.

Nevertheless, the machine or the SW logics of the C2 plays an essential role, because the most important requirement of such applications is the reaction time and no human is able to manage the amount of data/information and the complexity of the technology deployed in the short time (few seconds) needed to raise an alarm and counter act.

As an example, Figure 1.8 shows the typical range of effectiveness of the main sensors and shooters of a counter-drone solution, which gives an idea of the reaction time needed to guarantee a fit-for-purpose probability of success.

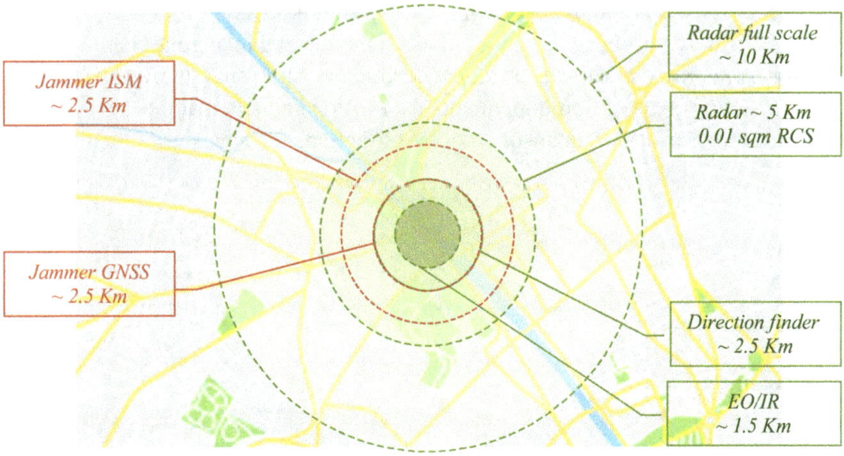

Figure 1.8 Typical range of effectiveness of a counter-drone solution

[§]This definition is reported in several US DOD Documents, as for example in JP-06 Joint Communications Systems available on the web as an open-source document.

Modern solution of SW logics implementing C2 functionalities in support of the decision-making process is based on the so-called OODA (observe–orient–decide–act) loop. Colonel John Boyd, a military strategist for the US Air Force, has theorized this in the 1990s, and it is often applied at the operational level during military campaigns. It is now also applied to understand commercial operations, learning processes, and many cases of decision-making processes.

Figure 1.9 summarizes the four steps of the original cycle, as Col. Boyd described them.[¶]

Modern views of the cycle have introduced the capability of the commander to improve the decision based on the experience (LEARN), which could be apparently obvious but requires methods, information storage and sharing, and related procedures of verification–validation when applied to operational context.

When applied to the mission of countering UAVs, SW logics in support to the decision-making process are typically organized in a function tree as depicted in Figure 1.10.

Behind every functional "box," complex algorithms could be required to contribute effectively to the success of the mission. In the last years, the availability of high computer power and the possibility to gather, store, and manage a large amount of data have boosted the application of many branches of AI, including the following:

Machine learning and data fusion. This allowed the reduction of overload of the human in the decision-making process, assigning to the machine (the computer) the execution of repetitive or computationally complex tasks, or those requiring times shorter than human capability.

In any case, every proposed solution, modular and scalable according to the operational scenarios and the needs of the final user, is several times more complex than the threat, requiring a plethora of assets (tangible and intangible) deployed. All the active and passive sensors and the effectors are integrated and connected in a local C2 station. This station implements data fusion and automatic procedures and rules in order to focus human operations on action, resources coordination, and

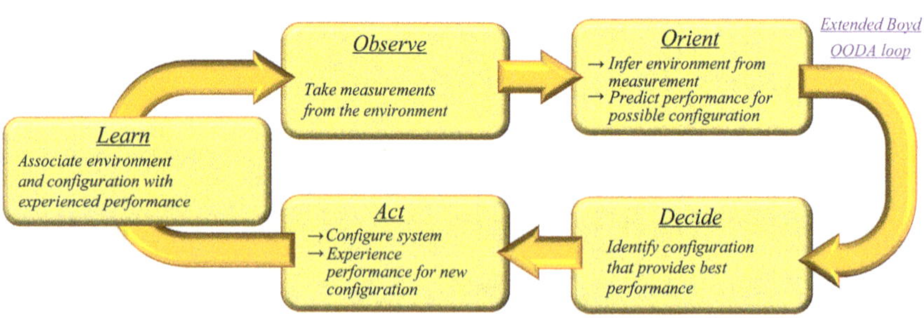

Figure 1.9 Extended OODA loop high-level description

[¶]Boyd, John, R., The Essence of Winning and Losing, June 28, 1995.

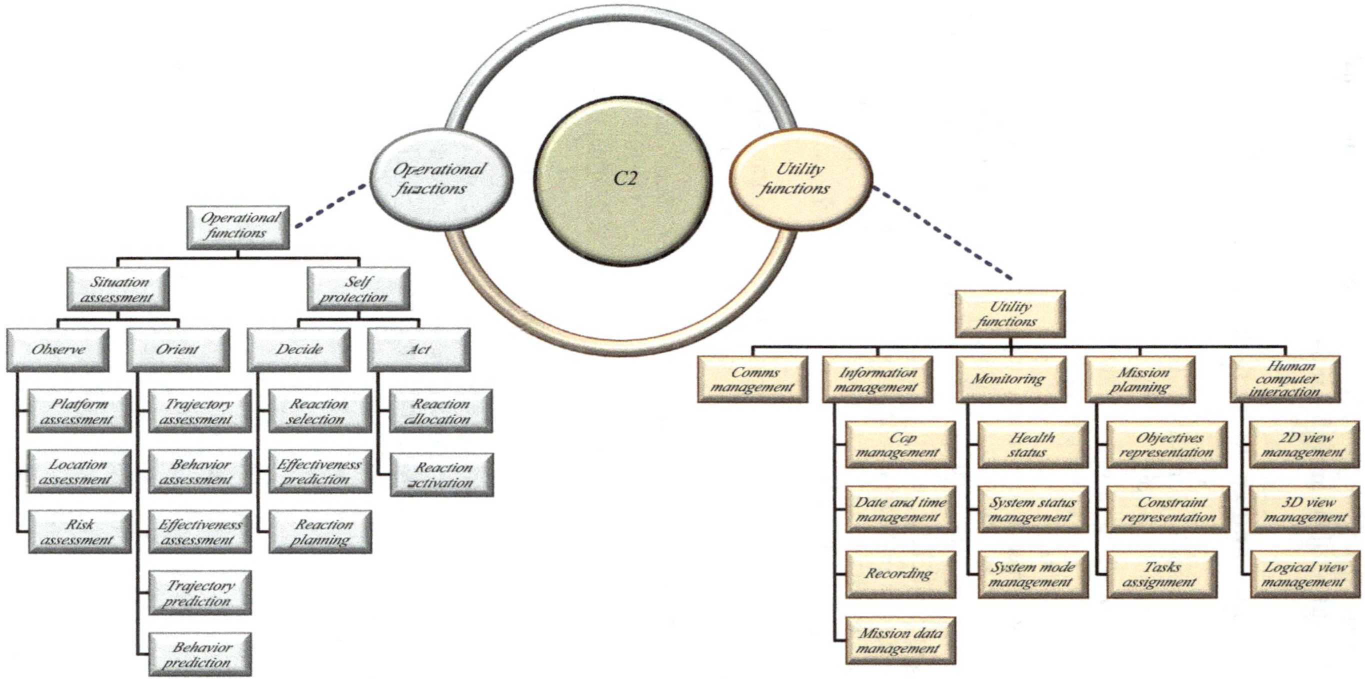

Figure 1.10 Typical functional tree of a C2 for counter-UAV solutions

cooperation. Controlling these assets will require highly qualified and best trained operators, whose mission preparation needs to be much more professional and complex compared to the relative simplicity of the threat.

1.5 Future scenarios and enabling technologies

As already anticipated, despite the advances in system architectures and sensor technologies, the game is far to be over.

The threat is rapidly changing and continuously evolving, taking advantage from the availability of COTS technologies and SW algorithms shared in the web even by smart young students.

Drones can be assembled from scratch, buying all parts via web, and programmed using open source SW codes or tools. More and more they will fly independently from the aids of GNSS, using inertial navigation or taking benefits from other sources of positioning, as cellular grid, for example.

Swarm intelligence is no more a fantasy of some smart fiction writer nor a theoretical mathematical discipline. "Soon the sky will be full of toy-like drones flying in formation over unidentified mountains in China," *Financial Times* has published in August 2017.[‖]

This unlikely spectacle could represent a revolution in military and security affairs. The June 11, 2017, demonstration of "Swarm" technology by China Electronic Technology Corporation (CETC), a state-owned high-tech company, included 119 drones. That made it the world's largest ever swarm, according to CETC, breaking US-held record. Nevertheless, on December 31, 2019, a spectacular exhibition of 500 drones took place in Singapore, at Marina Bay Sands, during the New Year celebrations: the figures created in the sky were "coordinated" and "programmed," while swarm intelligence algorithms were used to guarantee collision avoidance and correction during flight.

Each tiny aircraft is loaded with SW and sensors capable of communicating with other drones in the swarm. Developers are working toward a future where thousands could operate in accord, identifying and attacking targets. In theory, such swarms could feature drones fitted with any kind of payload, including warheads capable of overwhelming defenses with their sheer numbers. Swarms of drones, low-tech hardware knitted together with high-tech AI SW, will become a weapon of the future.

The gamble is that they can be effective both as lethal and nonlethal weapons. Thousands of cheap, 3D printed drones, for example, could swarm aircraft carriers or jets, which currently have no countermeasures for such attacks. They can also be effective without being lethal by crossing the line into a shooting war—a valuable form of deterrence, especially for weaker countries.

Many say that mature swarm technology is still a long way off and will require developing the necessary technology to boost communication between the drones,

[‖]https://www.ft.com/content/302fc14a-66ef-11e7-8526-7b38dcaef614.

methods to keep them in the air longer, and a modern military approach and culture capable of deploying the swarms effectively. However, before addressing swarms of thousands or even hundreds of drones, thinking to deploy swarms of few (10–20) is today already feasible and this opens to a number of scenarios in our domain of interest never imagined before.

Furthermore, trials are reported of few drones (in the order of the tenths) flying in swarm using 5G infrastructure as a per-opportunity communication link, against which no countermeasure is on the market at the moment.

1.5.1 What will be the future of counter-drone systems?

Sensors will need to improve in terms of detections-classification capability: the radar needs to detect weaker signatures, to recognize the presence of a swarm and to cue the reaction properly. The EM sensors need to extend their bandwidth to frequencies unconventionally used by the threat and being able to produce information related to the topology and hierarchy of the swarm. Cameras will face with the challenge to focus not just a single target and to continue to give the right support to the human in the loop in the decision-making process.

A complete paradigm change will rise for firing or any kind of reaction. From the concept of a "single bullet for a single target," a strategy of "sparse effect" should be put in place. Jamming remains the most effective, due to its capability to affect a wide volume simultaneously, and the most efficient, in terms of cost-per-shot. Challenges are related to the new frequencies unconventionally used by the threat, availability of power, and reaction time (i.e., being in the right place at the right time).

Behind the scenes, the real improvement will be required to the C2 chain. The overload of the human in the loop will dramatically increase and need to be strongly supported by SW logics much more than today. The idea is bioinspired or better, human-inspired and seeks to emulate a "cognitive" and "autonomous" approach to the environment in terms of observing, orienting, deciding, acting, and applying experience.

As is well known, the mathematical-engineering discipline that aims to create an intelligent agent that is a machine emulating the human behavior is commonly identified as AI. Nevertheless, under this terminology, many specific disciplines are categorized, each one typically dedicated to one specific aspect of the human behavior that it is assumed to be emulated.

Generally, the C2 environment is not the ideal place for "expert systems" because it is considered an intellectual exercise traditionally associated only with human intelligence.

There are, however, specific applications that fall within the bounds of control mechanisms that meet these prerequisites.

These expert systems are embedded usually into real-time military applications, typically as part of a weapon or weapons platform, and perform such military-specific functions as battle management, threat assessment, and weapons control.

As the battle-space becomes more complex and the requirement for faster decisions and reactions increases, there will be a growing need for automated expert systems for functions as sensor interpretation and "automatic target recognition–classification–tracking," based on anomalous behavior detection systems, trained on "standard" behavior data.

Other AI-based systems, known as "decision support systems," have broader applications in the C2 environment. This is because they assist in the organization of knowledge about ill-structured issues.

The emphasis is on effectiveness of decision-making, as this involves formulation of alternatives, analysis of their impacts, and interpretation and selection of appropriate options for implementation.

These systems aid humans in mission planning, information management, situation assessment, and decision-making.

A particularly relevant application is in the area of "Data Fusion," where AI advancements in the fields of natural language, knowledge discovery, and data mining are assisting in the analysis and interpretation of the vast quantities of unstructured/structured, synchronous/asynchronous, formatted/unformatted data collected by the ever-increasing number of intelligence, surveillance, and reconnaissance assets. This cognitive approach is rapidly overcoming the traditional approach based on rules, maximum likelihood, and correlation among elements.

In addition, AI contributions to "Game Playing" and simulation are also leading to better training for personnel who must perform within the demands of the C2 environment.

More in general, by exploiting its ability to interact with humans, AI-based C2 can learn diverse behaviors (off-line and online) and eventually, into the context of military operations, can perform a number of tasks oriented to support the decision-making process (cognitive decision-making).

As an example, the communicative ability of the human brain is limited by low-bandwidth input and output, and the thinking power of the brain, while impressive, is rather slow. In the C2 context, the amount of time required to train and develop human commanders is significant and yet they still are affected severely by such prevalent factors as stress and data saturation. In these areas, as well as in more recognized areas involving human physical and emotional vulnerabilities, AI technologies will offer significant advantages.

Since command, by nature, is an intellectual activity that requires decision-making in a dynamic environment, these developments are promising elements for the progressive incorporation of AI technology within the command element of C2.

Meanwhile this introduction can be supported by means a progressively decreasing supervision activity thanks to a human–machine–teaming (HMT) approach. AI advances in human/machine interface capabilities will then help one to close the gap between humans and the machines that support them. The new trend is "human–machine–teaming" where the machine learns from the past events, decision of the human, and increases its experience and probability of success in the future (proposed/autonomous) decisions.

The still controversial question lies in whether AI technology can advance to the point where this gap no longer exists, effectively allowing machines to replace humans and perform with human intelligence.

It is at this stage that AI will complete its entrance into the realm of command.

1.6 Conclusions

In 2016, Michael Blades, research director at the market research firm Frost & Sullivan, said that the counter-drone industry was too new to offer a market estimate.

However, things have changed, and quickly: the counter-drone business is worth between $500 million and a billion dollars right now in 2020 and the market projects growth rates as high as 25 percent, with market values hitting $1.5 billion by 2023.

"I think double-digit growth is a foregone conclusion," says Blades, "just because they're starting from almost zero right now."

Many companies worldwide are proposing solutions in this emerging field, even if a lot of research and development activities are permanently in place and no vendor is able to demonstrate the maturity of a "total weapon," due to the continuous evolution of the threat.

From the United States to Russia, including across Europe and Israel, announcements of new solutions and experimental results are published every day.

In any case, every proposed solution, modular and scalable according to the operational scenarios and the needs of the final user, is several times more complex than the threat, requiring a plethora of assets deployed.

Controlling these assets will require highly qualified and best trained operators, whose mission preparation needs to be much more professional and complex compared to the relative simplicity of the threat.

The costs of the defense could therefore be magnitudes higher than the cost of the attack. This is asymmetric warfare.

However, this is not the end of the story. The availability and the easy access to guidance and propelling technologies are leading to the proliferation of UAS for countries/forces that were not historically provided with autonomous capabilities.

The expected increase in the use of UAS by symmetrical and asymmetrical forces as well as their dissemination over all kinds of theaters has effectively occurred and extended.

Acronyms and abbreviations

AI	artificial intelligence
C2	command and control
COTS	commercial off-the-shelf
EM	electromagnetic

EMS	electromagnetic spectrum
EW	electronic warfare
GNSS	Global Navigation Satellite System
GPS	Global Positioning System
MANET	mobile ad hoc network
MTOW	maximum take-off weight
NATO	North Atlantic Organization
OODA	observe–orient–decide–act
RCS	radar cross-section
SW	software
UAS	unmanned aerial system
UAV	unmanned aerial vehicle

Bibliography

[1] Marshall, D.M., Barnhart, R.K., Hottman, S.B., Shappee, E., and Most, M.T. (Eds.). (2012). Introduction to Unmanned Aircraft Systems (1st ed.). Boca Raton, FL: CRC Press.

[2] McCurley, T.M. and Maurer, K. (2016). Hunter Killer: Inside the Lethal World of Drone Warfare. Allen & Unwin, Crows Nest, NSW.

[3] Van Brunt, L.B. (1978). Applied ECM: Volume 1. Dunn Loring, VA: EW Engineering.

[4] Van Brunt, L.B. (1982). Applied ECM: Volume 2. Dunn Loring, VA: EW Engineering.

[5] De Martino, A. (2012). Introduction to Modern EW Systems. Norwood, MA: Artech House.

[6] Electro Magnetic Spectrum Operation (EMSO), www.emsopedia.org.

[7] Joint Air Power Competence Centre. (2021). A Comprehensive Approach to Countering Unmanned Aircraft Systems. Kalkar: JAPCC.

Chapter 2
Systems design considerations

David Greig[1], Stephen Harman[2] and George Matich[1]

2.1 Introduction

To design a system which can counter a particular threat you must, first understand what the threat is, what measures can be taken to counter it and who will operate the system to do this. In this chapter, we will consider these points first of all and then discuss how the various possibilities can be translated into requirements for a radar system.

Small drones (or Unmanned Air Systems – UAS – to be more formal) have become widely used and easily obtained in recent years. Many of these are bought essentially as toys or by hobbyists, others are used commercially, mainly for aerial photography and surveying. As with any human invention, it is not long before a nefarious purpose was found for new technology.

Probably the most widely covered incident involving a misuse of drones occurred over Gatwick Airport, south of London in December 2018, when drones flown in the approaches to the airport saw flights cancelled for over 20 h causing a lot of disruption and with a huge financial impact. At the time, the immediate speculation was of this being a terrorist attack or deliberate disruption by climate change protestors. When no groups came forward, speculation moved onto the disruption being due to 'state actors' but months later the police investigating still had no idea who had carried out the flights or why [1]. While large companies such as Amazon have trialled drones to be used as delivery vehicles in congested cities and their use to deliver food and medicine in remote areas suffering natural disasters is also being looked at, the most widespread use of drones for delivery so far has been dropping contraband over prison walls [2].

In war zones, commercially available drones have been used for military purposes. A small drone equipped with a camera and a downlink to a tablet can be carried by ground troops and used to see what is over the next hill. Combined with larger, high flying military drones, it is possible for information across a battlefield

---

Dedication: To George Matich

[1]Leonardo UK Ltd, London, United Kingdom
[2]Aveillant Ltd, Cambridge, United Kingdom

to be immediately available to commanders on the ground. In the recent conflict in Ukraine, drones of many types were deployed making it very difficult for troops and vehicles to be deployed without being spotted well in advance of the position they were intending to take up. Commercially available drones can also be adapted to carry small but potentially lethal quantities of explosives (like a hand grenade) effectively turning them into guided weapons. Use in this way has been widespread by various parties in Syria and Iraq.

This latter threat has led to widespread concern of the use of drones for terrorist attacks on civilians and even political assassinations. While the former has not yet happened, in 2018 an assassination attempt was made on Venezuelan president Nicolás Maduro [3]. Targeting an individual in a crowd is not very easy; rather landing an explosive device randomly in that crowd is considerably easier. The possibility of an attack on a political rally, music festival or sporting event is one of the many concerns which modern police and security services have to think about.

The first attempt to limit the threat from drones has been a legislative and technological strategy to present misuse. In the United Kingdom, anyone responsible for a drone between 250 g and 20 kg has to register with the Civil Aviation Authority (CAA) as a user [4]. Drones which fly by GPS also now have pre-programmed software which prevents the craft from going above a certain height or flying over airports or sensitive buildings like nuclear power stations. These measures may prevent accidental misuse or casual attempts at causing a nuisance, and they will not stop more organised miscreants who can bring in drones from abroad which do not have the limiting software or reprogram the drones to remove its inhibitions (and certainly will not send a form off to the CAA). More to the point, as is clear from the previous text, drones currently pose most threat in regions where legislative control has effectively broken down.

The drone has flown out of Pandora's Box and it is not going to be easily piloted back in.

2.2 The systems design challenge

There is therefore a growing need for detecting and countering threats from small drones. Counter-UASs (C-UASs) can take many forms and are very much still emerging as a technology. At the core of any C-UAS is a sensor which can at least detect, most likely track drones and if possible, identify its type and purpose. To understand what a sensor must do to help counter a drone, we must first think about the system that it will be part of. The initial 'systems thinking' involves three questions:

1. Who is the customer/end user?
2. What are they trying to achieve?
3. How much can they afford?

The answer to question 3 will depend on question 1 and will likely influence question 2. It is possible to adapt high-grade military sensors for use in countering drones but this would likely be beyond the budgets of civilian agencies and the weaponry where such systems are designed to support would be overkill, possibly literally. If police forces, prison services and security contractors require a C-UAS, then they need to be affordable, reliable, easy to deploy and easy to use. The engineering challenge in producing a low-cost system which can be used with minimal training is greater than that for a specialised system used by experienced operators. This is why innovative research is still very much needed in determining how the drone threat can be reduced and eliminated.

The system design really needs to start at the endpoint and work backwards. To do this, we need to know exactly what will be done to counter the drone once it has been detected. There is no single answer to this and what countermeasures can be deployed will depend very much on the environment a C-UAS will operate in. In a war zone, the risk of collateral damage and death or injury has a level deemed acceptable under rules of engagement. In a crowded civilian environment, the potential harm the drone could inflict must be weighed up against the possibility of the countermeasure doing more harm than the drone would have. Various methods of countering drones have been tried or proposed. These will be summarised later, with discussion on their viability and what the sensor requirements to support the countermeasure would be.

Avoidance – this is the simplest case where once a drone is detected in an area where it poses a risk that area is avoided by anyone or anything at risk. For instance, this could be closing down airspace around an airport or evacuating a crowd of people for an area they are gathered in. While this approach minimises the risk of harm, it still allows the malicious operation of drones to be disruptive which may be the desired effect, and the publicity gained could encourage others. There is also the likelihood of legitimate drone use being wrongly reported as a threat and creating a panic out of nothing.

Tracking back to launch point – finding out where the drone came from and where it returned to allows the drone users to be found and questioned. This can minimise disruption by preventing repeats of the incursion and also determine whether the usage was malicious or ignorance. To do this, the sensor needs to be able to detect and track the drone and work out the trajectory to estimate the location of the launch point.

Jamming the UAS guidance – for drones with a remote 'pilot' the link from the operator console to the drone is either jammed or in a more sophisticated approach spoofed so that guidance of the drone can be taken over. For this countermeasure, the presence of the drone in an area where it poses a risk needs to be detected and the frequency of the link between operator and drone found to get up the jamming. The jamming signal can fill a large volume of airspace so does not need to be highly directional. If the drone can be directed away from the area at risk or if it crashing poses minimal risk then this approach is effective. If, however, the drone is carrying explosives, then allowing to lose control still risks death, injury or damage – just not where the terrorists intended. Moreover, many drones can have pre-programmed flightpaths that are followed by GPS. Jamming their guidance would require jamming GPS which would have wider consequences in the area of operation which may be unacceptable and still does not get round the problem of a weaponised drone going astray.

Directed energy weapons – increasing the energy of the jamming signal to be a Radio Frequency (RF) Directed Energy Weapon (RFDEW) temporarily disrupts or even can burn out the electronics in the drone causing it to crash. If the drone is weaponised, there is the still the risk of the drone exploding where it was not aimed and still causing harm unless the RFDEW can destroy the detonator also. The RFDEW will need to be directed much more accurately than a jamming signal which will require the drone to be tracked and very likely a linking of the pointing of the countermeasure to the sensor. Even then, there is still a risk of the RFDEW damaging other nearby electronic systems, which may be unacceptable. A high-power laser would produce a more directional beam which is less likely to cause collateral damage so long as there is nothing vulnerable immediately behind the drone. The narrower beam requires very accurate steering and therefore almost certainly high accuracy tracking coupled to the pointing of the laser.

'Soft' interception – these countermeasures make physical contact with the drone to divert it or bring it down. Methods include throwing a net to catch it in or directing a high-pressure jet of water or air to know the drone off course. The sensor would be required to detect the drone in a high-risk area then track it with enough accuracy to allow an operator to direct the countermeasure at it. The same proviso of collateral damage from a weaponised drone holds here too. Another interception method demonstrated has been to train large birds of prey to catch the drone. Other than providing a warning as to when to release the raptor, it is difficult to see how a sensor would be used in this case.

'Hard' interception – as the drones appeared over Gatwick, many asked 'why can't the police just shoot it down?'. The answer to that is that this has been tried with skilled marksmen who have found it very difficult to hit a small object which can go from moving to hovering almost instantly. A high-velocity rifle bullet would certainly shatter the drone while disintegrating itself; however, if it misses it can do a lot of damage as it comes back to the ground so risky to use in the urban environment. Spreading shot (like a clay pigeon shoot) has proven more effective in hitting a drone to bring it down but will not do the same level of damage to the drone and its payload. Using a sensor to aim the gun is generally considered to be necessary over relying on a human to find and shoot down the drone. Guided munitions have been shown to be able to intercept small drones [5] but using a $3 million missile to shoot down a drone which costs £24.99 from Argos is not a widely feasible solution. Smaller, cheaper guided rockets may be developed specifically for the task. If this happens, then the guidance system of the weapon needs to be matched to the accuracy of the track the sensor used to cue it can achieve.

Drone on drone – a system which has a guidance system and can match the speed and manoeuvrability of a drone is of course another drone. The best way to bring down a drone makes to guide another drone to the vicinity of the threat drone and then let it use a camera linked into its guidance system to either crash into the threat to destroy it (possibly with a small explosive charge), to divert it of course of possible even capture it in a net.

From the previous survey, we can see that the performance the sensor must achieve varies from detecting the presence of a drone to being able to track it accurately enough to point a gun at it. In designing the C-UAS, we must also consider the following:

- The range at which the drone needs to be detected.
- The volume of airspace which needs to be searched to find where the threat may come from.

The range is driven by two things as follows:

- The range at which the countermeasure is effective.
- The amount of warning time ahead of that the operator needs to make the decision to deploy the countermeasure or not.

The volume to be surveyed depends on what is being protected from drones.

- Battlefield deployment: if there is a frontline, the direction from which drones will appear can be anticipated narrowing the search volume.
- Perimeter protection of a sensitive facility (e.g. prison or sports stadium): the direction in which a drone may appear is less easily anticipated, especially when surrounded by buildings or trees which can provide cover for the launch point.
- Approaches to an airport: in theory the volume to be protected is any direction an aircraft can take off and land from up to the height which small drones can reach – potentially a very large volume. Multiple sensors may be needed to protect such large volumes of airspace which could rapidly become expensive unless innovative low-cost solutions can be found.

Electronic Support Measures (ESM), which intercept RF transmissions and can find the direction from which they come (and estimate range if triangulation by multiple intercepts is possible), can reduce the volume to be searched or at least the direction in which the search should be started. This is only possible for drones which are remote piloted via a data-link and rather than following a pre-programmed flightpath. Even if the search volume can be narrowed by ESM, it is still likely that the drone will need to be located and tracked more precisely and identified as a threat before a countermeasure can be deployed.

The range at which the sensors need to be effective and the volume of airspace to be surveyed are not the only challenges in developing sensors for a C-UAS. A small drone is a 'difficult target' hard to spot optically and with a low Radar Cross Section (RCS). Conventional detection methods developed for larger, more reflective targets are unlikely to work against drones, not at least until they are already close enough to pose a threat. Trying to process into noise and clutter to find small objects further away using established methods would likely result in a False Alarm Rate (FAR) which would be so high as to render the system unusable to the operator without novel detection methods to determine what is the true object of interest in a scene.

Potentially as much of a problem of as false alarms are false positives – genuine objects which will be detected but which are not of interest to the end user. Birds are the most obvious entities in the environment which can be confused with drones. Trees blowing in the wind can also cause false alarms. As drones can approach a low altitude, moving vehicles and even people could also produce confusing returns. The ability to determine not only where a target is and predict where is going, but also to be able to classify what that object is and thus if it is of interest to the end user. Even the broadest classification of drone/not drone is useful in informing the end user on where most attention should be paid. The finer the categorisation of drone types can be made, then the better informed the end user will be to make decisions with confidence.

Sensor fusion can significantly increase the confidence in identifying the drone as a threat and in eliminating false positives. If a drone is communicating to a remote operator via a data-link, it will have an RF signature which can be found using ESM, while birds should not be emitting at RFs. Similarly, a drone may have a similar RCS to a bird, but a different thermal signature. The application of sensor fusion requires an accurate estimation of direction and/or position of the possible threat from each sensor so the data produced by each can be associated with the same object in a scene.

Having confidence in the information provided is as important as the other attributes of the sensor performance such as range and volume coverage – detecting more objects at longer range without knowing what they may be only adds to the workload of the operator and reduces their ability to make quick decisions. In developing classification algorithms, it is vitally important to assess how well they work in theory and in a practical environment. Robust classification can aid the decision-making process by the end user significantly. As well as having confidence that what has been detected is very likely to be a drone, having an indication what the type of drone is and hence its size and potential payload allows the end user to assess what threat the drone may pose. If it can also be determined if the drone is carrying a payload and even what that payload may be, then this would make the possible threat clearer to the end user at an earlier stage of the operation.

The combination of detection, tracking and classification of the drone allows the end user to determine the intent of the drone. Intent is what the end user needs to work out to decide what action, if any, it is appropriate to take. A small drone with no payload flying where it should not be but not approaching a sensitive area is more likely down to nuisance or ignorance than posing a threat in which case trying to find the drone user to have a word with them is more appropriate than risking collateral damage with a more extreme countermeasure. For a larger drone, with a payload approaching a football stadium immediate action with some risk is more likely to be acceptable. If multiple or even swarms of drones are deployed then intent becomes even more important in determining which of the drones pose the greatest and most immediate threat and therefore need to be countered first.

Protecting facilities and airspace around airports will require continuous monitoring of the volume of airspace where a threat could appear. Having human operators doing this becomes expensive in terms of man hours and training and it is also difficult for an operator to monitor a system for hours on end without becoming fatigued and potentially finding false threats and missing real ones. An autonomous C-UAS becomes attractive in these scenarios. It would be more likely to be semi-autonomous where the system would alert an operator to potential threats for them to make the final decision on the appropriate response. Introducing autonomy in a system requires a low false alarm/false-positive rate; otherwise, the operator ends up being as busy as if the system were manually operated. For such systems, classification and determination of intent become essential in only alerting the operator as needed.

2.3 The role of radar

The usual benefits of using radar systems for wide area surveillance – high volume and long-range coverage, day/night operation and penetration of atmosphere, weather and obscurants – hold to some extent for drone detection but with some additional challenges also. The RCS of the drone can be very small so the ranges at which they can be detected will be shorter than conventional air surveillance radars typically operate – a few hundreds of metres to a few kilometres as opposed to tens or hundreds of kilometres. While this reduces the effects of the atmosphere, a relatively high peak power will still be needed to detect a small drone at a distance which gives enough time for a countermeasure to be decided upon and deployed. As wind and heavy rain can prevent drones from being able to fly, weather penetration is also less of a concern. Drones can be operated in the dark, or through mist or smoke especially if they are following a pre-programmed flightpaths so the operator does not need visual guidance.

It is possible to adapt a conventional pulse-Doppler radar to detect drones. In 2014, a Leonardo PicoSAR compact e-scan radar [6] designed for ground surveillance from medium sized unmanned aerial vehicles, helicopters and light aircraft was used in a ground-based experiment to determine if a small hobbyist drone (a DJI Inspire I quadcopter) could be detected at a useful range (Figure 2.1). The PicoSAR operated in Moving Target Indicator (MTI) mode with a waveform modified to have a shorter pulse to reduce the minimum range for detection. An example of the data collected from the experiment is shown in Figure 2.2.

Figure 2.1 The PicoSAR radar and van mounting for counter-drone experiment against DJI quadcopter

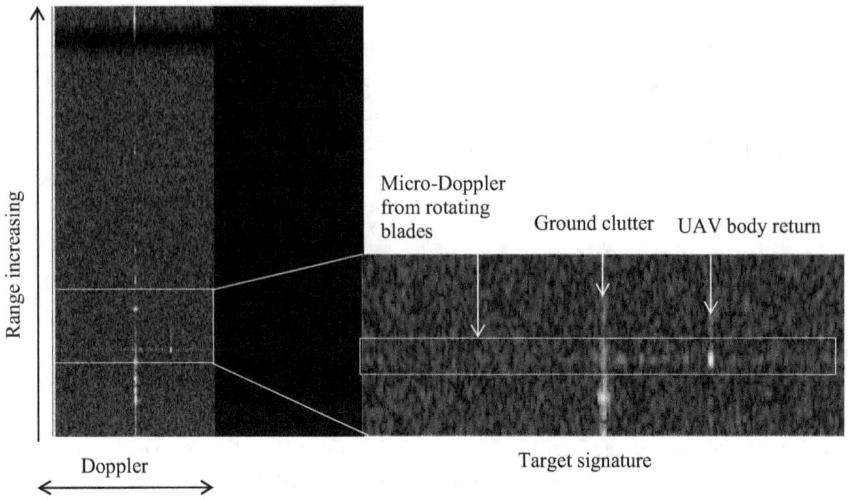

Figure 2.2 Example range-Doppler map from PicoSAR counter-drone experiment

This range-Doppler map is typical of what can be expected as a drone measurement from a stationary ground-based pulse-Doppler radar. The ground clutter is confined to a strong line at DC. The return from the body of a moving drone is separated from this line by the Doppler shift in frequency related to the speed of the drone and whether it is moving towards or away from the radar. The rotating blades also return energy to the radar which can be seen as a spectrum in Doppler related to the number of blades and their rotation rate known as micro-Doppler. During the experiment, the drone spent some time hovering at which time the return from the body became coincident with the ground clutter; however, the micro-Doppler spread out of the line at DC so the presence of the drone could still be detected.

While radars can clearly detect small drones, knowing that they are drones rather than birds or trees and what type of drone it is and what intent it has is another problem to be solved. While radars cannot produce the high-resolution images of a camera, other characteristics which cannot be seen can be measured, specifically the micro-Doppler which is a signature characteristic of particular drone types. The exploitation of this measurement for the classification of drones and the interpretation of their intent form a large part of what will be discussed in this book.

While conventional pulse-Doppler radars have been shown to be able to detect drones and measure their characteristics, they are generally optimised for long-range detection and tracking of larger objects in a sparse environment, that is relatively few targets in the volume scanned with a low FARs. Adapting such radars to provide short-range hemispherical coverage is not necessarily the most efficient, effective or economic solution. This has led to a number of concepts for novel solutions to RF surveillance systems for drones (which can still be broadly called radars). The Aveillant holographic radar (Figure 2.3) forms multiple

Figure 2.3 Aveillant holographic radar

simultaneous beams to achieve the hemispherical coverage needed to protect a large volume of airspace from incursions by drones in any direction.

In addition to design constraints, active radar systems may also be constrained by what frequency and bandwidth they can be licenced to operate at in the area they are positioned. The 'crowded spectrum' issue – the competing demand for access to the RF bands driven largely by the huge growth in mobile communications in recent years – has led to an increased interest in passive radar systems which can exploit persistent pervasive signals from other systems. This not only gets round the problem of claiming a section of the RF spectrum to operate in but can also lower the cost of owning and operating the system as it does not need to transmit reducing the complexity of the technology and the power required to operate it. The down side to this is that the power of illumination and selection of the waveform are no longer under the control of the system designer.

So what is the optimum system for countering drones? When there are so many diverse scenarios in which drones pose varying levels of threat, the idea that there is a single solution to the problem should probably be dismissed from the outset. What we can be certain of is that radar, or more broadly exploitation of the RF spectrum for detecting and identifying drones, can and will be a significant part of dealing with a growing problem. The following chapters will set out in detail concepts for how radar can be used as part of a counter-drone system. The system designer should consider each in detail and determine how they can be used to meet the requirements of their end user.

2.4 System design examples

2.4.1 Compact electronically scanned pulse-Doppler radar

Let us first consider a radar of the class of the PicoSAR radar which formed the range-Doppler maps in Figure 2.2 – a compact, electronically scanned rectangular array. The system parameters assumed for the radar we will analyse are given in Table 2.1.

Table 2.1 Typical parameters for a compact pulse-Doppler radar system

Parameter (symbol)	**Value(s)**
Operating band	X-band (9–10 GHz)
Wavelength (λ)	0.03 m (nominal)
Peak antenna gain (G_p)	25 dB
3 dB beamwidth in azimuth (unsteered)	5°
3 dB beamwidth in azimuth (unsteered)	12°
Electronic steering limits	±70°
Peak power (P_p)	200 W
Duty ratio (DR)	20% (maximum)
PRF	100 Hz to 100 kHz
Coherent processing interval (N_{CPI})	Up to 10,000 pulses
Receiver noise figure (F_N)	3 dB
System operating temperature (T)	290 K (nominal)
Losses (L)	3 dB (nominal)

These will allow us to design a waveform (or waveforms) for drone detection and classification using standard radar theory. Before doing that, we need to make some assumptions on the drone we are trying to detect. Let us assume that it is an openly available drone for the better off hobbyist or low-cost commercial use that can carry a camera size/weight payload and that it has a mean radar cross-section (σ) of 0.001 m^2. The DJI website [7] gives the maximum speed of this class of drone as 72 kph which is 20 m s^{-1}. If we assume that the drone could be carried by a strong wind of up to 10 m s^{-1}, then the range of speeds we need to be able to detect at is ± 30 m s^{-1} (with the drone opening or closing on the radar).

The starting point for waveform design is to consider adapting an existing waveform, as this has obvious benefits in both the amount of work required to develop it and the level of testing and verification which has to be applied. In turn, the cost of adapting the system could be significantly reduced making it more affordable to multiple agencies. Detecting drones is most similar to the MTI waveforms used in radar to detect ground vehicles, helicopters and increasingly people. The conventional approach in designing an MTI waveform is to assume that the pulse repetition frequency (PRF) is selected such that range rate measurement can be made unambiguously. In this case, we need the range rate ambiguity (v_a) to be 60 m s^{-1}, so

$$PRF = \frac{2v_a}{\lambda} = \frac{2 \cdot 60}{0.03} = 4000 \text{ Hz} \tag{2.1}$$

Radar systems were originally designed to detect objects as far beyond human sight as possible. It is thus conventional to operate the radar with the highest mean power it can output which means operating at the highest duty ratio than can be achieved. For this system that is 20% and the cycle time is defined by the PRF for a pulsed radar so the length of the pulse transmitted, τ, would be

$$\tau = \frac{1}{PRF} DR = \frac{1}{4,000} 0.20 = 50 \ \mu\text{s} \tag{2.2}$$

In a monostatic pulsed radar, the radar cannot receive while it is transmitting, which imposes a minimum detectable range, $r_{\min}$, given by

$$r_{\min} = \frac{c\tau}{2} \tag{2.3}$$

For a maximum duty ratio, this is 7,500 m. Where a drone with such a small RCS can be detected this far away is one consideration which must now be examined. The other is what minimum range is actually required for detecting and tracking the target. This will be driven by the minimum or optimum range at which the selected countermeasure to the drone can be deployed. There is a trade-off in a duty ratio that we need to assess.

An appropriate version of the range equation for this analysis is

$$SNR = \frac{N_{CPI} P_p \tau G_0^2 \sigma \lambda^2}{(4\pi)^3 r^4 k_B T N_F L} \tag{2.4}$$

where SNR is the signal-to-noise ratio at the receiver for a drone at range r and k_B is the Boltzmann constant. If we assume that an SNR of 10 dB is needed to detect the target, then this equation can be used to determine the maximum range at which the drone is detectable.

The selection of the length of the Coherent Processing Interval (CPI) is another factor in the trade off. The longer the CPI, the greater the SNR. The resolution in measuring the Doppler spectrum that also improves at CPI is increased, and the Doppler bin width being defined as

$$f_{bin} = \frac{PRF}{N_{CPI}} \tag{2.5}$$

This will become important in measuring the micro-Doppler spectrum of targets. Lengthening the CPI increases the time to measure a target which in turn increases the time needed to search a volume of space. We will examine that trade of later, but first of all we will assume N_{CPI}=128 which is typical for an MTI mode.

The minimum and maximum ranges for different transmitted pulse lengths are shown in Table 2.2. As expected, if the maximum duty ratio is selected, the drone can be detected furthest away; however, there is only an interval of 1,200 m before the minimum range is reached. For the shortest pulse considered, the maximum range is reduced to 3,300 m. Is this far enough away to be useful? If we consider this from the operator's point of view, an object, the size of a drone over 3 km away is well beyond what could be seen with the naked eye. At the maximum closing speed of the drone, it would still take around 2 min to reach where the system has been set up, which is a reasonable length of time for an experienced operator to decide whether the drone is a threat and what countermeasure should be deployed.

In a modern digital electronically scanned radar system, the ability to generate waveforms and instantly reposition the beam creates a great deal of flexibility which can be exploited in waveform design. To increase the time the operator has to deal with a drone, the optimal strategy may be to search the volume of concern with a longer pulse waveform with the minimum range matched to the maximum range of the countermeasure, for example 750 m, so the 5-μs pulse, which would then extend the search range and increase the time available for decision-making (in this case to around 5 km giving at warning time of nearly 3 min at maximum

Table 2.2 Variation of minimum and maximum detectable range for a drone with RCS of 0.001 m² for a pulse-Doppler MTI radar

Duty ratio (%)	Pulse width (μs)	Minimum range (m)	Maximum range (m)
20	50	7,500	8,700
10	25	3,750	7,300
5	12.5	1,875	6,200
2	5	750	4,900
1	2.5	375	4,100
0.4	1	150	3,300

closing speed). As the drone gets closer, the waveform which optimises the ability to detect and track is selected by the track estimate of position. This 'track update' waveform can be interspersed with the search waveform so that a check for further threats can be made.

If hemi-spherical coverage is needed for the counter-drone system, then three of the radar antennas can be positioned in equilateral triangle configuration as shown in Figure 2.4. Each would have its own receiver and signal processor (Rx/Px) controlled by a central processor which also includes a tracker taking inputs from each array and forming the interface to the operator display and control. As the bandwidth of the MTI waveform is around 1 MHz, each array can operate within the 1 GHz coverage of the system in X-band with suitable separation in frequency and isolation between the arrays in the construction of the system.

The duration of the CPI is

$$N_{CPI} \times \frac{1}{PRF} = 128 \times \frac{1}{4,000} = 0.032 \text{ s} \tag{2.6}$$

For an electronically scanned array, the beamwidth broadens as the cosine of the electronic steering applied. For this system, the maximum electronic steering angle needed is 60° so at the end of the scan, the beamwidth is 5°/cos(60°) = 10°. If we take an average beamwidth of 7.5°, then to cover the 120° sector, each array needs 120°/7.5°=16, beam positions. In practice, to minimise beam-shape loss and reduce FAR, we would allow something like a 50% overlap of consecutive beams, so 32 positions would be needed. The time to perform a search is then

$$32 \times 0.032 \text{ s} = 1.024 \text{ s} \tag{2.7}$$

A scan every second provides a sequence of opportunities to detect and measure the drone, which should be sufficient to form and maintain a track.

The final task before deciding on if and how to counter the drone threat is to identify the drone and assess its intent. Using the micro-Doppler spectrum of the drone to do this is the basis of most identification methods. To be able to achieve the measurement of the spectrum, a longer CPI is needed to enhance the SNR of the individual components, have high resolution in frequency to be able to separate these

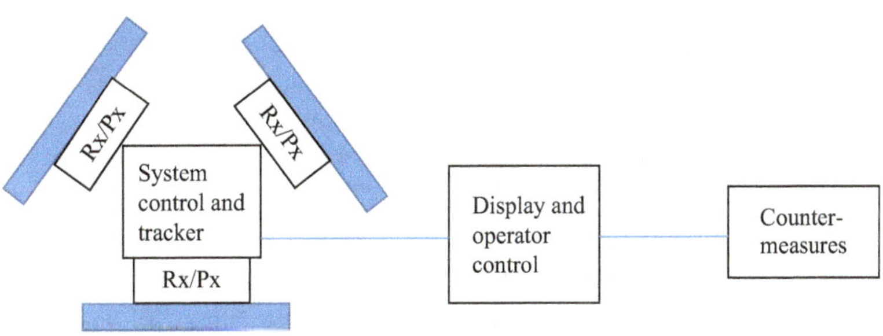

Figure 2.4 Pulse-Doppler radar based counter-drone system

components and allow time to measure the periodicity of the returns. A measurement period of around a second is generally needed to achieve this, for which there should be sufficient time after the track has been established.

This quick analysis tells us that adapting an MTI mode in a compact electronically scanned radar can provide a counter-drone capability. We are not yet in a position to go off and start marketing the radar for that purpose, the only conclusion we can make is that the radar is not unsuitable for the task. Whether it is optimum in terms of performance and cost for customer requirements now needs more detailed thought. Some points for further analysis would be as follows:

- Beam-broadening: when the antenna is scanned to 60° the broadening of the beam also causes a loss in peak gain on transmit and receive. The normal method of compensating for this loss is to extend the CPI as the electronic steering angle increases, which in turn makes the time to search the volume longer. There are trade-offs in beam overlap, beam-shape loss, CPI, etc. which need to be considered here to find the optimum waveform and control set-up to counter the classes of drones the customer has asked for.
- Angular measurement accuracy: given the wide beam-width, it is likely that monopulse measurement in azimuth would be needed to give the required angular accuracy in tracking the target. To achieve this, two channels are needed, which increases hardware and processing costs.
- Height coverage and measurement: similarly, if there is a need to determine the height of the drone to deploy the countermeasure, then elevation monopulse is also required. Two more channels separated in elevation are needed to include this. If we consider the lower part of the main beam positioned on the horizon, then the boresight will be pointing up by 6°. At a 1-km slant range, the 3-dB beamwidth in elevation covers 200 m in height. Given that the legal limit for flying a drone is 400 ft, this would seem to be adequate. Drones of the type we are considering here can fly much higher [8], which is likely for malicious use. Scanning to higher altitudes is feasible but extends the time to search the required volume if the height at which a drone approaches cannot be covered in a single scan.
- Size, weight, power and cost: the solution proposed before requires three simultaneously transmitting radar antennas with receiver–processor units. While this achieves more rapid coverage of the volume of space being protected, it will require more power, be heavier and considerably more expensive than an arrangement of three antennas which switch on and off in sequence to provide the coverage (at the expense of increasing search time by a factor of 3). Whether the lower cost system would have the capability to counter the threats, the customer is most concerned with is a matter to discuss with those purchasing the system.

Performance improvement and/or cost reduction can be achieved through innovative system designs and processing techniques. Some of these may be enhancements to a pulse-Doppler radar of the type discussed here, while others may be entirely novel concepts. The aim of this book is to show how novel approaches to the counter-drone problem can be exploited in a system. Let us first look at a different system concept.

2.4.2 Holographic radar

Now let us consider the design parameters for a holographic (staring) radar such as the Aveillant Gamekeeper 16U radar [9], which was designed primarily for use as a drone detection radar.

Ground-based surveillance radars such as Gamekeeper are required to provide high coverage extending from short to long range with high dynamic range. High dynamic range is needed for a drone surveillance radar for the following reasons:

- The low RCS return of a drone's body and even lower micro-Doppler components from rotor blades must be sensed in the presence of high stationary clutter and both ground and air targets with high RCS, which are often present where drones fly. For a typical installation, stationary clutter from buildings can be as high as 50 dBm^2 (at L band), with drone micro-Doppler components being 90 dB smaller.
- The surveillance objective is to persistently sense drones over the whole of the radar's instrumented range, which for ground-based drone surveillance radar can represent a significant span. The radar range equation tells us that received power varies as $1/R^4$, so the power received from a target at 200 m is 80 dB greater than the same target at 20 km.

Therefore, low duty, low PRF waveforms (where only targets with a range delay of less than one PRI are considered) must be used owing to such a high radar system dynamic range requirement, which is up to $90 + 80 = 170$ dB for sensing drones.

The Aveillant Gamekeeper radar operates at a lower frequency radar than the PicoSAR radar, in which such operation benefits from a waveform response with greater combined unambiguous range and velocity space, since Doppler frequency shift is proportional to the illumination frequency. Therefore, at L band, a low PRF waveform that can cover a long-range swath can also unambiguously sense the entire range of likely Doppler frequencies returned from drones.

For example, a PRF of 8,000 Hz provides a maximum unambiguous range of

$$r_a = \frac{c}{2PRF} = \frac{3 \times 10^8}{2 \cdot 8{,}000} = 18{,}750 \text{ m} \tag{2.8}$$

which consequentially provides unambiguous Doppler frequency coverage (noting complex sampling in radar) of

$$f_d = \pm\frac{PRF}{2} = \pm\frac{8{,}000}{2} = \pm 4{,}000 \text{ Hz} \tag{2.9}$$

and which relates to maximum unambiguous velocity of

$$v_a = \frac{\lambda f_d}{2} = \frac{0.24 \pm 4{,}000 \text{ Hz}}{2} = \pm 480 \text{ m s}^{-1} \tag{2.10}$$

where ± 480 m s^{-1} is a more than adequate radial velocity range covering all types of targets encountered by a ground-based surveillance radar. Given this information, choosing a lower PRF may sound attractive to provide greater maximum unambiguous range for protection against the risk ambiguous returns from large targets (e.g. airliners) that could alias into and appear to be smaller targets within the closer unambiguous coverage range. However, a greater maximum unambiguous range comes at the cost of either increased minimum range or a reduced duty ratio, for example given a 200-m minimum range requirement the maximum duty ratio achievable drops from 1% to 0.5% when the PRF is reduced from 8,000 to 4,000 Hz, since the maximum achievable duty ratio relates to the minimum range:

$$DR = r_{\min}\frac{2\,PRF}{c} = \frac{r_{\min}}{r_a} = \frac{200}{18,750} = 1.07\% \tag{2.11}$$

Therefore, since duty ratio affects the sensitivity of a radar, it is more advantageous to use signal processing, tracking and target classification techniques for the mitigation of ambiguous target returns.

Holographic radars differ to other surveillance radars in that they do not scan, their transmitters illuminate their entire 3D coverage sector persistently and multiple simultaneous receive beams are formed to cover the entire 3D coverage sector, see Figure 2.5. This affords three key advantages for drone detection, tracking and recognition:

- Persistent coverage provides very low minimum detectable velocity, enabling slow moving target detection. This is achieved because long coherent dwell times in all 3D sectors simultaneously provides very high velocity resolution, which in turn allows moving targets to be detected against a background of high levels of stationary ground clutter encountered by all 'near-to-ground' surveillance radar.
- The high-resolution Doppler frequency provided by long coherent dwells allows for high-resolution sensing of the micro-Doppler signature returned by a drone's rotor blades, thereby enabling effective recognition.
- Persistent coverage provides high detection rates on all targets, which is essential for effective tracking of drones given their ability to follow agile trajectories. Scanning radars are fundamentally limited in this aspect by their scan rate: such that when scanning radars switch to modes using long coherent dwells to provide the preceding two properties, the scan rates of such radars can be very slow, so detections from any one target will be infrequent.

The disadvantage that holographic (staring) radars have with respect to scanning radars is that transmitter antenna gain is considerably less than a scanning equivalent, given that the transmitter beam is required to persistently illuminate the entire 3D coverage space, whereas a scanning radar can focus its beam on the narrow angular sector which is being sampled any one point in time. In theory, this disadvantage is counteracted by holographic radars being able to coherently integrate over a time period equivalent to the comparative scanning radar scan time; therefore both radars receive the same energy over a defined period. This does,

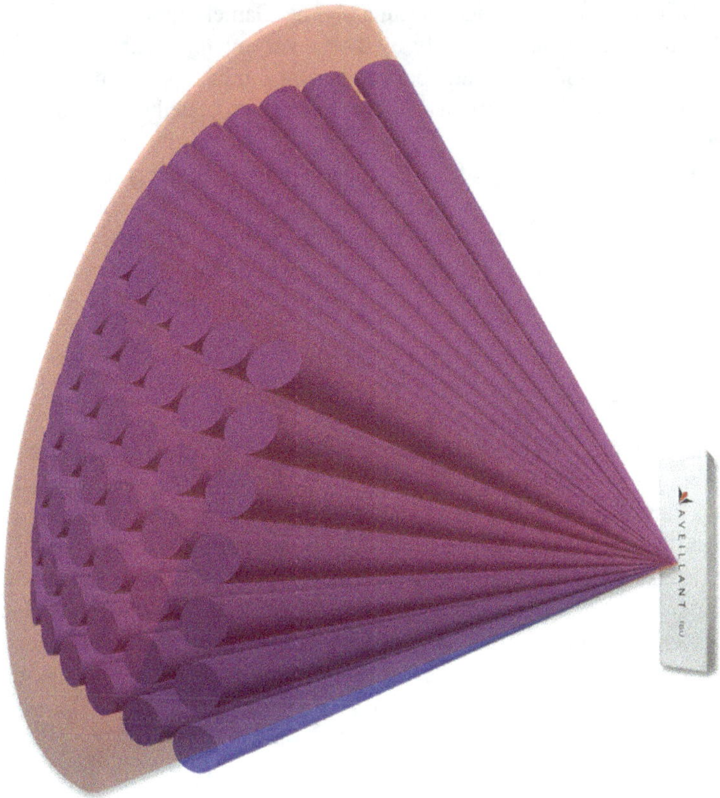

Figure 2.5 Transmit beam pattern (red) and multiple simultaneous receiver beam patterns used by a holographic radar

however, lead to the processing loads of holographic radars being greater, since long coherent dwells require long Fourier transforms which need to operate on multiple receiver beams simultaneously.

Holographic radars necessarily have array receiver antennas so that multiple receive beams can be formed, allowing angular direction to be measured. A greater number of receiver antennas provide greater sensitivity owing to the greater total antenna area and greater angular accuracy; however, the greater the number of receivers, the greater the size, the number of receiver channels and hence cost of a radar. Holographic radars are all built from similar modular components allowing flexibility in configuration of a radar to meet the required application. For the drone detection problem, the most important requirement of the antenna array, in addition to sensitivity, is that it provides good elevation accuracy. Good elevation accuracy provides the capability to determine whether a target of interest is airborne or not, particularly given the low altitudes at which drones fly. Therefore, the Gamekeeper radar has four times greater elements vertically that it does horizontally, prioritising elevation accuracy over azimuth accuracy. The Gamekeeper 16U radar has 16-receiver antenna elements vertically by 4-receiver antenna elements horizontally, in which the '16' refers to the number of receiver modules, each of which consists of two-times-two receiver antenna elements and all associated independent receiver chains and digitisers.

To review the sensitivity of the Gamekeeper 16U holographic radar, we use a similar version of the range equation as used in Section 2.4.1 but adapting it for holographic radar in which the receiver and transmitter gains differ and by adding an Array factor to account for beamformer gain when summing across the Array of receiver antenna elements:

$$SNR = \frac{N_{CPI} P_p \tau G_{tp} G_{rp} A \sigma \lambda^2}{(4\pi)^3 r^4 k_B T N_F L} \tag{2.12}$$

The approximate system parameters assumed for the holographic radar for this analysis are given in Table 2.3.

For a surveillance radar, we assume that a threshold SNR of at least 16 dB is required to detect a target, which provides an acceptable FAR given many simultaneous detection decisions are made independently over multiple ranges, azimuth, elevation and Doppler frequency cells. This threshold though atypically high also provides some margin for the inhomogeneity in background noise encountered when detecting against a complex background containing high clutter and potential EMC sources, which can be encountered in environments in which drones fly.

Table 2.3 Approximate system parameters for the Gamekeeper 16U radar

Parameter (symbol)	Value(s)
Operating band	L band (1,215–1,350 MHz)
Wavelength (λ)	0.24 m (nominal)
Peak transmitter antenna gain (Gt_p)	13 dB
Transmit antenna 3 dB azimuth beamwidth	>90°
Effective* transmit antenna elevation beamwidth	±30°
Peak receiver element antenna gain (Gr_p)	6 dB
Receiver antenna element 3 dB azimuth beamwidth	>90°
Receiver antenna element 3 dB elevation beamwidth	>90°
Number of receiver elements	64 (16×4)
Peak power (P_p)	2 kW
Typical duty ratio (DR)	1%
Typical PRF	8 kHz
Typical coherent processing interval (N_{CPI})	4,096 pulses
Receiver noise figure (F_N)	4 dB
System operating temperature (T)	290 K (nominal)
Losses (L)	8 dB (nominal)

*Note the transmit antenna is shaped in elevation to provide greater sensitivity at elevations closer to horizontal.

Comparatively high losses for the Gamekeeper radar (with respect to other radars) are quoted in Table 2.3. This loss covers a number of worst case losses that may be encountered over the entire surveillance coverage volume, which is as follows:

- Beam loss with respect to the peak gains achieved at the boresight of both the transmitter and receiver arrays; this is common to all non-mechanically scanned radars.
- Signal-processing losses, such as those caused by windows applied to time series and antenna aperture and losses in the necessary constant false-alarm rate detector, which is common to all array antenna pulse Doppler radars.

In addition to the 16 dB surveillance operation threshold, here, we also consider a lower SNR detection threshold of 10 dB, which is more appropriate for the detection of the micro-Doppler components in the return of a drone. This lower threshold is appropriate due to the following reasons:

- Further discrimination processes are applied to recognise multiple micro-Doppler components in the Doppler frequency spectrum, which reduces the risk of false alarms.
- Micro-Doppler recognition is often only used a following detection and hence such processes only operate on focussed 3D resolution cells of interest, so a surveillance assumption is not appropriate.

The parameters in Table 2.3 and the equation for SNR quoted in this section determine the maximum range at which a drone's radar return is detectable given different SNR threshold levels. Table 2.4 provides these ranges for common RCSs. For context, at L band the RCSs of readily available quadcopters are typically 0.01 m^2, whilst micro-Doppler components can be up to 20 dB below a drone's dominant return, whilst larger drones do offer higher returns.

The ranges calculated in this exercise show that Gamekeeper can provide long-range surveillance despite the small size of drones and the environments in which they are encountered. In addition, L band radar operation provides a consistency of performance, in particular being little affected by attenuation and backscatter from rain and other weather-related affects, which lends its use to applications that require higher levels of assurance.

Table 2.4 Derived range performance from the Gamekeeper 16U radar parameters

RCS (m^2)	Range ($SNR = 16$ dB) (m)	Range ($SNR = 10$ dB) (m)
0.1	13,900	19,700
0.01	7,800	11,100
0.001	4,400	6,200
0.0001	2,500	3,500

References

[1] https://www.bbc.co.uk/news/uk-england-sussex-49846450, accessed March 2021.
[2] https://www.bbc.co.uk/news/uk-england-45980560, accessed March 2021.
[3] https://www.bbc.co.uk/news/world-latin-america-45073385, accessed March 2021.
[4] https://www.caa.co.uk/Consumers/Unmanned-aircraft/Our-role/Drone-and-model-aircraft-registration/, accessed March 2021.
[5] https://www.bbc.co.uk/news/technology-39277940, accessed March 2021.
[6] https://www.leonardocompany.com/en/products/picosar-1, accessed March 2021.
[7] https://www.dji.com/uk/products/compare-consumer-drones, accessed March 2021.
[8] https://www.technik-consulting.eu/en/analysis/drone.html, accessed March 2021.
[9] https://www.aveillant.com/products/gamekeeper/, accessed March 2021.

Chapter 3

Applications of millimetre wave radar for UAV detection and classification

Samiur Rahman[1] *and Duncan A. Robertson*[1]

Millimetre wave radars operate in the frequency range of 30–300 GHz and offer several advantages over lower frequency radars: (1) high-range resolution due to the availability of wider bandwidths, (2) narrow beamwidths for a given antenna aperture, (3) high-Doppler resolution in a given integration time and (4) system compactness, mainly due to the smaller antenna dimensions. Their principal drawback is that they have shorter range coverage (usually sub-km) compared to their lower frequency counterparts, which is mainly due to the greater free-space path loss, the lower available transmitter power and the general increase in receiver noise figure that all arise at shorter wavelengths. Nonetheless, in the context of unmanned aerial vehicle (UAV) detection and classification, millimetre wave radars are attractive since their good Doppler sensitivity and resolution can yield richly detailed micro-Doppler signatures from the rotating propeller blades, which provide a strong basis for robust target classification. The emerging availability of miniaturised, highly integrated millimetre wave radars, including single-chip devices, is also enabling new UAV-mounted radar systems for sense and avoid and 'hunter-killer' applications.

A successful UAV countermeasure system must detect and track UAVs over usefully long ranges and wide areas then discriminate them from static ground clutter and other small, slow moving confusers. Whilst humans and vehicles may move at similar speeds, they are ground-based, so birds are usually the primary airborne confusers. The radar cross section (RCS) of UAVs is comparable to that of birds, as is discussed in Section 3.2. As a result the magnitude of the radar return is not a reliable discriminant; hence micro-Doppler features are more widely used for target classification. Radar micro-Doppler signatures are intrinsic characteristics of any UAV due to the propeller rotation. The concept of the micro-Doppler signature of an object and its effect on radar signal returns is thoroughly discussed in [1–3] along with the mathematical approaches for extracting micro-Doppler information from radar data.

[1]School of Physics and Astronomy, University of St Andrews, St Andrews, United Kingdom

This chapter presents a comprehensive survey of published work on millimetre wave radar applied to UAV detection and classification, covering the reported radar systems, UAV RCS characteristics and micro-Doppler signatures. Recent work on the specific signatures of UAVs equipped with a variety of threat payloads is then described. The chapter concludes with a review of classification methods that are applied to millimetre wave radar data in order to discriminate UAVs.

3.1 Millimetre wave radar systems for UAV detection

In recent years, the availability of low-cost components at millimetre wave frequencies has increased, mainly driven by the markets for automotive radar (at 77 GHz) and free-space communications (e.g. 60 GHz). Additionally, as future commercial applications of the millimetre wave spectrum emerge (e.g. next generation car radar, 5G and 6G communications), more semiconductor components and chipsets are becoming available at even higher frequencies. This has facilitated the development of radar systems for a variety of applications at these frequencies with reduced cost and greater integration. Within the wider context of UAV detection radar research [4], a number of research groups in academia and industry have specifically been exploring millimetre wave radars for detecting UAVs, and this section presents an overview of these systems. It should be noted that only radars operating at frequencies within 30–300 GHz are included in this survey. Whilst there are quite a few 24 GHz systems reported for this application, those are excluded here as this frequency is not technically millimetre wave.

Prior to the advent of consumer UAVs, the US Army Research Laboratory reported the Doppler signatures of an MI-24 Hind-D helicopter in 1998 using a 92 GHz continuous wave (CW) radar [5]. Whilst not a UAV, this was one of the first published reports of the millimetre wave micro-Doppler signatures of rotor blades. The radar, which had a transmit power of +23 dBm and used 24 dBi standard gain horns in vertical polarisation (Figure 3.1), was Doppler calibrated before measurement and the helicopter engine was modified for remote operation. Doppler spectra up to ±150 kHz were measured with a resolution of ~1 Hz.

The same institute later developed a fully polarimetric 216 GHz frequency modulated continuous wave (FMCW) radar for micro-Doppler measurements, initially for human targets [6]. The radar had a transmit power of +17 dBm for vertical polarisation and +2.6 dBm for horizontal polarisation and a chirp bandwidth of 1.6 GHz. In 2014, this radar was used for micro-Doppler measurements of a toy helicopter – the first known millimetre wave radar measurements of a UAV [7]. It is believed that this remains the highest reported frequency radar measurement of a UAV.

Warsaw University of Technology presented the use of a 35 GHz FMCW staring mode radar in 2016 [8]. The radar, known as DANIEL-35, was originally designed as a synthetic aperture radar and the front end was developed with the support of the Fraunhofer Institute for High Frequency Physics and Radar Techniques (FHR). The radar operated with +33 dBm transmit power and 1 GHz

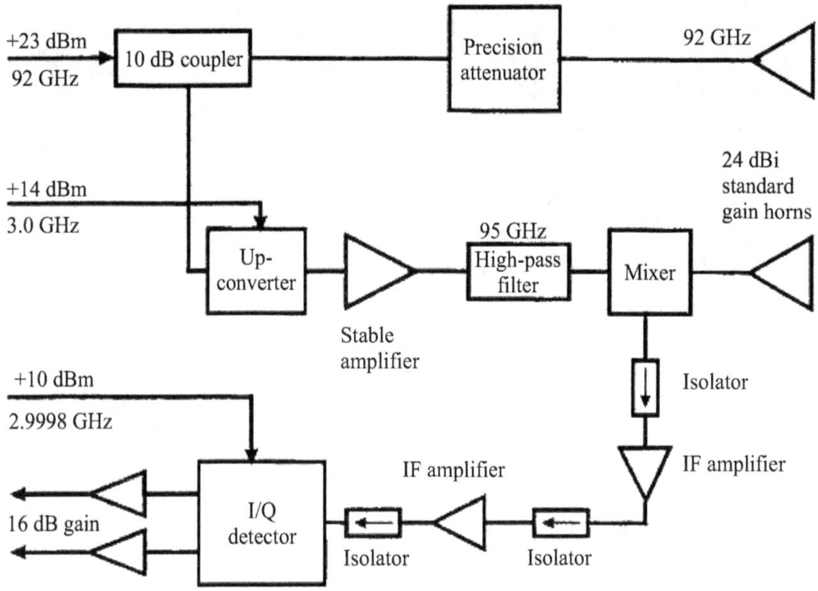

Figure 3.1 Block diagram of the 92 GHz CW radar built at the US Army Research Laboratory for Doppler measurement of an MI-24 Hind-D helicopter [5]

bandwidth, obtaining fine range resolution. Results obtained showed the radar detecting a DJI Phantom 2 quadcopter and a DJI S1000 octocopter at ranges up to 50 m, with range Doppler plots showing the micro-Doppler signatures. Whilst ISAR capability was mentioned, no such results were reported. There appears to have been no further publications arising from the use of this radar to detect UAVs.

The Fraunhofer Institute for High Frequency Physics and Radar Techniques (FHR) has been one of the most prolific institutions in this field, and they have developed multiple millimetre wave radar systems for UAV detection. Their designs frequently benefit from using state-of-the-art millimetre wave semiconductor devices, such as power amplifiers and low noise amplifiers, developed by the Fraunhofer Institute for Applied Solid State Physics (IAF). In 2015, they introduced two radar systems for perimeter surveillance and detection of UAVs [9].

The first system, named Scanning Surveillance Radar System (SSRS), is a 94 GHz FMCW mechanically scanned radar operating with +20 dBm transmit power and 750–1,000 MHz bandwidth for high-range resolution [10,11] – Figure 3.2 (left). SSRS uses slotted waveguide sectoral horn antennas with a narrow azimuth beamwidth of 1.25° and an elevation beamwidth of 13° which rotate at 8 Hz for 360° target localisation and tracking. The preliminary results showed a 2D range tracking based detection of a −20 dBsm target using RCS values, as seen in Figure 3.2 (right). The target was detected up to 150 m. The radar was also used with an optical high-resolution camera to augment target identification.

Figure 3.2 (Left) Fraunhofer Institute for High Frequency Physics and Radar Techniques (FHR) 94 GHz FMCW Scanning Surveillance Radar System (SSRS) and (right) PPI plot of real-time detection and tracking of a quadcopter using SSRS. The colour code represents the RCS of the detected UAV [9]

The second system, named Multi-Channelled Radar for Perimeter Surveillance (MuRPS), is a staring 94 GHz FMCW radar with four receive antennas in an azimuth-elevation monopulse configuration, covering 30°–60° in azimuth depending on the antenna choice – Figure 3.3 (top). The system, which was developed for detecting UAVs up to several hundred metres, operates with +20 dBm transmit power and 1 GHz bandwidth [10,11]. The monopulse architecture was selected to achieve 3D localisation of targets without the need for scanning antennas – a network of up to six units would provide 360° azimuth coverage. In Figure 3.3 (bottom), RCS-based detection and tracking results show that MuRPS can detect small UAVs out towards 100 m. The signal-to-noise challenge of detecting low-RCS targets with a millimetre wave radar out to ranges greater than 100 m using low gain, wide beamwidth antennas is highlighted in [12].

Fraunhofer FHR considered a 'radar fence' multi-sensor network for UAV detection in urban environments in 2018, with the goal of achieving 200 m range [13]. To reduce costs and maximise target observation time, two non-moving antenna concepts were considered: frequency scanning antennas and monopulse (e.g. MuRPS). An electronically steered 60 GHz substrate integrated waveguide meander antenna design is presented and used to detect UAVs, but only at very short ranges. For the monopulse approach, highly integrated, single chip, radar front ends at 30, 60 and 94 GHz are proposed and transmit–receive leakage is shown to hamper this architecture. A real-time FPGA-based detection and classification architecture is proposed. Another highly integrated 60 GHz FMCW monopulse radar sensor suitable for a radar fence, with +15 dBm transmit power and 2 GHz bandwidth, is reported in [14]. Low phase noise chirp generation, as required for high clutter urban environments, is achieved using a low-cost voltage controlled oscillator and phase locked loop architecture.

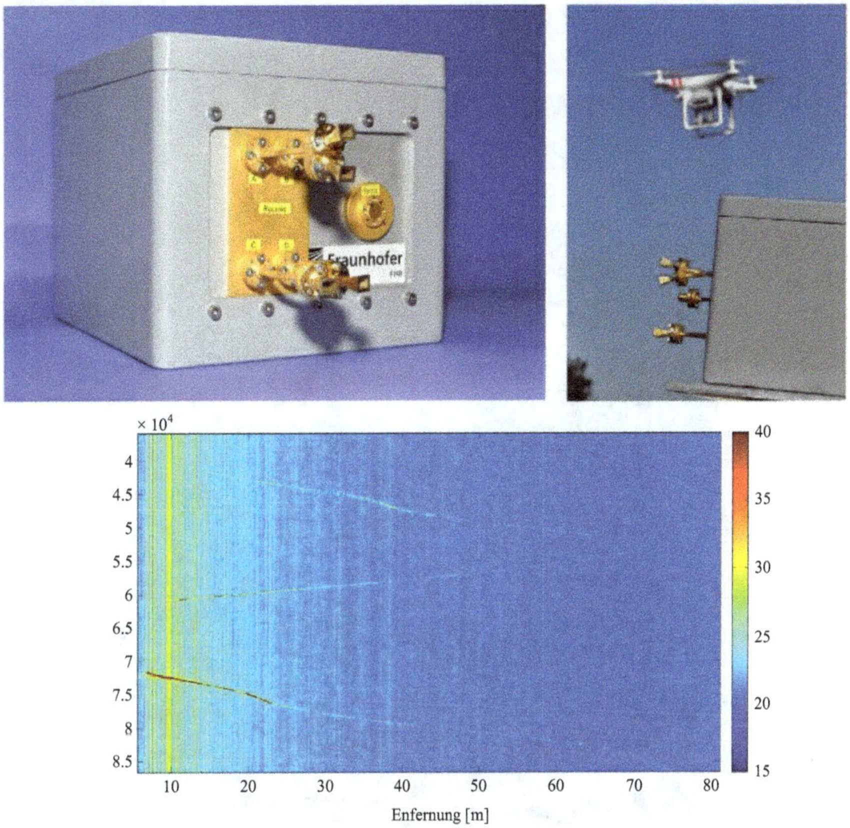

Figure 3.3 (Top) Fraunhofer FHR 94 GHz MuRPS system used for UAV detection and (bottom) range-time plot of UAV detection and tracking using MuRPS [9]

Another radar used for UAV detection by Fraunhofer FHR is MIRA-CLE Ka which is a 36.2 GHz pulsed MIMO radar with time multiplexed 16 transmit and 16 receive antennas forming 256 virtual elements (Figure 3.4) [15]. It was originally developed to detect slowly changing subcentimetre deformations in hillsides or walls. In 2016, this radar was modified for UAV detection which requires the rapid measurement of fast motions [16,17]. Only 8 transmit antennas were used (128 virtual elements) and the radar operated with 400 MHz bandwidth. The wide beam, staring MIMO radar yielded azimuth-range-Doppler data from a hexacopter at ranges of 100–150 m with a signal-to-noise ratio (SNR) of ~20 dB. The challenge of detecting slowly moving UAVs in the presence of non-stationary clutter (e.g. wind-blown trees) is noted. Doppler analysis of the targets was also done for classification purposes. In [18], MIRA-CLE Ka was used alongside optical and acoustic sensors within a multi-sensor UAV detection trial. Multi-copter targets were detected out to ranges of 200–300 m.

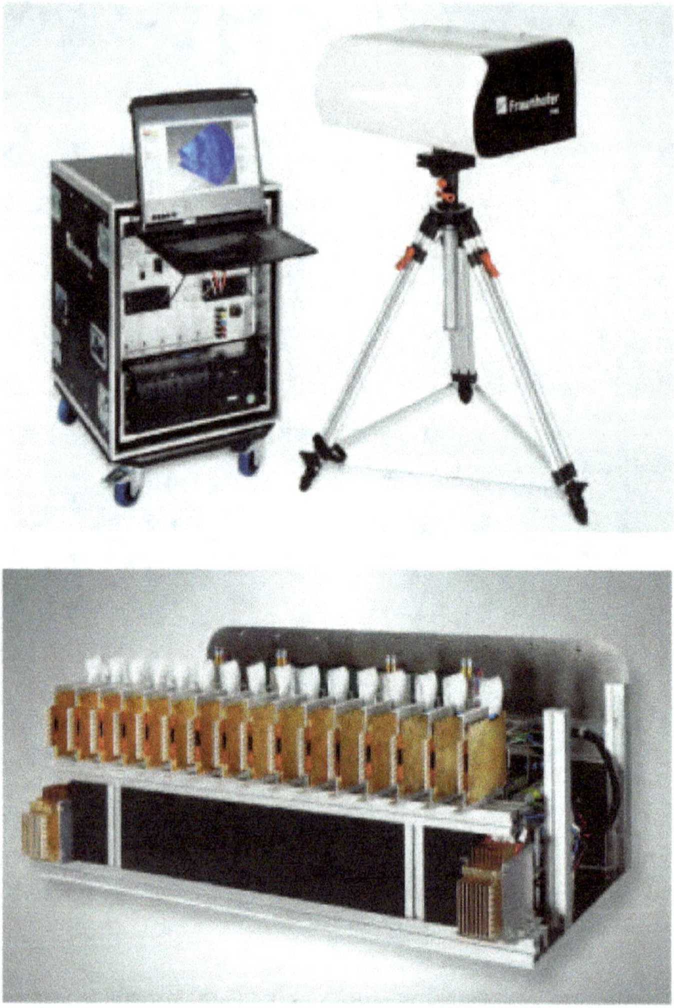

Figure 3.4 Fraunhofer FHR MIRA-CLE Ka radar system [15]

In 2017, 35 GHz radar measurements of UAVs are reported by the University of Siegen and Fraunhofer FHR [19]. The single-channel radar was assembled using parts from a 3D imaging MIMO radar within the Radar Warning and Information System (RAWIS). It operates in CW mode with +30 dBm transmit power and uses two horn antennas with 20 dBi gain. The radar was used to gather micro-Doppler data with a maximum Doppler of ±25 kHz from four different types of quadcopters, one octocopter, one small helicopter and one fixed-wing plane. The gathered data was processed to obtain spectrogram, cadence velocity diagram (CVD) and cepstrogram plots. Singular value decomposition (SVD) was used for feature extraction and

Figure 3.5 University of St Andrews W-band radar system T-220 [21]

then support vector machine (SVM) for classification training. The classification accuracy range for different UAVs was 96.2%–100%.

The University of St Andrews reported using several 94 GHz radars to measure UAVs in 2017. A basic CW radar based on a 94 GHz Gunn oscillator was used to collect blade flash micro-Doppler signatures of a single-propeller blade in the laboratory [20]. The same paper presented bulk and propeller micro-Doppler signatures of a DJI Phantom 3 UAV flying at a range of 120 m, collected with a 94 GHz FMCW radar called NIRAD, using 600 MHz bandwidth and a 6.9 kHz chirp repetition frequency (CRF). In [21], the authors used another 94 GHz low phase noise radar called T-220 (Figure 3.5), to collect CW data of a flying UAV and a bird-like biomimetic UAV. A sampling frequency of 100 kHz was used to collect unambiguous micro-Doppler data, allowing individual blade flashes to be obtained from a single propeller on a Phantom 3 at a range of ~20 m. Very recently, the same authors have reported a low noise, compact 77 GHz FMCW-Doppler radar, designed specifically for drone detection, which uses elevation fan beam antennas that rotate 360° in azimuth [22]. The radar is capable of detecting a Phantom 3 out to ~300 m based on SNR and classifying as a UAV out to ~150 m.

Details of the classification are not given for commercial reasons, but it is based on micro-Doppler and is reported to work even for drones with zero bulk Doppler, i.e. hovering.

A lab-based homodyne CW radar was built at Gwangju Institute of Science and Technology and used to collect data on 1, 2 and 4 propeller blades in 2018 (Figure 3.6) [23,24]. The system comprises MMIC transmit, receive and down converter modules and operates at 92.16 GHz with +25 dBm transmit power and uses 23 dBi horn antennas. The micro-Doppler signatures obtained were processed to extract blade length and rotation rate.

A prototype UAV detection system was developed by KAIST to investigate target fluctuation characteristics [25]. This is a 34.5 GHz chirp-pulse radar using three pulse widths for range ambiguity resolution. The radar transmit power is +23 dBm and the bandwidth is 75 MHz. A Cassegrain antenna is used and whilst the gain is not specified, it appears to be of order 30 cm in diameter. The radar easily detected a large hexacopter out to 850 m. Doppler from the propeller blades was measured out to 240 m with the blades rotating in the conventional horizontal plane and out to 1,000 m when the drone was orientated so the blades rotated in the vertical plane, presenting a larger surface area. The report proposes that the fuselage of a UAV is a non-fluctuating Swerling 3 target that would benefit from coherent integration whilst the blades are fluctuating Swerling 4 targets that would benefit from non-coherent integration. However, no statistical analysis of experimental data is presented to confirm the proposed Swerling types. The chapter concludes that for better detection performance, both coherent and non-coherent integrations should be implemented due to the low SNR and rapid RCS fluctuation of the UAVs.

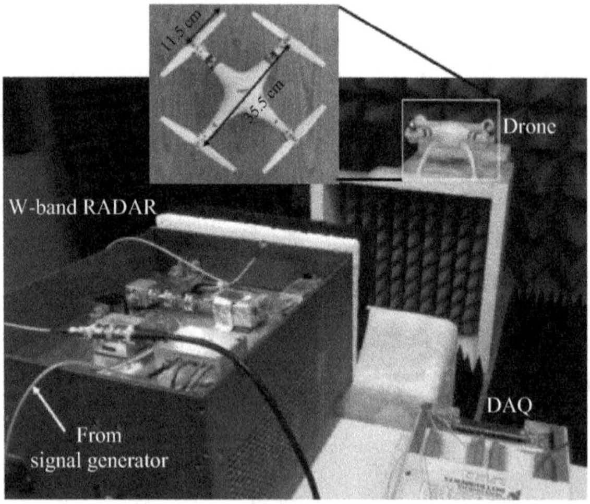

Figure 3.6 Gwangju Institute of Science and Technology W-band lab-based CW radar for propeller blade signature analysis [23]

The Technische Universität Ilmenau reported a 34–35 GHz single-input–multiple-output bistatic radar configuration in 2018 [26]. A 12-bit pseudo-random binary sequence was used as the modulating signal and a maximum-likelihood parameter estimation method was used to determine target range and bearing. The low link gain available limited the measurements of various UAVs to less than 20 m.

XY-Sensing Ltd, a spin-off company from Warsaw University of Technology, reported using a 122 GHz radar system for UAV detection, named XY-DemoRad in 2019 (Figure 3.7) [27]. The system is based on the single-chip radar transceiver developed by Silicon Radar GmbH [28] which was integrated with a DSP unit. This FMCW radar operates with −3 dBm transmit power and maximum bandwidth of 2 GHz and uses a lens to augment the integrated patch antennas. The demonstrator system comes with a built in GUI for operation and trial results show the UAV being detected at ~11 m with negligible micro-Doppler signature. Some very close range measurements (~2 m) were also performed with this system and the micro-Doppler signatures were reported in [29].

A system level analysis of a proposed 37.5 GHz GaN-based FMCW 2×2 MIMO radar for UAV detection was published in 2019 by the University of Florence – Figure 3.8 (top) [30]. Design and simulated performance are presented for the power amplifier, low noise amplifier and zero-IF mixer GaN ICs. The proposed radar design would have +36.8 dBm transmit power, receive antenna gain of 18 dBi and transmit antenna gain of 12 dBi and a bandwidth of 1.5 GHz. System level simulations are shown for different numbers of MIMO virtual elements (4×16 and 2×2), indicating that the performance would increase with more virtual elements. The critical aspect of minimising transmit–receive leakage is emphasised – Figure 3.8 (bottom) – and can be mitigated by using more MIMO elements. No measurements of UAVs are reported.

Recently, the University of Coimbra proposed a slightly different approach to detecting UAVs: using a small UAV-mounted radar to detect and track other UAVs [31]. They selected a 76–81 GHz Texas Instruments AWR1642 2D radar with two

Figure 3.7 (Left) XY-Sensing Ltd XY-DemoRad radar hardware and (right) radar hardware enclosed in cover [27]

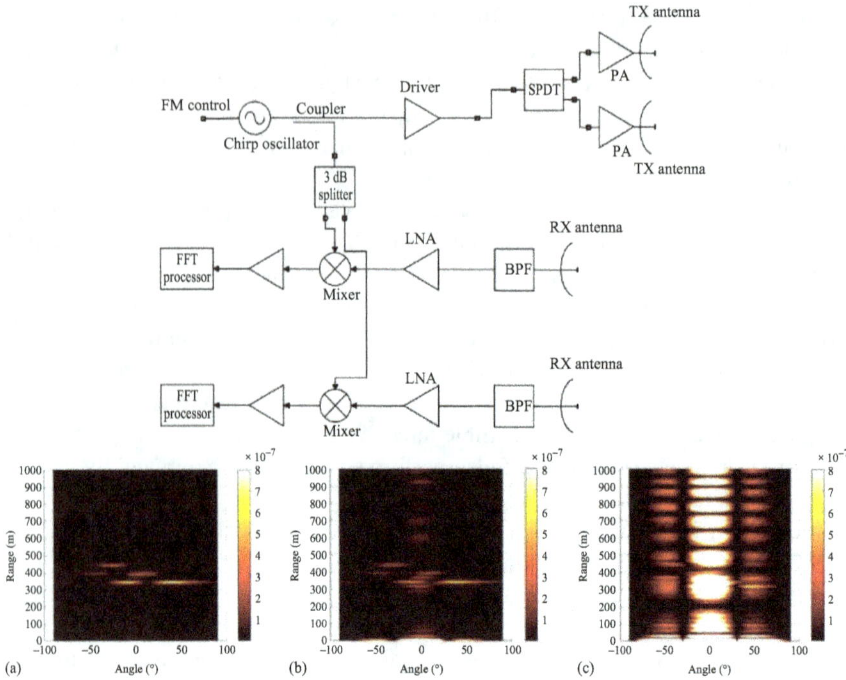

Figure 3.8 (Top) Block diagram of a 2×2 GaN-based FMCW MIMO radar from University of Florence and (bottom) a 2×2 MIMO radar image related to three identical UAVs, with an average RCS of −10 dBsm, with different levels of TX-to-RX leakage: (a) −70 dB, (b) −50 dB and (c) −40 dB [30]

Tx and four Rx antennas, which yields range and azimuth information, and mounted it on a Sky Hero Spyder X4 UAV. An Extended Kalman Filter algorithm, which makes a use of the changing radar position, was used to track and localise the target UAV. During the experimental trial, the target UAV was a smaller Sky Hero Little Spyder and the system was able to detect it up to 25 m. The AWR1443 3D radar version (range, azimuth, elevation) was also used but was not sensitive enough for detection. In [32], a 3D position estimation method based on successive 2D measurements by the manoeuvring pursuer UAV is proposed and validated with experimental results obtained at similar range.

Several companies have been marketing millimetre wave radar systems for UAV detection. ELVA-1 Ltd offers a 76 GHz radar [33], Texas Instruments are promoting their 76–81 GHz single-chip-based evaluation modules [34] and Plextek have developed a miniature 60 GHz radar for UAV-mounted situational awareness [35]. Russian firm RTI recently announced that they are developing a submillimetre wave (or terahertz) radar for detecting miniature UAVs [36].

3.2 Millimetre wave RCS of UAVs

Knowing the RCS of a given type of target is critical information for radar system designers. Detailed characterisation is needed to determine the static RCS as a function of polarisation, aspect angle and bistatic angle plus the dynamic statistics of how it fluctuates as the target moves. The RCS of UAVs is known to be low at microwave frequencies, principally due to their small size and typically low metal content – they are constructed mostly from plastics and carbon fibre which reflect weakly [37]. Their complex shape confers strong variations in RCS with angle. To date, there are only a few reports on RCS modelling and measurement of different types of UAVs at millimetre wave frequencies and an overview of those studies is given below.

In 2017, Ilmenau University of Technology and Saab Medav Technologies GmbH performed wideband measurements of Ka-band static RCS of three types of unspecified UAVs (quadcopter, hexacopter and helicopter, in that order of increasing size) [38]. The data collection was done in an anechoic chamber for full polarimetric backscatter behaviour investigation with respect to different UAV shapes, azimuth angle and bistatic angle. The frequency range was 30.4–37.1 GHz, azimuth angle range was full orientation (0°–360°) and the bistatic angle range was 30°–150°. The results were analysed to find out the influences of these factors. In Figure 3.9, the effects of the orientation angle and frequency are clearly observed and the RCS values of the UAVs vary from −50 to −20 dBsm. The results confirm the general trend of greater RCS with increasing UAV sizes. The effect of polarisation is strongly evident in the bistatic data. Unlike monostatic measurements where the scattering matrix is typically symmetric, for larger bistatic angles, distinct differences in the cross-polar components become visible. In Figure 3.10, it can be seen that backscattering along the side arms of the rotors produces cross-polar components, which are useful features for target detection. Hence, it was proposed that a polarimetric joint processing of co- and cross-polar data is required for robust UAV detection.

Comprehensive quasi-monostatic RCS measurements of nine different rotary wing UAVs using a 26–40 GHz VNA based radar were reported in 2020 by Université Catholique de Louvain, VTT, Aalto University and NYU Wireless [39]. As seen in Figure 3.11, the antennas were mounted on a mast at one end of the anechoic chamber and their RCS measured in the lower hemisphere, mimicking the view from below as if seen by a ground-based radar. Note that the measurements are reported as a function of the positioner angles of azimuth (θ) and roll (φ). Since the UAVs were mounted sideways on the positioner, these axes correspond to the elevation and azimuth angles respectively of the UAV with respect to the in-flight vertical axis, as viewed by a radar. The transmit antenna was a standard gain horn whereas the receive antenna was a dual polarised Vivaldi horn. The UAVs were divided into two groups: (1) small UAVs made of plastic materials (DJI F450, Phantom 4 Pro, DJI Mavic Pro Quadcopters and Kyosho helicopter) and (2) large UAVs made mainly of carbon fibre (DJI Matrice 100, custom built hexacopter,

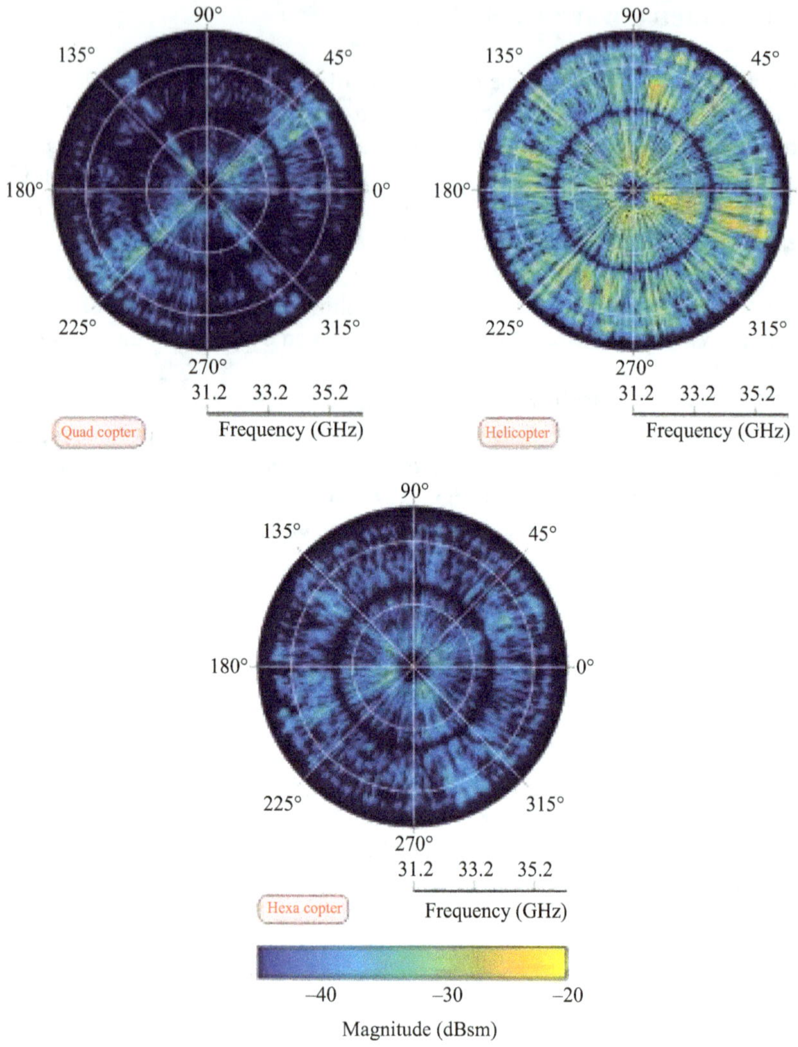

Figure 3.9 Static RCS of different UAVs versus incidence angle and frequency across 30.4–37.1 GHz for 30° bistatic angle and VV polarisation. The dark ring in the middle of the frequency axis is due to a hardware filter that splits the wideband measurement into two sub-bands [38]

HMF Y600 and Walkera Voyager 4). A 10,000 mAh 6-cell LiPo battery was also measured. Figures 3.12 and 3.13 illustrate the RCS characteristics with respect to the positioner azimuth and roll angles at 38 GHz for the group 1 and group 2 UAVs, respectively. Statistical analysis of the RCS values over the measured angles shows that the mean RCS of group 2 UAVs was ~7 dB higher than for group 1 and the maximum value is 10–20 dB higher for group 1, which is expected as those are

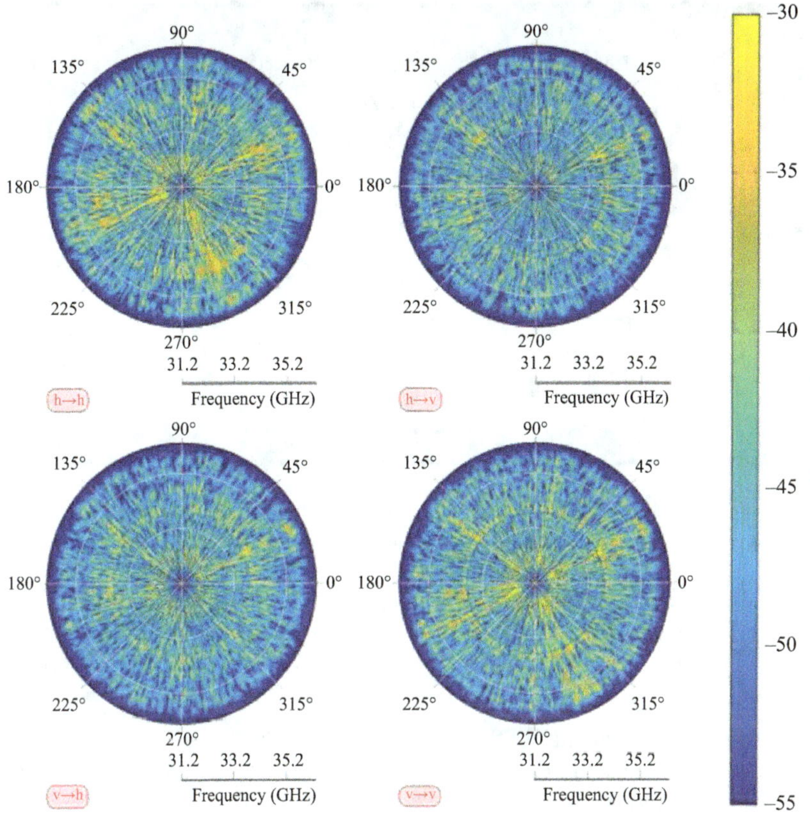

Figure 3.10 Static RCS in different polarisations of the hexacopter versus aspect angle and frequency across 30.4–37.1 GHz for 30° bistatic angle [38]

larger and made of carbon fibre which reflects more strongly than plastic. The mean RCS at 40 GHz of the group 1 UAVs varied from −15 to −11.9 dBsm at 40 GHz and from −13.6 to −6.1 dBsm for the group 2 UAVs. The mean RCS increases slightly with frequency (at ∼0.25 dB/GHz) irrespective of UAV model and the standard deviation is consistently ∼6 dB irrespective of frequency or UAV model. Also, no significant variation was found between HH and VV measurements. The report presents a very comprehensive table of the mean, standard deviation and maximum RCS values of all the UAVs for all frequencies and polarisations (mainly co-polar). Cross-polar values are only available for the M100, showing ∼7 dB lower mean RCS than the co-polar values. The measurement results are available as an open database, which is very valuable to researchers in this field.

For system performance modelling and real-time classification, if RCS values are to be used as one of the features, it is important to know the dynamic RCS characteristics of UAVs as they fly and manoeuvre, plus those of any confusers

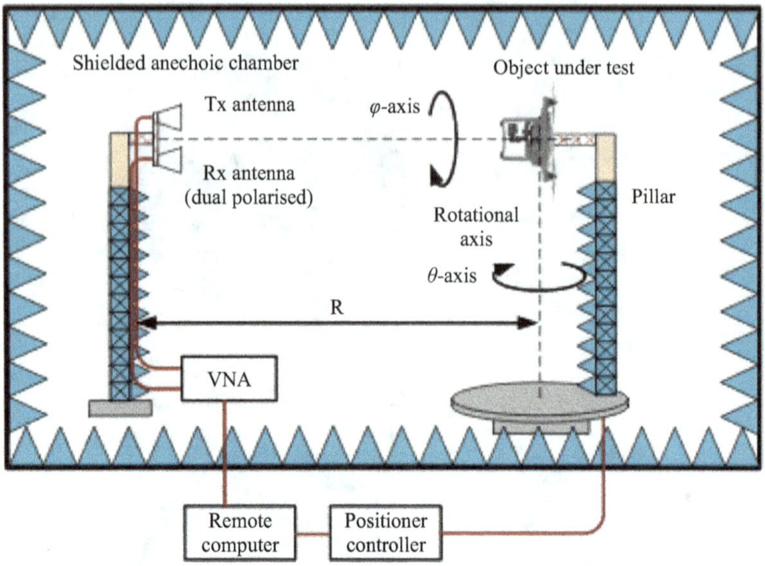

Figure 3.11 Schematic view of the RCS measurement set-up based on a 26–40 GHz VNA based in an anechoic chamber from [39]. Note that θ and φ are the positioner azimuth and roll angles, respectively

such as birds. The dynamic RCS fluctuation of UAVs at high microwave frequencies can be quite significant as they are electrically large, complex structures of distributed scattering elements.

Detailed measurement and analysis of the in-flight RCS of UAVs and birds at W-band were reported by the University of St Andrews in 2019 [40]. Two coherent 94 GHz radars, previously used for vibrometry and micro-Doppler studies of various targets, were used during the experimental trials and were amplitude calibrated to within ±1 dB. The first system is a low phase noise, heterodyne 94 GHz FMCW radar called NIRAD [41]. The transmit power is +20 dBm. It has a circular polarised (CP), 42.5 dBi gain single pencil beam antenna (0.74° azimuth and 0.8° elevation beamwidths). Owing to the very narrow beamwidth in both planes, NIRAD was only used for hovering target measurements. The second system is a very low phase noise, homodyne 94 GHz FMCW/CW radar called T-220 [42]. The transmit power is +18 dBm and its noise figure is 6 dB. This radar has a dual fan beam antenna (0.9° azimuth and 3° elevation beamwidths) with 40.5 dBi gain and operates in single-bounce CP. Both radars were configured to use 150 MHz bandwidth for 1 m range bins. Three different sizes of UAVs were used, with the following details:

- *DJI Phantom 3 Standard* (weight 1.216 kg, width 35 cm, blade length 13 cm)
- *DJI Inspire 1* (weight 2.845 kg, width 58 cm, blade length 34.5 cm)
- *DJI S900* (3.3 kg, width 90 cm, blade length 38.1 cm)

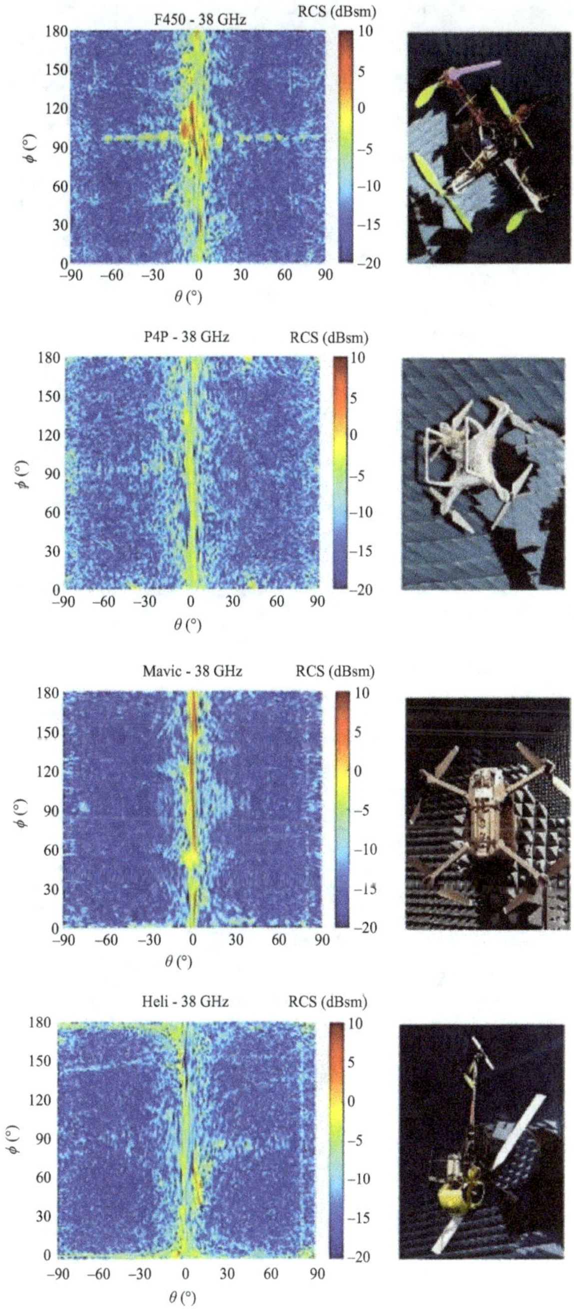

Figure 3.12 RCS values of UAVs from group 1 with respect to positioner azimuth (θ) and roll (φ) angles at 38 GHz [39]

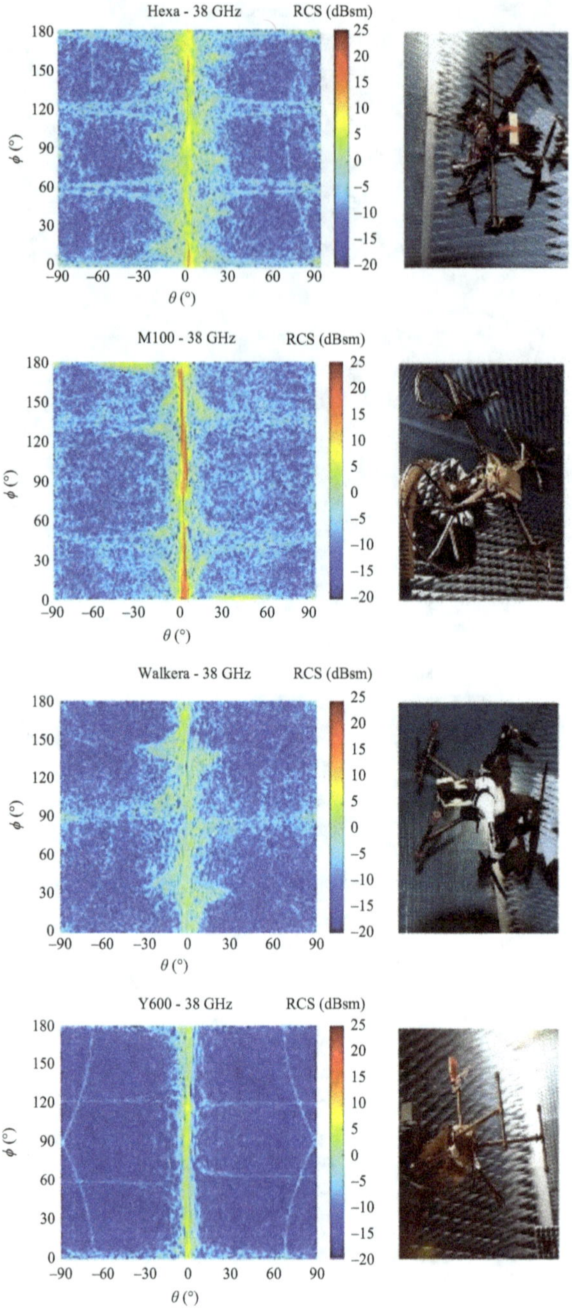

Figure 3.13 RCS values of UAVs from group 2 with respect to positioner azimuth (θ) and roll (φ) angles at 38 GHz [39]

The UAVs were always flown at least 7 m above the ground and no further than 150 m range to minimise clutter returns from the ground and distant trees. Care was taken to ensure that only those data were considered where the targets were well within the radar beam. On average, a typical measurement run was about 5–7 s long, accumulating more than 20,000 chirps in a single dataset for any given radar. The data was then processed to generate RCS histogram plots along with the corresponding empirical cumulative distribution function (ECDF) plots being overlaid.

Example RCS results for the UAVs at 94 GHz are shown in Figure 3.14. In Figure 3.14(a), the Phantom 3 was hovering 102.5 m away from the NIRAD radar and 12 m above the ground. The RCS modal value is −23 dBsm and the maximum is −15 dBm. A secondary peak in the histogram is observed at around 6–8 dB higher than the modal value. This phenomenon is often observed in the case of the Phantom but not all the time and not for the other UAVs. It is suggested that this is because the propeller blades of the Phantom, which are quite close to each other and to the fuselage, sometimes line up and increase the overall return signal. Figure 3.14(b) shows RCS values obtained with T-220 when the Phantom was flying radially back and forth between approximately 60 and 80 m and at least 7 m above the ground. The RCS modal value is −20 dBsm, the distribution is wider and extends to higher values and contains subsidiary peaks. This is attributed to the greater range of static RCS values available when viewed from a diversity of aspect angles, as seen in the anechoic chamber measurements described earlier. Figure 3.14(c) shows example RCS distribution plots of the hovering DJI Inspire 1 at 94 GHz measured with T-220 in which the modal RCS is −16 dBsm. Figure 3.14(d) shows example RCS values for the flying S900 at 94 GHz measured with T-220 in which the modal RCS is −9 dBsm. These results confirm the trend reported in the static RCS publications in [38,39] that the RCS increases with increasing size of the UAV – this is illustrated with overlaid histograms and ECDF plots for the three UAVs in Figure 3.15(a). In Figure 3.15(b), two example RCS distribution plots are shown for the DJI Inspire 1 at 94 GHz measured with T-220. The UAV was hovering at 85 m range and at 8 m altitude. The histograms compare the Inspire hovering with the body stationary and when it was rotating slowly through 360° in azimuth whilst hovering. The RCS modal value increases by about 2 dB and the histogram width increases by about 5 dB when the whole fuselage is rotating on the spot. The Inspire has a relatively asymmetrical geometrical shape compared with the other two UAVs (it has a long, thin fuselage) and so has a greater variation in RCS with azimuth angle than the other two UAVs.

These results clearly illustrate the fact that in real scenarios the RCS values will fluctuate substantially due to the dynamics of these electrically large, complex targets. Table 3.1 provides a summary of the 94 GHz in-flight RCS values obtained from multiple measurement runs for the three UAVs. Whilst these results were obtained with odd-bounce CP, the values are expected to similar for VV and HH with little polarisation dependence, as reported in [39]. A remaining open question which has yet to receive definitive analysis is which Swerling model best describes the RCS fluctuation statics of UAVs at millimetre wave frequencies? Whilst not

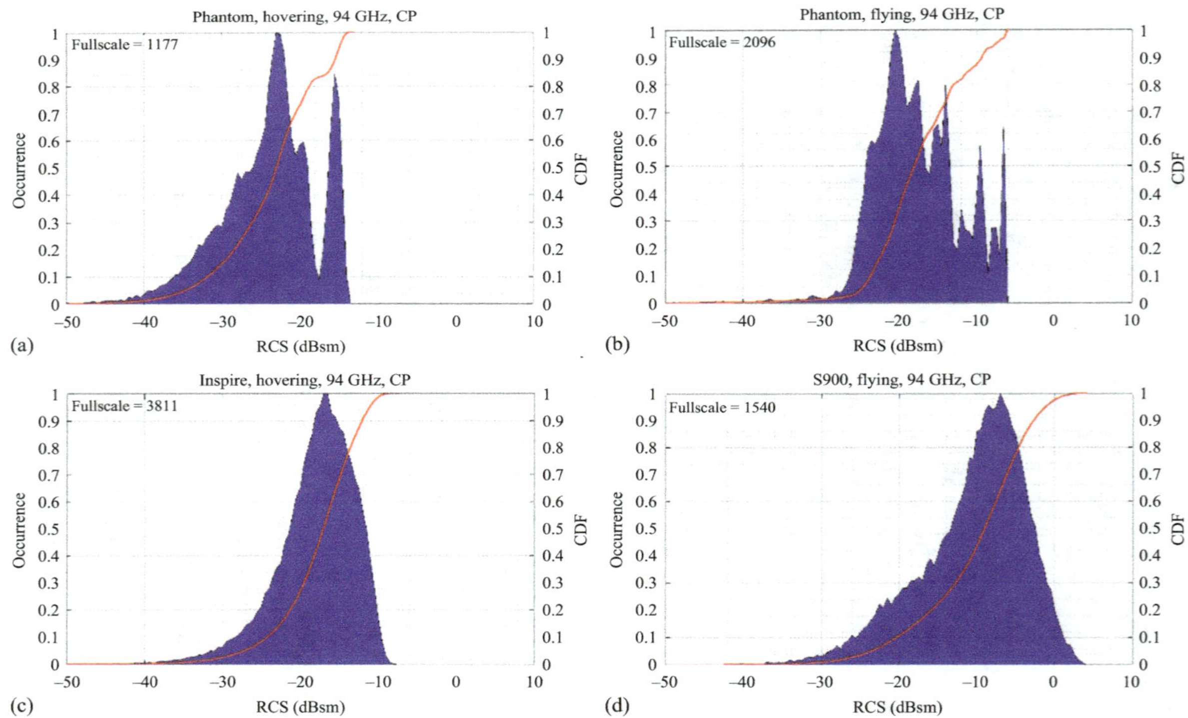

Figure 3.14 *(a) 94 GHz RCS histograms and the corresponding ECDF plots of hovering DJI Phantom 3 Standard, (b) flying DJI Phantom 3, (c) hovering DJI Inspire 1 and (d) flying S900 [40]*

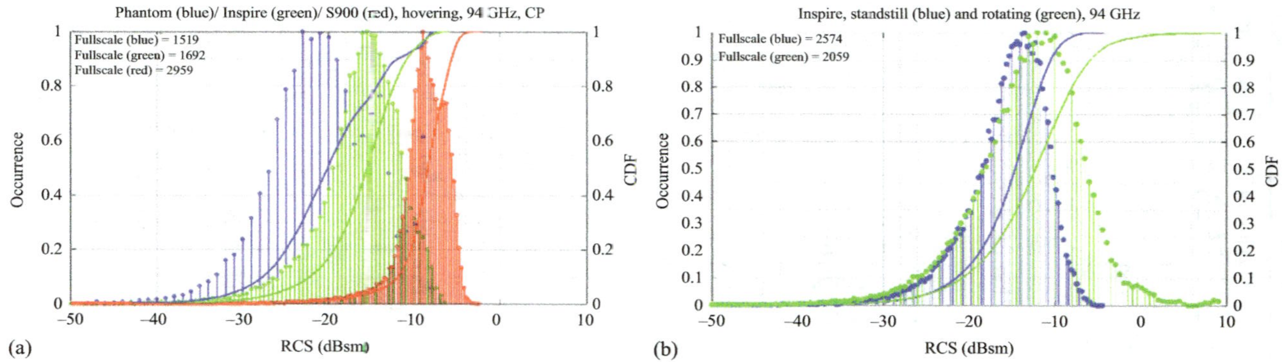

Figure 3.15 94 GHz RCS histograms and the corresponding ECDF plots of UAVs: (a) comparison of Phantom 3, Inspire 1 and S900, (b) comparison of Inspire hovering stationary versus hovering whilst rotating slowly on the spot [40]

Table 3.1 Summary of 94 GHz RCS statistic for three UAVs (odd-bounce circular polarisation) [40]

Target	Average modal RCS (dBsm)	Maximum RCS (dBsm)	Average CDF 10% and 90% points (dBsm)
DJI Phantom 3 Standard	−20	−6	−27.5, −12
DJI Inspire 1	−15	−2	−23, −11
DJI S900	−8	2	−21, −2.5

discussed in detail here, [40] confirms that the millimetre wave RCS distributions of birds in flight overlap closely with those of UAVs of similar size. This confirms that using RCS as a feature for discriminating between UAVs and birds is unlikely to be successful so other features such as micro-Doppler signatures must be used.

3.3 Millimetre wave micro-Doppler signatures of UAVs

The micro-Doppler effect in radar has direct frequency dependence. Considering L as the radial length of a rotating propeller blade from its centre and Ω as the rotation rate in revolutions/second, the blade tip velocity is $V_{tip} = 2\pi L\Omega$. The maximum Doppler shift can then be readily calculated as

$$\{f_d\}_{\max} = \left(\frac{2V_{tip}}{\lambda}\right)\cos\beta \tag{3.1}$$

where β is the elevation angle of the radar boresight with respect to the propeller axis of rotation and λ is the radar wavelength [3]. It can be seen from (3.1) that the maximum Doppler frequency is proportional to the radar operating frequency. This means that using a higher frequency radar yields a comparatively higher frequency Doppler signal for a given target velocity which may then be measured with a finer resolution in a given integration time. Additionally, the Doppler spectrum may be more distinct from zero-Doppler clutter and hence more easily allow the measurement of very slow moving targets. These attributes are advantageous for a millimetre wave radar when measuring low-speed UAVs and for producing high-quality micro-Doppler signatures of the propeller blades, which is one of the most essential features for target discrimination. Simple simulations of the micro-Doppler signatures of a UAV can visually illustrate the effect. In Figure 3.16, spectrogram plots, generated using a short-time Fourier transform (STFT), are shown of simulated radar return signals at 10 and 94 GHz from a DJI Phantom 3 Standard. The model is based on the UAV's geometry (four rotors, each with double-ended propeller blades of radial length 13 cm and spaced in a square on a diagonal of 41 cm from the fuselage centre). The propellers are modelled as rotating sticks whose RCS is around 30 dB lower than the fuselage which itself is modelled as a single point target at the centre of the UAV [20]. The rotors usually

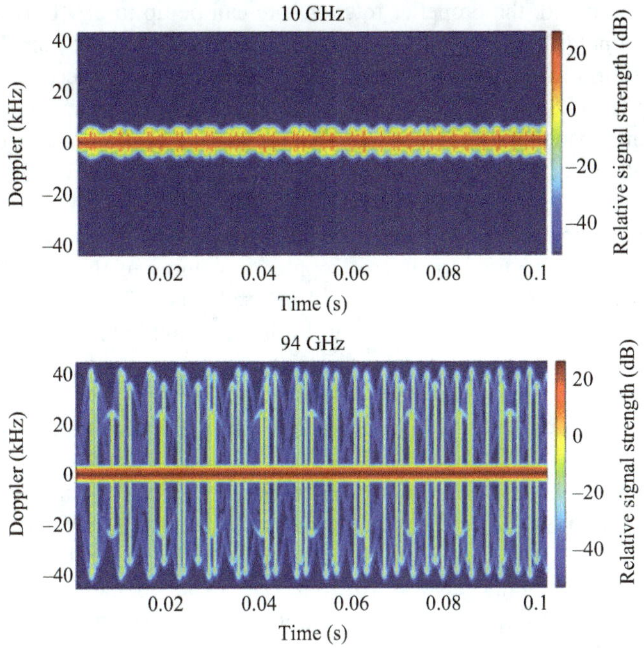

Figure 3.16 Simulated spectrograms of a hovering DJI Phantom 3 Standard at 10 GHz (top) and at 94 GHz (bottom) showing better resolved micro-Doppler blade flashes in the latter

rotate at different rates to maintain required orientation; hence different rotation rates were allocated (80, 47, 69 and 77 Hz). The sampling rate was defined to be 100 kHz, to avoid Doppler aliasing. In Figure 3.16 (top), the spectrogram is generated for 10 GHz centre frequency and the blade flash signatures are quite hard to detect. In contrast, the spectrogram at 94 GHz in Figure 3.16 (bottom) shows a much wider Doppler excursion and very clear blade flash signatures, demonstrating the advantage of millimetre waves in the context of UAV micro-Doppler. Whilst operation at higher frequencies requires a higher Doppler sampling rate, this is a worthwhile trade-off as it significantly increases the fidelity of the micro-Doppler signatures obtained, which can improve the performance of classification algorithms.

3.3.1 FMCW micro-Doppler signatures

In practice, a radar-based UAV detection and classification system will mostly operate in pulsed or FMCW mode with the ability to measure range as it enables localisation of the UAV for tracking, along with classification, which cannot be done using CW radar. One issue with using an FMCW radar for micro-Doppler-based UAV detection is that the CRF must be quite high to achieve unambiguous Doppler sampling. For instance, according to the product specification of the DJI

Phantom 3 Standard, the propeller rotation rate can be up to 280 Hz leading to a maximum Doppler frequency of 143 kHz at 94 GHz from (3.1). To comply with the Nyquist criterion for unambiguous Doppler sampling, the radar needs to operate at a CRF of at least 286 kHz. This not only puts strain on the hardware but also on the computational load, requiring extremely fast chirps ($<3.5\ \mu s$) which are difficult to generate without distortion, very high frequency digitisers to sample the resulting high baseband frequencies (likely to be in the 100s MHz) and rapid processing of extremely large amounts of data. Hence, in most cases, FMCW radars operate with sub-Nyquist CRF and produce micro-Doppler signatures of the propeller blades that are aliased. Nonetheless, they can still be used very effectively for classification as very few other targets (especially birds, the main confusers) produce signatures at such high Doppler values. The mere presence of a micro-Doppler signature beyond a certain Doppler threshold can be a very good indication of the presence of a UAV. A few example FMCW micro-Doppler signature plots are discussed later.

In Figure 3.17, millimetre wave spectrogram plots of a flying DJI Phantom 3 Standard are shown [20]. The data was obtained by using the 94-GHz NIRAD radar [41]. The chirp duration was 102.4 μs and the chirp repetition interval was 144 μs and the sampling rate was 10 MHz. The radar operating bandwidth was 600 MHz, corresponding to a range resolution of 25 cm. The maximum unambiguous Doppler frequency can be readily calculated from the chirp interval and is 3.47 kHz. Figure 3.17 (top) illustrates a spectrogram (with STFT window length 73.7 ms) for a single-range bin. The UAV was flying at a range of approximately 120 m from the radar, at an angle to the radar beam, and passed through the beam between 1 and 2.1 s. The bulk Doppler in bright red is in the range of 2–3 kHz (3.2–4.8 m/s radial velocity) and the change in bulk Doppler frequency is clearly evident, as are the fluctuations in the magnitude due to the changing aspect angle on the UAV. The micro-Doppler features can be seen around both sides of the bulk Doppler; the under-sampled propeller micro-Doppler is spread across the Doppler spectrum. A very good SNR of the UAV was observed at this range: ~50 dB for the fuselage and ~25 dB for the propellers. In Figure 3.17 (bottom), the UAV flew on a radial flight path within the radar beam and the spectrogram was produced from all the range bins it passed through. As can be seen in the first 0.5 s of the image, the bulk Doppler exceeds the maximum unambiguous value and Doppler aliasing has occurred.

The frequency resolution of the spectrogram can be increased by using a longer STFT window, which reveals Helicopter Rotor Modulation (HERM) lines for periodically rotating targets [43]. Figure 3.18 (top) gives an example of such a spectrogram for a quadcopter showing the HERM lines (STFT window length 294.9 ms) and Figure 3.18 (bottom) is a Doppler slice taken from that spectrogram for better visualisation of the HERM line spacing [20]. In theory, the HERM line spacing is directly related to the rotation rate of the blades [43]. The HERM lines in Figure 3.18 are quite complicated as the signal is modulated by the returns from four propellers on four rotors rotating at different speeds and the radar return signal is under-sampled in the Doppler domain.

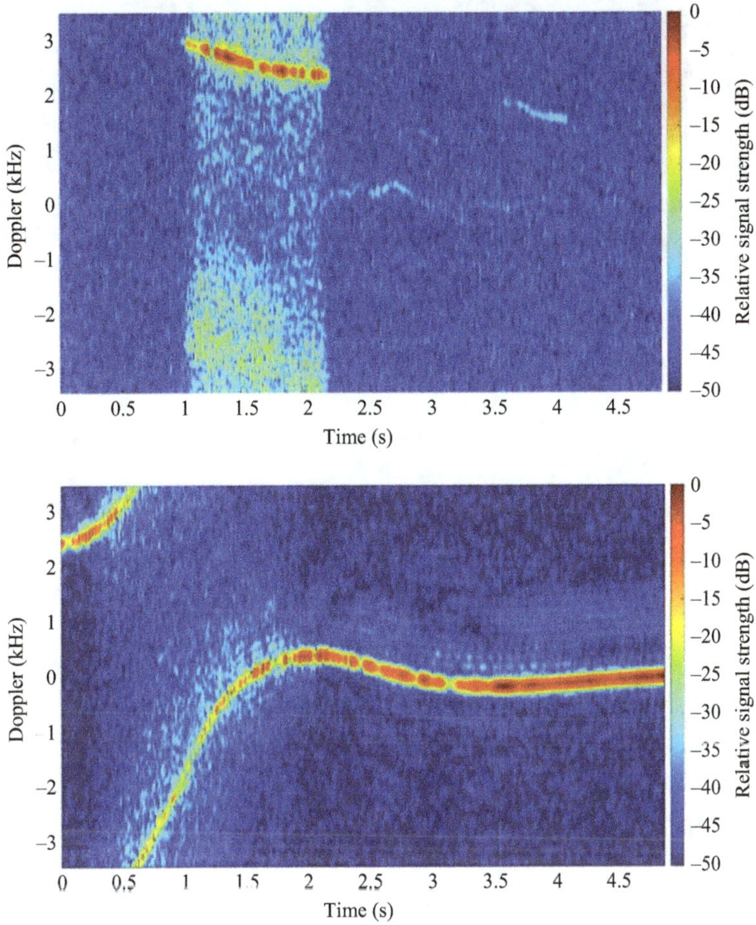

Figure 3.17 (Top) A 94 GHz spectrogram from a single range bin of a flying DJI Phantom appearing in the radar beam and then disappearing again and (bottom) a spectrogram obtained from multiple range bins along the UAV's flight path. In both spectrograms the STFT window length was 512 samples/73.7 ms [20]

It is also possible to reveal the rapid temporal modulations in the Doppler spectrum caused by propeller blades, known as blade flashes, if a short window is used in the STFT and the Doppler spectrum is fully sampled. The NIRAD radar was used to measure a hovering DJI Phantom 3 Standard at a range of ~45 m. A very high CRF of 50 kHz (chirp repetition interval of 20 μs) was used to achieve a maximum unambiguous Doppler frequency of 25 kHz. Figure 3.19 shows the resulting spectrogram which clearly reveals the fully sampled individual blade flashes [44]. The rotation period (two blade flashes) is ~13 ms so the rotation rate

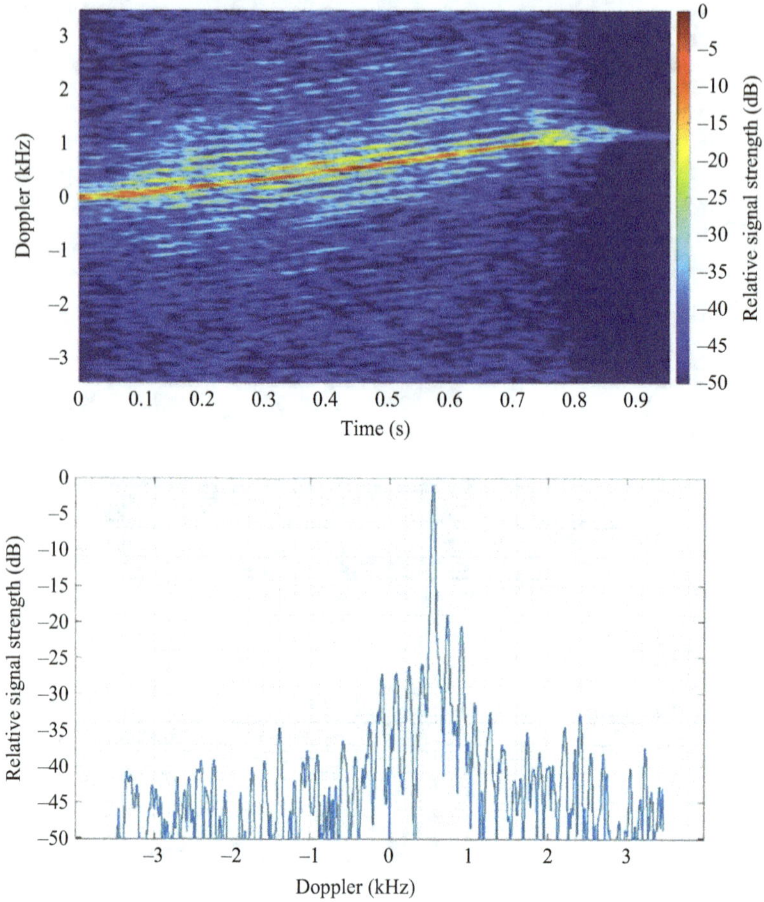

Figure 3.18 (Top) A 94 GHz spectrogram of the DJI Phantom using long STFT window (2,048 samples, 294.9 ms) and (bottom) Doppler frequency slice obtained from the spectrogram with long STFT window showing the HERM lines [20]

is ~77 Hz, which is consistent with the typical rotor speed for a Phantom in normal flight.

The micro-Doppler signatures of various UAVs have been investigated using the 94 GHz T-220 radar [44]. In Figure 3.20 (top), a spectrogram derived from FMCW data of two UAVs (DJI Phantom 3 Standard and DJI Inspire 1) flying simultaneously at ~75 m range is shown. The UAVs flew radially along the radar beam at different heights and in opposite directions. For both UAVs, both the bulk Doppler and micro-Doppler signatures are quite prominent. The micro-Doppler signature of an S900 hexacopter shown in Figure 3.20 (bottom) is even stronger,

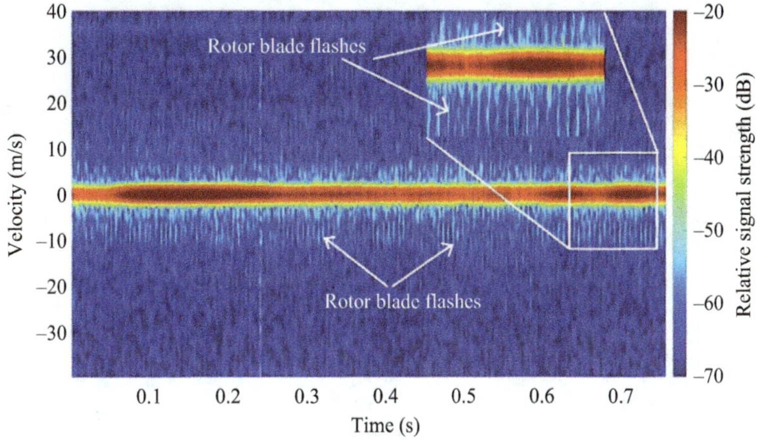

Figure 3.19 A 94 GHz spectrogram of DJI Phantom hovering at ~45 m range obtained with a very high chirp repetition frequency of 50 kHz and short window STFT (512 samples, 10.2 ms) revealing individual blade flashes [44]

likely due to the increased size of the propeller blades compared with the other two UAVs.

To demonstrate the distinctiveness of millimetre wave micro-Doppler signatures from UAVs compared with other moving targets, Figure 3.21 shows a spectrogram of a Tawny Eagle flying and approaching the perch 30 m away from the radar [44]. The radar set-up was the same as for the results shown in Figure 3.20. A strong micro-Doppler signature due to the wing flapping is observed around the bulk Doppler, which is aliased in the first few seconds. When compared with UAV spectrograms, the differences in micro-Doppler characteristics are clearly obvious. Whilst UAVs and birds can fly at similar speeds and hence have similar bulk Doppler, the spectrograms reveal significant differences in the micro-Doppler signatures. Bird wing flaps occur on a significantly longer timescale (100s ms) than propeller blade flashes (few ms) and this difference allows them to be differentiated reliably for classification.

Another way to display micro-Doppler information is the range-Doppler plot processed from a single coherent processing interval (CPI). Figure 3.22 shows some example range-Doppler plots of different UAVs obtained by various millimetre wave radars operating at different centre frequencies. The return from the UAV fuselage appears as a bright spot at the range and bulk Doppler coordinates measured in that CPI. If the micro-Doppler returns from the propellers are visible to the radar, they appear as a wide spread in the Doppler axis, centred on the bulk Doppler. When FMCW modulation is used, as discussed earlier, the Doppler is often under-sampled so the micro-Doppler energy is spread right across the Doppler axis.

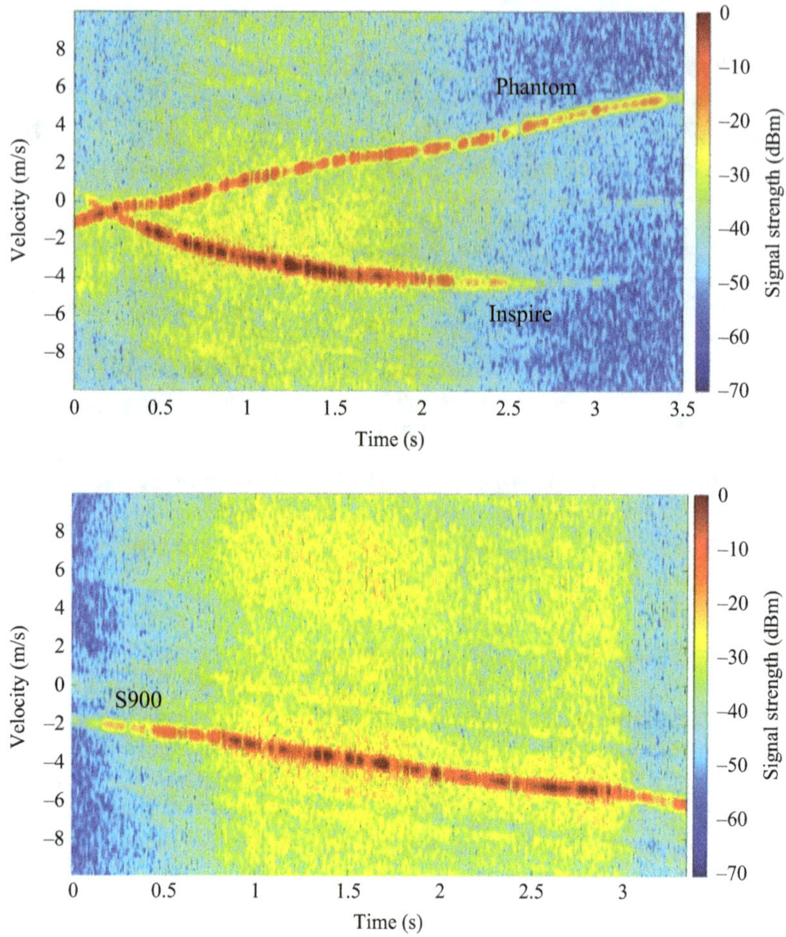

Figure 3.20 94 GHz spectrograms of (top) DJI Phantom 3 Standard (~75 m range) and DJI Inspire 1 (~85 m range) flying together and (bottom) a flying DJI S900 Hexacopter at ~95 m range [44]

3.3.2 CW micro-Doppler signatures

CW radars have the advantage of more easily sampling Doppler without ambiguity, which is very pertinent for analysing the micro-Doppler signatures of the very fast rotating propeller blades of a UAV. Due to the larger Doppler spread and fine Doppler resolution achievable at millimetre wave frequencies, detailed information on the propeller blades can be obtained.

The micro-Doppler signals from UAVs contain signatures from multiple blades rotating at different rates. To observe the propeller signature of a single blade, data was collected from a Phantom held static with only one propeller blade fitted and the motor idling [45]. Measurements were made using the 94 GHz T-220

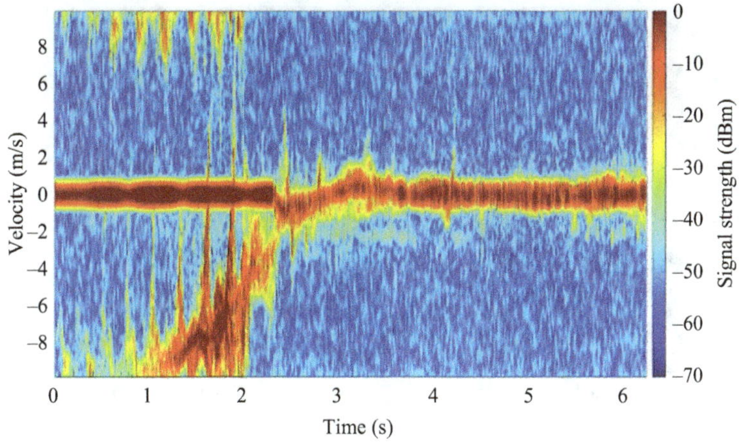

Figure 3.21 A 94 GHz spectrogram plot of a Tawny Eagle approaching the perch, showing distinct wing flaps and a very different micro-Doppler signature from a UAV [44]

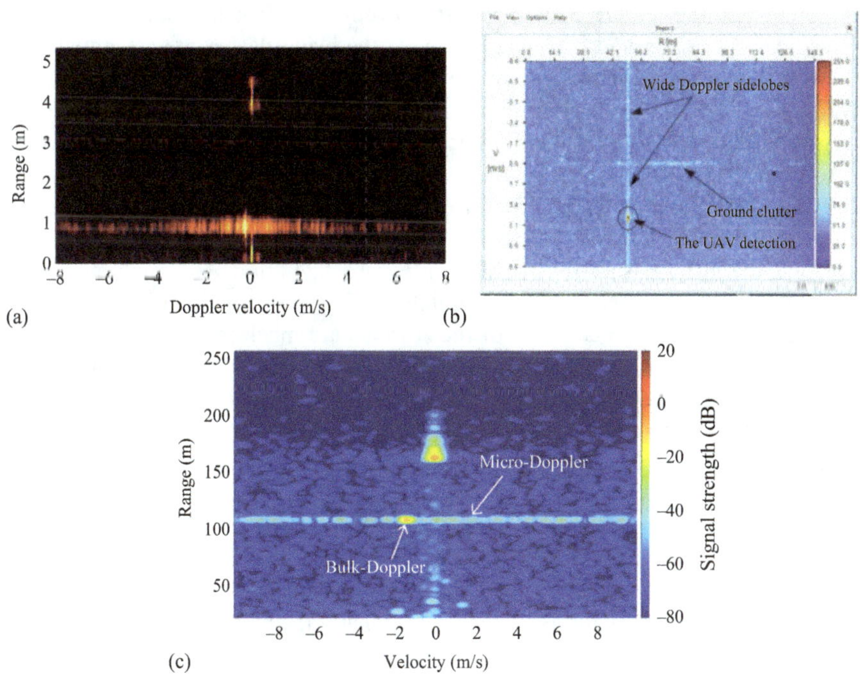

Figure 3.22 (a) Range-Doppler plot of a mini helicopter in flight at ~1 m at 216 GHz [7], (b) range-Doppler plot of DJI Phantom 2 at ~40 m range at 35 GHz [8], (c) range-Doppler plot of a DJI S900 in flight at ~110 m at 94 GHz [44]

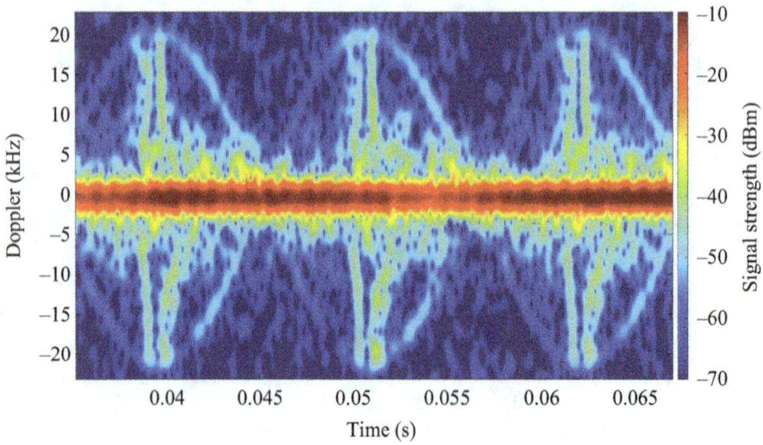

Figure 3.23 A 94 GHz CW spectrogram of DJI Phantom held by hand with only one propeller blade rotating and the other blades removed [45]

radar operating in CW mode with a 100 kHz sampling rate to ensure unambiguous Doppler sampling. Figure 3.23 shows the spectrogram which clearly reveals the sinusoidal tip motion and blade flashes. Note the blade flashes are asymmetric in Doppler due to the curved profile of the blades and the viewing angle of the radar. The periodicity and maximum tip velocity can be used to extract the propeller length. In Figure 3.24, 94 GHz CW spectrograms are presented for three different UAVs, hovering at close range (~15–30 m), measured with T-220 in CW mode at a 200 kHz sampling rate [46]. The difference in the micro-Doppler signatures is visually noticeable, which is due to the different sizes and shapes of the blades. For instance, the blade flashes for the S900 in Figure 3.24 are quite straight and have distinct Doppler maxima due to the straight leading edges and square tips of the blades. The micro-Doppler traces of the Joyance are much more curved which is consistent with the more curved blade profile. The low-frequency Doppler components differ between the UAVs as well due to the shape of the roots of the blades and the motor bodies. The presences of multiple blades rotating at different frequencies make the spectrograms more complex than the idealised single propeller case. The information present in these spectrograms is very rich and can even be used for determining the type of UAV [46]. Figure 3.25 shows the spectrogram plots and extracted peak frequencies of single, double and quad rotor blades obtained with a W-band CW radar [23]. The peaks from the blade flashes were used to determine the rotation rate and subsequently the blade length, which could be characteristic of the type of UAV.

3.3.3 Alternative time–frequency analysis methods

Whilst the STFT is the most commonly employed time–frequency transform, other methods to analyse millimetre wave micro-Doppler signatures have been reported.

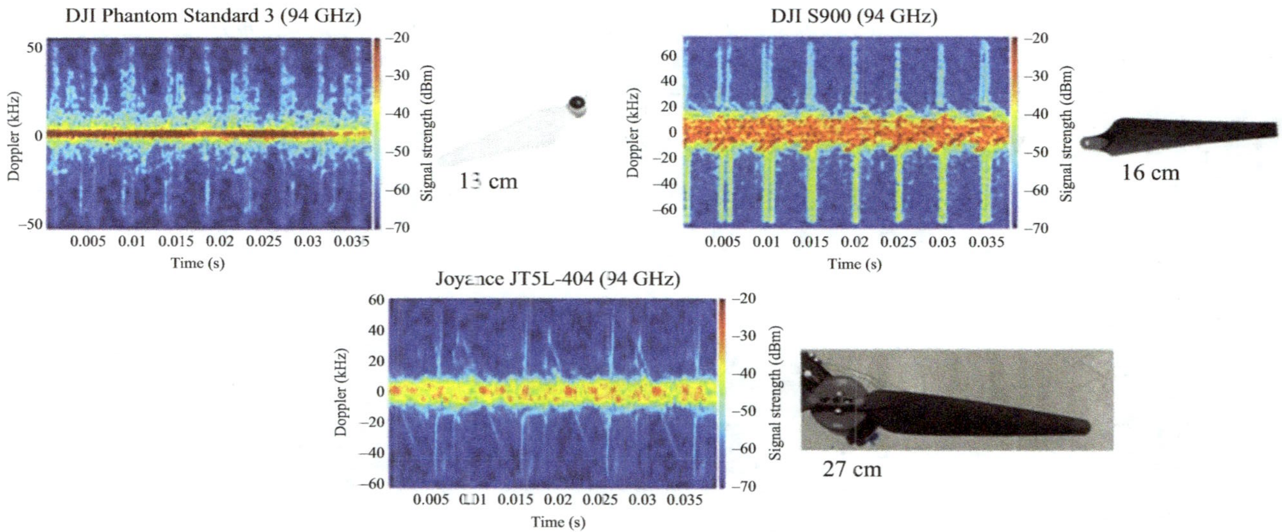

Figure 3.24 94 GHz CW spectrograms of a DJI Phantom 3 Standard, DJI S900 and Joyance JT5L-404 and the corresponding propeller blades [46]

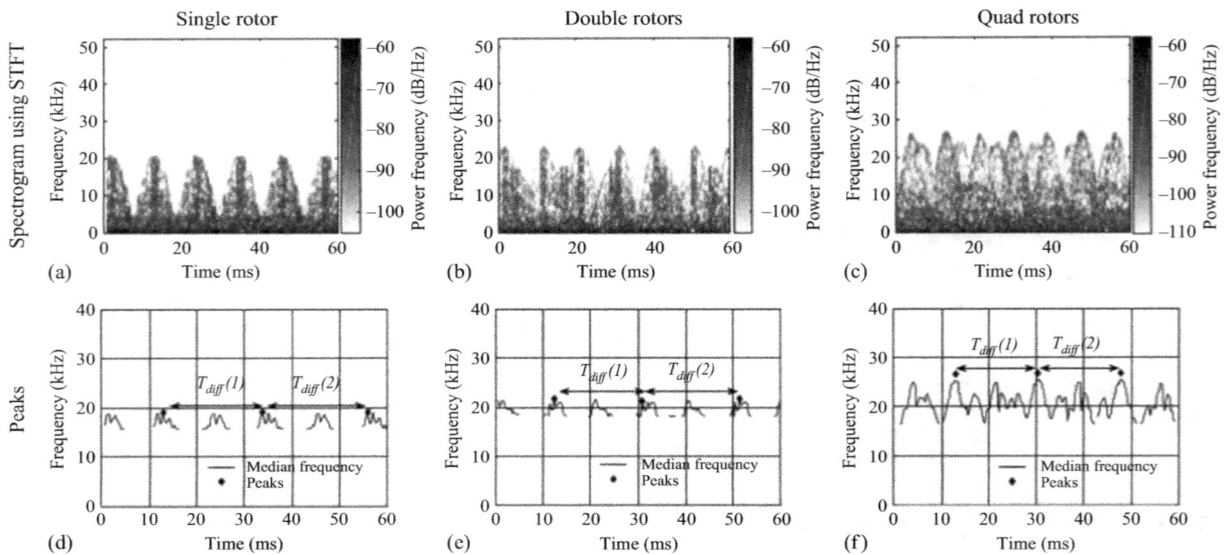

Figure 3.25 Analysis of received data from a 92.16 GHz CW radar: spectrograms using STFT for (a) single, (b) double and (c) quad rotors; (d)–(f) median frequency with markers on alternate peaks [23]

In [19], 35 GHz CW micro-Doppler measurements of UAVs were analysed using STFT, CVD and cepstrogram. CVD was used to obtain information on the cadence of repeating frequencies over time and the cepstrogram was used to obtain an estimate of occurring periodicities in the UAV spectrograms, i.e. the blade flash frequency and hence rotation rate of the rotors. The rotation rate and blade length were extracted successfully from the data. Different feature vectors were extracted from these three types of time–frequency diagrams and used for classification (discussed in Section 3.5). CVD and cepstrogram analysis were also applied to UAV data measured with the MIRA-CLE Ka radar in [47]. It was noted that the cepstrogram consistently indicates the possible rotor flash frequencies (two blade flashes per revolution). A wavelet transform analysis of UAV Doppler data has also been done, as presented in [21]. It was shown that wavelet transforms can be used to segment the high-frequency components (corresponding to propeller blade rotation) from the low-frequency components (fuselage return), which then can be further used for classifying the target. No apparent advantage of using wavelet transforms over spectrograms was observed apart from their lower computational load ($O(n)$) compared to fast Fourier transform ($O(n \cdot \log 2(n))$). However, in practice, very efficient FFT library functions are more readily available than similarly effective fast wavelet transform functions.

3.4 Signatures of UAVs equipped with payloads at a millimetre wave

Whilst there is a growing body of literature on the in-flight micro-Doppler signatures of drones in their standard configurations, there is comparatively little work reported on the radar signature of UAVs equipped with payloads. The ease with which UAVs can be modified to carry payloads has made this a tangible security threat, where the UAVs can be used to carry out malicious acts like spreading bio chemical weapons, carrying a firearm, grenades or other contraband. Millimetre wave radars can be very useful to sense such actions due to their high range and Doppler sensitivity and, in the first example, stronger backscatter from liquid droplets. This section presents work considering the following three different scenarios:

1. UAV equipped with liquid spray payload
2. UAV equipped with simulated recoil
3. UAV equipped with heavy payload

3.4.1 UAV equipped with liquid spray payload

In [48], the analysis of the radar signature characteristics of small water droplets sprayed from a UAV, mimicking a chemical weapon attack, was performed. A commercially available crop spraying UAV named Joyance JT5L-404 [49] was used as the basis for liquid droplet radar backscatter modelling and for experimental data acquisition. The spray nozzle droplet parameters were used to model the RCS and the SNR of the liquid droplets at X-, K- and W-bands, using the

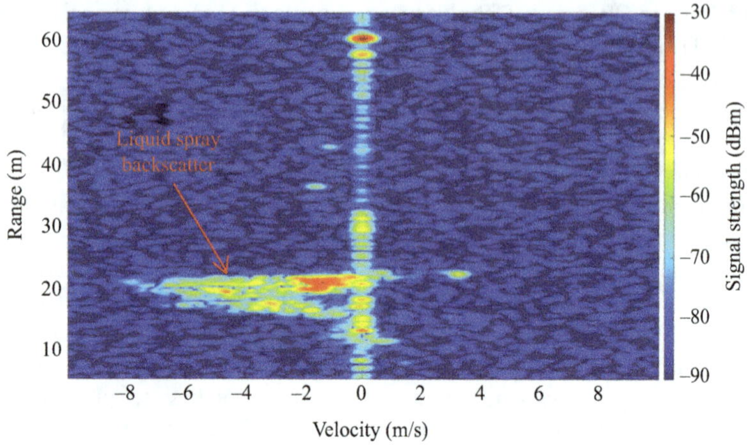

Figure 3.26 A 94 GHz range-Doppler plot of crop spraying UAV at ~22 m range and the radar antenna being bore sighted at the spray below the UAV [48]

Rayleigh approximation. Additionally, experimental data was obtained simultaneously with 24 and 94 GHz radars. The processed results show very good agreement with the model. It is clearly demonstrated that at W-band (94 GHz), the liquid spray produces strong micro-Doppler signatures observed from the range-Doppler plots (Figure 3.26), whereas no such detection was possible at K-band (24 GHz). The experimental results validate the hypothesis that a millimetre-wave radar offers superior sensitivity than lower frequency bands to reflections from liquid spray droplets of $\ll$0.5 mm size and the behaviour is consistent with Rayleigh scattering. Figure 3.27 illustrates this quite clearly as the measured SNRs from the liquid droplets at 94 GHz match quite well with the theoretical values. In contrast, at those ranges (20–50 m), the predicted SNR at 24 GHz was below 0 dB, so no backscatter was measured in this case. Hence, a millimetre wave radar system could potentially be used for detecting and classifying a UAV with a liquid spray payload.

3.4.2 UAV equipped with simulated recoil

In [50], it was investigated whether inertial forces due to a dynamic payload impart characteristic changes on the UAV's radar signature which can be used to identify the particular payload. The hypothesis is that a UAV reacting to some payload-induced inertial force, such as a UAV-mounted gun, is likely to show a characteristic change in its kinematic behaviour and position as it jerks about then re-stabilises. This will correspond to recoil and counter-recoil, respectively. The focus was only on FMCW data collection, as the recoil and counter-recoil effects will be more apparent from the bulk Doppler signature than the micro-Doppler. The lower Doppler sampling rate of the FMCW radar gave a relatively long integration time and hence a high Doppler resolution. The 94 GHz experimental data was collected

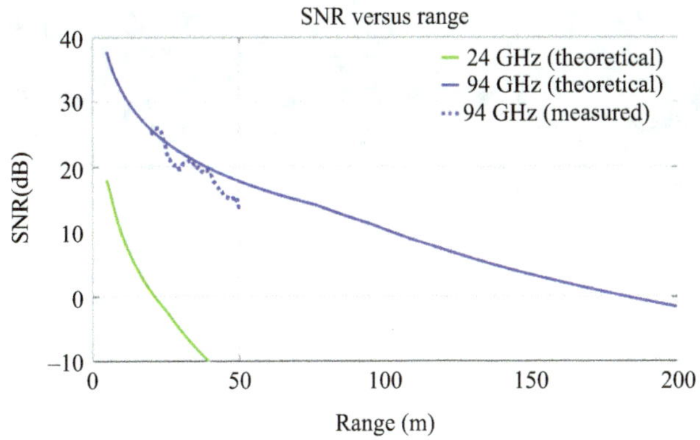

Figure 3.27 SNR of liquid spray versus range showing theoretical curves for the 94 and 24 GHz radars and the measured data at 94 GHz. No spray was detectable at 24 GHz [48]

on a DJI Phantom 3 Standard UAV subject to a sudden inertial force. As it was not possible to use a real gun due to logistical and safety reasons, the impulses were applied to the UAV by pulling it sharply several times using a wire. The wire was a few metres long with one end attached to the main body of the UAV and the other end to the tip of a wooden pole. This pole was held upwards and by sharply pulling it backwards, a horizontal impulse was applied to the UAV fuselage, thus simulating the recoil effect. Due to the UAV's internal stabilisation mechanism, it then returned to its previous position, which is the counter-recoil. The Doppler signature of the experimental data is seen in Figure 3.28 (top), where the recoil and counter-recoil effects are observed from the bulk velocity trace. A velocity–time profile can be produced by tracing the bulk velocity peaks from the spectrogram and then subsequently a force–time profile can be created by using basic force–velocity relations. Figure 3.28 (bottom) shows a close-up of impulse 4 in Figure 3.28 (top) to illustrate the details of this signature. The impulse (labelled 'recoil') takes place over a few hundred milliseconds, characteristic of the stick and wire jerk technique, and is followed by a longer, damped oscillation, stabilisation and recovery process (labelled 'counter-recoil') which is characteristic of the UAV's flight controller properties. This exhibited a Recoil Impulse of 1.71 N s which is nearly within a factor of 2 of the value for a Glock pistol (3.5 N s) [51]. The characteristic shape of the response could be used for classification.

3.4.3 UAV equipped with heavy payload

In [50], 94 GHz CW data was collected for a DJI S900 hexacopter with four different weights attached as heavy payloads (1, 1.6, 2, 2.5 kg). The payloads could be released remotely, hence providing the opportunity to analyse the effect of the radar signature change due to the presence or absence of the payload. The drop in the rotation rates of the propeller blades after the release was observed in all cases. This

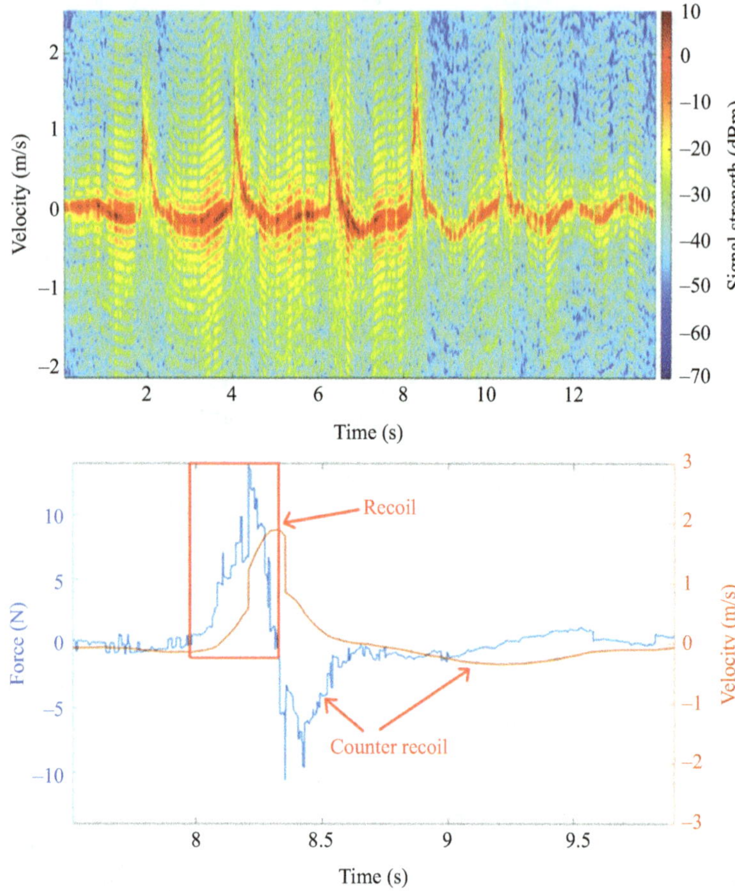

Figure 3.28 (Top) A 94 GHz FMCW-Doppler spectrogram of Phantom 3 subject to five impulse events and (bottom) close-up of impulse 4 force–time and velocity–time profiles showing recoil and counter-recoil regimes [50]

is to be expected since a heavier payload will make the UAV rotors spin faster to generate more lift. The spectrograms contain contributions from more than one blade; hence it is not always straightforward to calculate the rotation rate. Instead, SVD was used to determine the dominant rotational frequency within a spectrogram. Figure 3.29 (bottom) shows a clear increasing trend in the rotational frequency with heavier weights. It was also observed that the frequency value saturates at 200 Hz and whilst the reason is not fully understood; it may be due to the motor reaching the maximum rotational speed.

Even though there is a clear change in micro-Doppler signatures due to the presence of a heavy payload, this does not provide a robust basis for classification. First, these measurements are relative and it is unlikely that the rotational rate of a particular UAV without payload will be known beforehand. Second, other factors

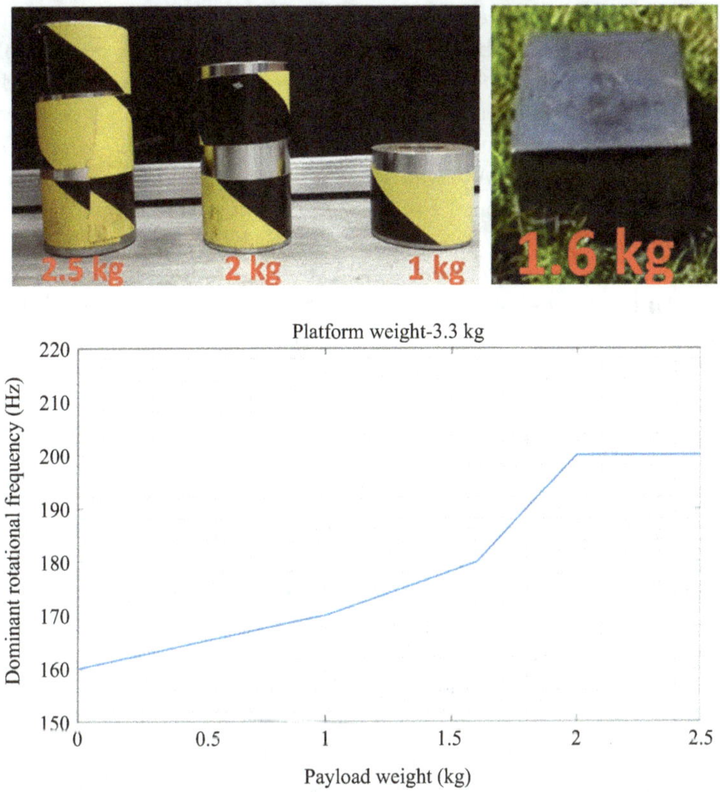

Figure 3.29 (Top) Steel payloads used for the experimental trials which were attached to the S900 with remote controlled electromagnet and (bottom) propeller dominant rotational frequency for S900 versus attached payload weight, measured at 94 GHz [50]

like wind or platform manoeuvres can generate the same effect. Hence, for reliable detection of the presence of heavy payloads, further information is required, perhaps from a different sensor providing information on the wind speed in the area or passive RF sensor providing information on the UAV type. It is possible that multiple successive classifications could average out the effects of wind or manoeuvres and potentially improve payload classification.

3.5 Classification methods of UAVs using millimetre wave data

The eventual goal for the radar measurement of UAVs is to use the data to classify the target. Different classification methods have been applied to millimetre wave radar data of UAVs, as discussed in this section.

Feature extraction based classification is quite widely used for radar data of all frequencies, and there have been a few reports of the approach being applied to millimetre wave radar data. In [19], 35 GHz measurements of UAVs and birds were analysed using STFT, CVD and cepstrogram, from which various features were extracted and then classified using an SVM. A classification accuracy of 100% was achieved when classifying UAVs versus birds with randomly selected training and test data. The accuracy dropped to 96.2% when classifying small UAVs versus medium-sized UAVs.

In [45], the following features were extracted from 94-GHz FMCW radar Doppler data: (1) micro-Doppler spread width, (2) micro-Doppler spread weight, (3) micro-Doppler strength and (4) micro-Doppler periodicity. The last feature was extracted by performing a Fourier transform on the first row vector of the SVD unitary matrix. The other features were obtained by calculating feature values directly from the spectrogram plots, in order to avoid computationally expensive operations such as SVD as much as possible. The targets used for the feature extraction were three UAVs (Phantom 3, Inspire 1 and S900) and four birds of different sizes (Northern Hawk Owl, Harris Hawk, Indian Eagle Owl and Tawny Eagle). To verify the performance, the feature values were used to train two well-known classifiers, linear discriminant (LD) and SVM. It was clearly shown that the features are very distinctive for birds versus UAVs, enabling an identification of the targets, with both the classifiers showing 100% prediction accuracy during training – see Figure 3.30. It was also shown that the proposed method can differentiate the three different types of UAVs, achieving accuracies of 100% for LD and 91.7% for SVM. Much lower accuracy (<25%) was observed for birds alone as their features overlapped with each other when all four birds were considered. However, the prediction performance improved significantly by considering only the largest and smallest birds (87.5% with LD, 100% with SVM). Overall, it was demonstrated that the proposed algorithm has very good potential to be used in a UAV detection FMCW radar system, to classify UAVs and birds in a dynamic environment. The method described here uses thresholds very close to the noise floor to extract the micro-Doppler information, which requires high fidelity Doppler measurements, best achieved by using a very low phase noise radar with excellent spectral purity.

One issue with the feature extraction based classification is the computational time. The raw radar data first needs to be time–frequency analysed for Doppler information, then further computation is done to extract feature values. For real-time operation, this becomes a compromise between processing time and classification accuracy. Alternatively, a neural network based classification has the potential to reduce the computational load for real-time operation as the computationally heavy work is done in advance when the model is trained. Then, the processed data can be fed directly to the trained model for rapid target prediction, eliminating the time required for feature extraction. In [52], a convolutional neural network (CNN)-based classification based on spectrogram images was reported. A large dataset was created to train a neural network using mainly 94 GHz FMCW data but to achieve diversity, 24 GHz FMCW data was included as well. Supervised training was done by labelling the targets into four classes initially

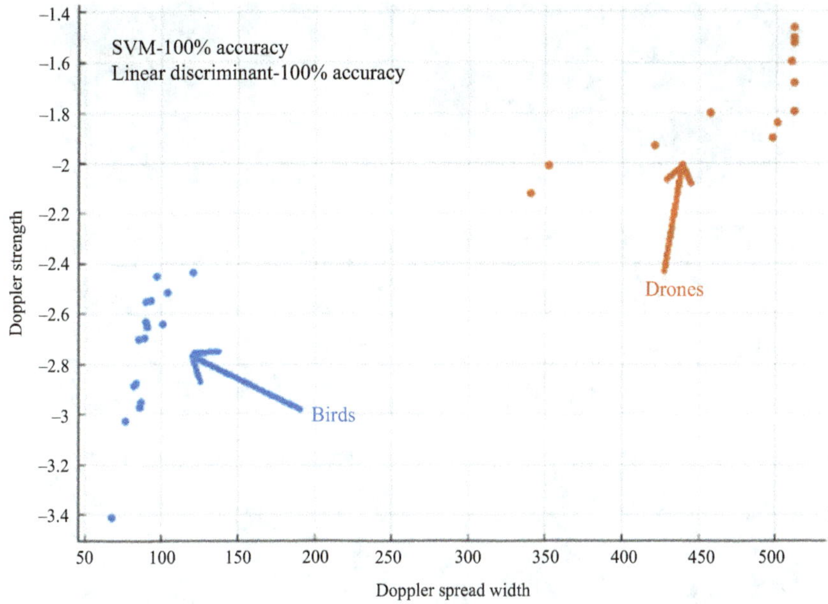

Figure 3.30 Screenshot of the Classification Learner App in MATLAB® while performing the training for two classes (birds and drones) using 94 GHz radar data, showing 100% accuracy with both SVM and linear discriminant classifiers [45]

(UAV, bird, clutter, noise) and then two classes (UAV, non-UAV). For 4-class training, each class had 600 spectrogram images, each 0.4 s long – see Figure 3.31, for examples. The number was limited by the availability of bird images. For 2-class training, each class had 1,000 images. One of the very popular and effective CNN models, GoogLeNet [53], was used as a pre-trained network. The advantage of using this is that the first few convolutional layers can be frozen as those correspond to low-level features (line, curve, etc.). During training, the first 10 layers were frozen. As GoogLeNet requires 3-channel RGB images and radar spectrograms are false colour images, the same dataset was produced in greyscale and was trained with a newly developed series network for comparison. During training, 80% data was used for training and 20% for validation. The validation and testing accuracy for the developed series network were found to be 99.6% and 94.4%, respectively, for four classes and 99.3% and 98.3%, respectively, for two classes. The GoogLeNet-based model showed both validation and testing accuracies to be around 99% for all the cases. Both models have been tested with previously unseen data and have again shown very good accuracy. GoogLeNet performs better but is more time-consuming compared to the series network as it deals with 3-channel images instead of monochrome images in the series network.

In [46], 94 GHz CW data was used for GoogLeNet-based training to classify different types of UAVs. In Figure 3.24, it is shown that unambiguously sampled

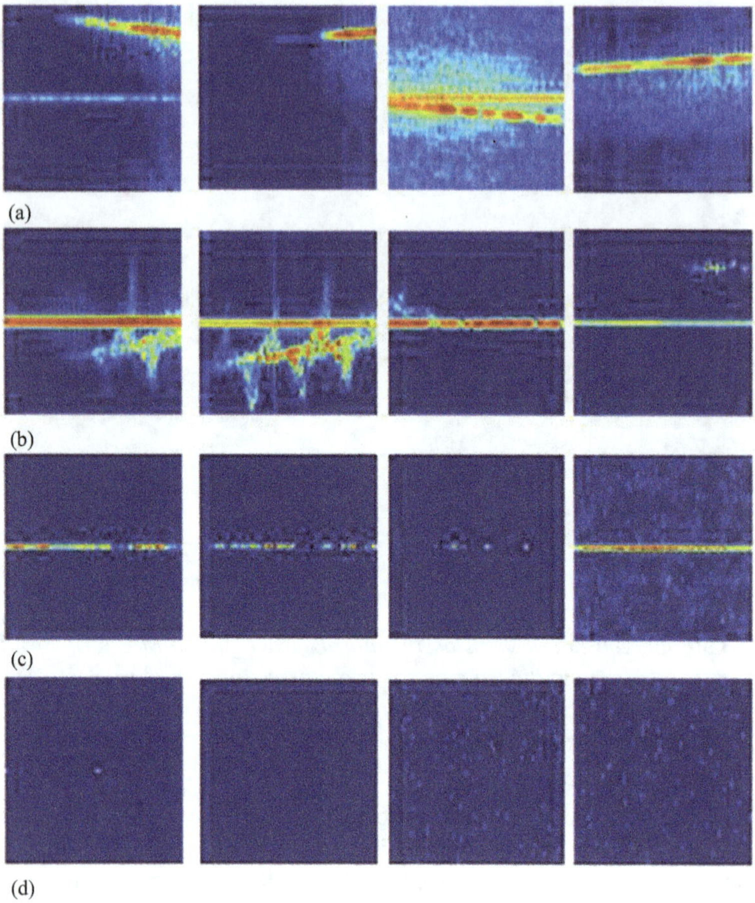

Figure 3.31 Example 94 GHz FMCW training data for a GoogLeNet classification of four classes: (a) UAV, (b) bird, (c) clutter and (d) noise [52]

micro-Doppler spectrogram plots obtained from millimetre wave radar data reveal the different blade shape features quite clearly. A dataset of spectrograms of this type, suitable for classification by CNN, was created using CW data for three different classes (Phantom, S900 and Joyance). Each class consisted of 650 spectrogram images and the dataset was used to train GoogLeNet. Over 99% validation accuracy was achieved. The success of the method was then demonstrated by feeding images unknown to the model, which resulted in 100% classification accuracy in the case of the S900 and Joyance and 88% accuracy for the Phantom. The CNN-based classification outperformed a feature extraction and SVM-based classification of the same data that only achieved 75% accuracy. Additionally, the CNN classification was faster than the feature extraction. These are very promising results which illustrate that a large and diverse experimental radar dataset cannot only be used for classifying UAVs from other targets, but for classifying different models of UAVs as well.

Several practical issues need to be considered when using spectrogram image based neural network training. The more diverse the dataset is, the more robust it will be for general use. In [46], only a single-radar system with a fixed polarisation was used for data collection. More variation in the dataset can be achieved by utilising radar systems with different operating frequencies, polarisations, antenna beamwidths, sampling rates, etc. Data taken at different places will also create changes in the surrounding clutter. Variation in the signal processing can also be achieved (e.g. by varying STFT window parameters, intensity range, etc.), but careful consideration should be taken when doing so. It should be kept in mind that, fundamentally, radar spectrograms are not optical images, so changing parameters like noise floor threshold, STFT length and colour scale will alter the spectrograms significantly and may affect the CNN-based classification process. When developing a CNN-based model, careful decisions should be made regarding the extent of the versatility of the model being used in practice.

3.6 Conclusions

This chapter has presented an overview of the state of the art in a millimetre wave radar for UAV detection and classification. A literature survey of millimetre wave radar systems that have been applied to this application has been presented, highlighting the international breadth of the research groups involved. To date, the range of frequencies reported spans of 30–216 GHz. A small body of work has yielded RCS information for a few types of UAVs, measured both statically in anechoic chambers and dynamically in flight, and example values are presented. It is confirmed that RCS generally increases with UAV size and that the RCS of birds, which are key confusers, overlaps with UAVs, rendering RCS a poor discriminant. It is the high Doppler sensitivity afforded by millimetre wave radars which makes them of particular interest for revealing the micro-Doppler characteristics of UAVs, due to the high-speed propeller rotation. The nature of UAV micro-Doppler signatures obtained with an FMCW radar, which tends to be sub-Nyquist Doppler sampled, and a CW radar, which is more easily fully sampled, is compared. Researchers report processing these signals using a variety of time–frequency analysis techniques, but the STFT-based spectrogram remains the most common. FMCW data tends to allow the measurement of HERM lines, whilst CW data is able to reveal individual blade flashes; from both these forms of data, it is possible to extract details of the propeller rotation rate and blade length, potentially aiding classification.

As counter-UAV systems evolve, there is desire to move beyond simply confirming the presence of a UAV to wanting to know what type it is, whether it is carrying a payload or what might its intent might be, to determine the level of threat and guide the appropriate response. Recent results investigating the millimetre wave micro-Doppler signatures of drones equipped with payloads (liquid spray, heavy objects and inertial recoil) are summarised. A millimetre wave radar is uniquely placed to be able to detect the tiny droplets in a liquid spray dispensed by a UAV, benefitting from the frequency to the fourth power dependence of Rayleigh scattering. Whilst no reliable or unique signature is reported for heavy payloads, a drone-mounted gun would produce a distinct and classifiable recoil/counter-recoil signature in the bulk motion of the UAV.

The classification of drones is the ultimate goal when developing a counter-UAV system, be that drone versus bird or drone versus drone. A range of work has been reported applying different classification methods to millimetre wave radar data of UAVs. Feature extraction-based classification can achieve high performance but often at the expense of a heavy computational load. However, this approach remains favourable as the decision process is explainable, which is relevant for a security system which has to consider the legal implications of the way it operates. In contrast, neural network-based classification is attracting ever more interest and has been shown to achieve extremely high performance and comes with a speed benefit at the classification stage. However, neural networks are often seen as ‘black boxes’ whose inner workings are not easily interpretable. The main practical drawback is the requirement for large amounts of training data, which can be difficult to collect experimentally, or challenging to simulate with high fidelity.

There remain many opportunities for research in the field of a millimetre wave radar for UAV detection and classification. RCS statistics are only available for a small set of UAV types, little RCS modelling has been done at these frequencies, and the statistical fluctuation behaviour has not yet received sufficient attention. In terms of micro-Doppler, which is one of millimetre wave radar’s key advantages, there is potential for higher fidelity measurement and modelling of signatures, potentially leading to the identification of UAV types.

Looking forward to how millimetre wave radars might become used operationally in counter-UAV scenarios, it is likely that these will be in more specialist roles. The relatively limited range capability of millimetre wave radars for wide area surveillance makes it unlikely that they will be used as primary counter-UAV sensors. However, they are more relevant for shorter range (sub-km) scenarios or applications which require physically very compact sensors. The ability to gather richly detailed and high fidelity micro-Doppler signatures using a millimetre wave radar means that they could be used as secondary sensors for robust classification of targets detected and tracked by a lower frequency primary radar. Hunter-killer UAVs equipped with on-board, highly integrated, miniature radars are already being developed. The reduction in hardware costs arising from the commercialisation of millimetre wave semiconductor devices for mass market communications like 5G and 6G and single chip radars are enabling new opportunities for millimetre wave radars in the field of UAV detection and classification.

References

[1] V. C. Chen and S. Qian, “Joint time-frequency transform for radar range-Doppler imaging,” *IEEE Trans. Aerosp. Electron. Syst.*, vol. 34, no. 2, pp. 486–499, 1998.

[2] V. C. Chen, F. Li, S.-S. Ho, and H. Wechsler, "Analysis of micro-Doppler signatures," IEE Proc. Radar Sonar Navig., vol. 150, no. 4, p. 271, 2003.

[3] V. C. Chen, *The micro-Doppler effect in radar*. Norwood, MA: Artech House, 2011.

[4] J. S. Patel, F. Fioranelli, and D. Anderson, "Review of radar classification and RCS characterisation techniques for small UAVs or drones," *IET Radar Sonar Navig.*, vol. 12, no. 9, pp. 911–919, 2018.

[5] R. J. Wellman and J. L. Silvious, *Doppler signature measurements of an MI-24 Hind-D Helicopter at 92 GHz*. Adelphi, MD: Army Research Laboratory, 1998.

[6] A. S. Hedden, J. L. Silvious, C. R. Dietlein, J. A. Green, and D. A. Wikner, "Extremely high-frequency micro-Doppler measurements of humans," in *Radar Sensor Technology XVIII*, 2014, vol. 9077, no. 29, p. 90771C.

[7] D. Tahmoush, "Detection of small UAV helicopters using micro-Doppler," in *Radar Sensor Technology XVIII*, 2014, vol. 9077, p. 907717.

[8] J. Drozdowicz, M. Wielgo, P. Samczynski *et al.*, "35 GHz FMCW drone detection system," in *Proceedings International Radar Symposium*, 2016-June.

[9] M. Caris, W. Johannes, S. Stanko, and N. Pohl, "Millimeter wave radar for perimeter surveillance and detection of MAVs (Micro Aerial Vehicles)," in *Proceedings International Radar Symposium*, 2015-August, pp. 284–287.

[10] M. Caris, S. Stanko, W. Johannes, S. Sieger, and N. Pohl, "Detection and tracking of micro aerial vehicles with millimeter wave radar," in *European Microwave Week 2016: "Microwaves Everywhere", EuMW 2016 – Conference Proceedings; 46th European Microwave Conference, EuMC 2016*, 2016, pp. 1553–1555.

[11] D. Noetel, W. Johannes, M. Caris, A. Hommes, and S. Stanko, "Detection of MAVs (micro aerial vehicles) based on millimeter wave radar," in *Millimetre Wave and Terahertz Sensors and Technology IX*, 2016, vol. 9993, p. 999308.

[12] M. Caris, W. Johannes, S. Sieger, V. Port, and S. Stanko, "Detection of small UAS with W-band radar," in *Proceedings International Radar Symposium*, 2017.

[13] D. Nüßler, A. Shoykhetbrod, S. Gütgemann *et al.*, "Detection of unmanned aerial vehicles (UAV) in urban environments," in *Emerging Imaging and Sensing Technologies for Security and Defence III; and Unmanned Sensors, Systems, and Countermeasures*, 2018, vol. 10799, no. 4, p. 34.

[14] S. Engelbertz, C. Krebs, A. Küter, R. Herschel, R. Geschke, and D. Nüßler, "60 GHz low phase noise radar front-end design for the detection of micro drones," in *16th European Radar Conference (EuRAD)*, 2019, pp. 25–28.

[15] J. Klare, O. Saalmann, and O. Biallawons, "The MIMO radar MIRA-CLE Ka," in *Conference Record – Asilomar Conference on Signals, Systems and Computers*, 2013, pp. 1418–1422.

[16] J. Klare, O. Biallawons, and D. Cerutti-Maori, "UAV detection with MIMO radar," in *2017 18th International Radar Symposium (IRS)*, Prague, 2017, pp. 1–8.

[17] J. Klare, O. Biallawons, and D. Cerutti-Maori, "Detection of UAVs using the MIMO radar MIRA-CLE Ka," in *Proceedings of EUSAR 2016: 11th European Conference on Synthetic Aperture Radar*, 2016, pp. 1–4.

[18] U. Böniger, B. Ott, P. Wellig *et al.*, "Detection of mini-UAVs in the presence of strong topographic relief: a multisensor perspective," in *Target and Background Signatures II*, 2016, vol. 9997, p. 999702.

[19] L. Fuhrmann, O. Biallawons, J. Klare, R. Panhuber, R. Klenke, and J. Ender, "Micro-Doppler analysis and classification of UAVs at Ka band," in *Proceedings International Radar Symposium*, 2017.

[20] S. Rahman and D. A. Robertson, "Millimeter-wave micro-Doppler measurements of small UAVs," in *Proc. SPIE 10188, Radar Sensor Technology XXI*, 2017, vol. 10188, p. 101880T.

[21] S. Rahman and D. A. Robertson, "Time-frequency analysis of millimeter-wave radar micro-Doppler data from small UAVs," in *2017 Sensor Signal Processing for Defence Conference, SSPD 2017*, 2017-January, pp. 1–5.

[22] S. Rahman and D. A. Robertson, "FAROS-E: a compact and low-cost millimeter wave surveillance radar for real time drone detection and classification," in *Proceedings International Radar Symposium*, 2021, June 21–22, Germany.

[23] A. K. Singh and Y. H. Kim, "Automatic measurement of blade length and rotation rate of drone using W-band micro-Doppler radar," *IEEE Sens. J.*, vol. 18, no. 5, pp. 1895–1902, 2018.

[24] A. K. Singh and Y. H. Kim, "Accurate measurement of drone's blade length and rotation rate using pattern analysis with W-band radar," *Electron. Lett.*, vol. 54, no. 8, pp. 523–525, 2018.

[25] B. K. Kim, J. Park, S. J. Park *et al.*, "Drone detection with chirp-pulse radar based on target fluctuation models," *ETRI J.*, vol. 40, no. 2, pp. 188–196, 2018.

[26] H. Stefan, M. Roeding, G. Sommerken, R. Mueller, R. S. Thomae, and G. Del Galdo, "Contribution to drone detection by exploiting parameter estimation for a prototype mm-wave radar system – VDE conference publication," in *WSA 2018; 22nd International ITG Workshop on Smart Antennas*, 2018.

[27] P. Samczynski, K. Stasiak, D. Gromek, K. Kulpa, and J. Misiurewicz, "XY-DemoRad – Novel K- and mm-band radar demo kit for educational and commercial applications," in *Proceedings International Radar Symposium*, 2019-June.

[28] "Silicon Radar GmbH – Manufacturer of Radar Front Ends, MMICs, ASICs." [Online]. Available: https://siliconradar.com/. [Accessed: 26-Jan-2021].

[29] K. Stasiak, M. Ciesielski, P. Samczynski, D. Gromek, and K. Kulpa, "Preliminary results of drone's propellers detection using K-band and mm-Wave FMCW radar," in *Proceedings International Radar Symposium*, 2019-June.

[30] A. Cidronali, M. Passafiume, P. Colantonio *et al.*, "System level analysis of millimetre-wave GaN-based MIMO radar for detection of micro unmanned aerial vehicles," in *Progress in Electromagnetics Research Symposium*, 2019-June, pp. 438–450.

[31] S. Dogru, R. Baptista, and L. Marques, "Tracking drones with drones using millimeter wave radar," in *Advances in intelligent systems and computing (AISC)*, 2020, vol. 1093, pp. 392–402.

[32] S. Dogru and L. Marques, "Pursuing drones with drones using millimeter wave radar," *IEEE Robot. Autom. Lett.*, vol. 5, no. 3, pp. 4156–4163, 2020.

[33] "76 GHz Drone Detection Radar." [Online]. Available: http://elva-1.com/news_events/a40120. [Accessed: 05-Mar-2020].

[34] D. Barrett, D. Wang, A. Ahmad, and V. Mahimkar, "Using mmWave sensors to enhance drone safety and productivity," Texas Instruments (whitepaper) [Online]. Available: https://www.ti.com/lit/wp/spyy001/spyy001.pdf [Accessed: 02-Jul-2021].

[35] "Millimetre-wave Micro-Radar for Persistent Situational Awareness – Plextek DTS." [Online]. Available: https://plextek-dts.com/portfolio/mmwave-radar/. [Accessed: 05-Mar-2020].

[36] "Russian Firm Developing Submillimeter Radars to Detect Small Drones." [Online]. Available: https://www.defenseworld.net/news/26215/Russian_Firm_Developing_Submillimeter_Radars_to_Detect_Small_Drones#.Xwa8IOfTXIV. [Accessed: 09-Jul-2020].

[37] A. Schröder, M. Renker, U. Aulenbacher *et al.*, "Numerical and experimental radar cross section analysis of the quadrocopter DJI Phantom 2," in *2015 IEEE Radar Conference*, 2015, pp. 463–468.

[38] M. Röding, G. Sommerkorn, S. Häfner *et al.*, "Fully polarimetrie wideband RCS measurements for small drones," in *2017 11th European Conference on Antennas and Propagation, EUCAP 2017*, 2017, pp. 3926–3930.

[39] V. Semkin, J. Haarla, T. Pairon *et al.*, "Analyzing radar cross section signatures of diverse drone models at mmWave frequencies," *IEEE Access*, vol. 8, pp. 48958–48969, 2020.

[40] S. Rahman and D. A. Robertson, "In-flight RCS measurements of drones and birds at K-band and W-band," *IET Radar Sonar Navig.*, vol. 13, no. 2, pp. 300–309, 2019.

[41] D. A. Robertson and S. L. Cassidy, "Micro-Doppler and vibrometry at millimeter and sub-millimeter wavelengths," in *Proc. SPIE 8714, Radar Sensor Technology XVII*, 2013, vol. 8714, p. 87141C.

[42] D. A. Robertson, G. M. Brooker, and P. D. L. Beasley, "Very low-phase noise, coherent 94 GHz radar for micro-Doppler and vibrometry studies," in *Proc. SPIE 9077, Radar Sensor Technology XVIII*, 2014, vol. 9077, p. 907719.

[43] D. Blacknell and H. Griffiths, *Radar automatic target recognition (ATR) and non-cooperative target recognition (NCTR)*. Institution of Engineering and Technology, London, 2013.

[44] S. Rahman and D. A. Robertson, "Radar micro-Doppler signatures of drones and birds at K-band and W-band," *Sci. Rep.*, vol. 8, no. 1, p. 17396, 2018.

[45] S. Rahman and D. A. Robertson, "Millimeter-wave radar micro-Doppler feature extraction of consumer drones and birds for target discrimination," in *Radar Sensor Technology XXIII*, 2019, p. 28.

[46] S. Rahman and D. A. Robertson, "Multiple drone classification using millimeter-wave CW radar micro-Doppler data," in *Radar Sensor Technology XXIV*, 2020, vol. 11408, no. 23, p. 15.

[47] O. Biallawons, J. Klare, and L. Fuhrmann, "Improved UAV detection with the MIMO radar MIRA-CLE Ka using range-velocity processing and TDMA correction algorithms," in *Proceedings International Radar Symposium*, 2018-June.

[48] S. Rahman, D. A. Robertson, and M. A. Govoni, "Radar signatures of drones equipped with liquid spray payloads," in *IEEE Radar Conference (RadarConf20)*, Florence, 2020, pp. 1–5.

[49] "5L Economic Drone Agriculture Sprayer for Beginner | Drone Agriculture Sprayer | Joyance Tech." [Online]. Available: http://www.foggerdrone.com/html/sprayer/products/low_price_drone_agriculture_sprayer_for_b.html. [Accessed: 26-Jan-2021].

[50] S. Rahman, D. A. Robertson, and M. A. Govoni, "Radar signatures of drones equipped with heavy payloads and dynamic payloads generating inertial forces," *IEEE Access*, vol. 8, pp. 220542–220556, 2020.

[51] "'Recoil' Entry in Wikipedia." [Online]. Available: https://en.wikipedia.org/wiki/Recoil.

[52] S. Rahman and D. A. Robertson, "Classification of drones and birds using convolutional neural networks applied to radar micro-Doppler spectrogram images," *IET Radar Sonar Navig.*, vol. 14, no. 5, pp. 653–661, 2020.

[53] C. Szegedy, W. Liu, Y. Jia *et al.*, "Going deeper with convolutions," in *2015 IEEE Conference on Computer Vision and Pattern Recognition (CVPR)*, 2015, pp. 1–9.

Chapter 4

Detection and tracking of UAVs using an interferometric radar

Xiangrong Wang[1] and Xianbin Cao[1]

4.1 Background

As mentioned in the first chapter of this book,[*] it is urgent to develop effective countermeasures, such as unmanned aerial vehicle (UAV) detection, localization, tracking, recognition and interception systems, to eliminate the potential risks caused by UAVs [3,4]. A radar exhibits unique advantages for the detection and tracking of UAVs and lends itself well to a potentially effective countermeasure for UAVs [5–7].

However, the detection and tracking of UAVs are not that easy as they seem, as UAVs are usually small-sized and flying relatively slow at a low altitude compared with other air targets [8–10]. Small UAVs are flexible in manoeuvring various motions and swift course change, such as hovering, turning or circling; thus it is difficult to track the UAV due to lack of sufficient motion information and feasible mathematical models. As the radar cross section (RCS) of UAVs is comparable to that of flying birds, the distinguishing of UAVs from other air targets is also a challenging task [11,12]. Conventional Doppler radars may fail when dealing with UAV detection and tracking, as they only provide one-dimensional (1-D) motion information along the radial direction. When UAVs are circling around the observing radar, their radial movement is too small to be detected by Doppler radar. We have shown in [13,14] that the motion information along radial and angular directions are complementary: when the motion along one direction is small; conversely, the other is obvious. For two-dimensional (2-D) UAV trajectory tracking, the most concerned information is the angle of the moving target relative to the observing radar. In addition to that, angular velocity is also necessary for UAV tracking. Although a timely update of angle estimation is a substitute of angular velocity, the low update rate is not adequate for real-time tracking. Subsequently, some new methods for UAV detection and tracking are proposed in the literature, such as multi-static radars in [15] and multiple-input–multiple-output (MIMO) radars in [16–18]. However, it is difficult to maintain accurate synchronization for

[1]School of Electronics and Information Engineering, Beihang University, Beijing, China
[*]Portions reprinted, with permission, from References [1,2].

multi-static radars in practical implementation, while MIMO radar suffers from the compromise between prohibitive cost and high spatial resolution.

An interferometric radar is capable of providing both radial and angular velocity information in addition to the 2-D position calculated from range and angle [1,2,13,14,19,20], where radial velocity and angular velocity are proportional to Doppler and interferometric frequency, respectively. Comparing with multi-static radars, the signal acquisition and processing of all receiving channels of interferometric radars are conducted referring to the same clock. While different from MIMO radars, the application of interferometry renders the spanning of a larger aperture with a very small number of antennas, in turn achieving high spatial resolution. This chapter investigates the problem of UAV detection and tracking using an interferometric radar, equipped with one transmitting antenna and two widely spaced receiving antennas. UAV detection and tracking are completed by extracting the information of 2-D position and 2-D velocities from interferometric radar measurements. The range and radial velocity of UAVs can be estimated by performing respective fast-time and slow-time fast Fourier transform (FFT) to the received data of each channel. As mentioned earlier, 1-D information is far below the requirement when it comes to detecting and tracking UAVs, which are flexible in manoeuvring various motions and swift course change. Fortunately, we can estimate the angle and angular velocity of UAVs from the interferometric measurement. We then propose to track the UAV by predicting the next position with the information of range, angle and 2-D velocities, namely radial and angular velocities, in this chapter. Extensive simulation and experimental results are presented to validate the effectiveness of the method.

The remaining part of this chapter is organized as follows. An introduction of interferometry theory is provided in Section 4.2. The UAV detection is solved in Section 4.3. The proposed UAV tracking method is delineated in Section 4.4. Simulation and experimental results are presented in Sections 4.5 and 4.6, respectively.

4.2 Theory of interferometry

Interferometry is a processing technique which correlates the signals from different samples or sensors to calculate the phase difference and then extract useful information. By widely spacing the sensors and forming a large baseline, a high accuracy of transversal velocity estimation is achieved. Next, we elaborate on the basic interferometry principle in detail.

Consider a radar with one transmitter and two nominally identical receiving antennas observing a far-field monochromatic point scatterer, with the beam covering the interested field of view (FoV). The baseline, which is the distance between two receivers, is D, and the transmitter is in the middle of two receivers. The configuration of the interferometric radar system is illustrated in Figure 4.1.

Assume that the transmit signal at time instant t is denoted as $s_T(t)$. For a UAV target located in the far field, where its range relative to the transmitter is $R_T(t)$, and its angle relative to the radar boresight is $\theta(t)$. The transmit signal propagates in the

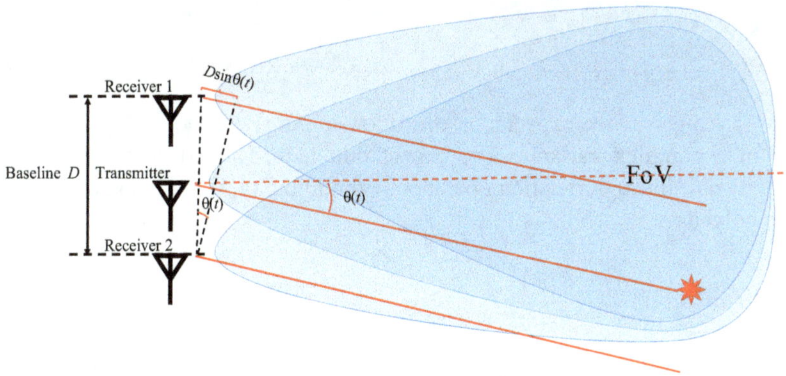

Figure 4.1 Interferometric radar system with one transmitter and two receivers

space and reaches the target. A small portion of the transmit energy is then back-scattered to two receivers, and the two received signals are denoted as $s_1(t)$ and $s_2(t)$, respectively. The range of the target relative to the receivers can be derived from a simple geometric relationship as follows:

$$R_1(t) = R_T(t) + \frac{D \sin\ \theta(t)}{2},$$
$$R_2(t) = R_T(t) - \frac{D \sin\ \theta(t)}{2}. \tag{4.1}$$

The time delay corresponding to the wave propagation path can be expressed by

$$\tau_k(t) = \frac{R_T(t) + R_k(t)}{c}, \qquad k = 1, 2, \tag{4.2}$$

where c is the speed of light. The difference of time delay further induces the phase difference between two receivers, which is determined by

$$\Delta\phi(t) = \phi_1(t) - \phi_2(t) = \frac{2\pi c[\tau_1(t) - \tau_2(t)]}{\lambda} = \frac{2\pi D \sin\ \theta(t)}{\lambda}, \tag{4.3}$$

where $\lambda = c/f_0$ corresponds to the wavelength of carrier frequency. Hence, we can estimate the angle $\theta(t)$ of the target by

$$\theta(t) = \arcsin \frac{\Delta\phi(t)\lambda}{2\pi D}. \tag{4.4}$$

The phase difference can be obtained by taking the phase of the interferometric measurement $s_I(t)$, which is the expectation of correlation between two receiving signals. That is,

$$s_I(t) = \left\langle s_1(t)s_2^*(t) \right\rangle = \left\langle s_t(t - \tau_1(t))s_t^*(t - \tau_2(t)) \right\rangle, \tag{4.5}$$

and

$$\Delta\varphi(t) = \arg\{s_I(t)\}, \tag{4.6}$$

where $\langle\cdot\rangle$ represents the expectation operation and $\arg\{\cdot\}$ denotes the argument or phase of a complex variable. The expectation here can reduce the noise level caused by non-linear correlation operation. Combining (4.4), (4.5) and (4.6), $\theta(t)$ is finally solved as

$$\theta(t) = \arcsin\frac{\arg\{s_I(t)\}\lambda}{2\pi D}. \tag{4.7}$$

In order to obtain high spatial resolution, the baseline of the interferometer is normally far larger than the wavelength, thereby introducing the phase ambiguity in the measurable range of $[0, 2\pi]$. A number of phase unwrapping methods are proposed in the literature to estimate the ambiguity number, such as second-order difference array (SODA), SODA-based interferometry, correlative interferometer, common angle search and phase line length matching [21–29]. However, all these methods require additional receivers to resolve the phase ambiguity, which is not applicable to the configuration of one transmitter and two receivers discussed in this chapter. Therefore, the angle of the UAV target cannot be measured reliably from the phase difference per (4.7) and we are going to present another angle estimation approach in Section 4.4.

Nevertheless, angular velocity, which characterizes the change rate of the angular displacement of the UAV target relative to the radar boresight, can be obtained by taking the derivative of the two sides of (4.3) with respect to time. We have that

$$\begin{aligned}\frac{1}{2\pi}\frac{d}{dt}\Delta\varphi(t) &= \frac{d}{dt}\frac{D\sin\,\theta(t)}{\lambda},\\ \frac{1}{2\pi}\frac{d}{dt}\arg\{s_I(t)\} &= \frac{D}{\lambda}\frac{d}{dt}\sin\,\theta(t),\\ f_I(t) &= \frac{D}{\lambda}\omega(t)\cos\,\theta(t),\end{aligned} \tag{4.8}$$

where $\omega(t)$ is the angular velocity, and $f_I(t)$ is the instantaneous frequency of the interferometric signal, which can be obtained applying time–frequency analysis methods, such as short-time Fourier transform (STFT), wavelet transform and Hilbert–Huang transform, to the interferometric measurement $s_I(t)$.

Under the small angle assumption, that is the UAV target is moving within the area around the boresight direction, it can be regarded that $\cos\theta(t) \approx 1$. And thus the angular velocity can be simply estimated as

$$\omega(t) = f_I(t)\frac{\lambda}{D}. \tag{4.9}$$

For those UAV targets with a large moving area, the effect of the angle term $\cos\theta(t)$ on interferometric frequency $f_I(t)$ cannot be neglected. In this case, the angular velocity should be estimated using the exact formula as follows:

$$\omega(t) = f_I(t)\frac{\lambda}{D\cos\,\theta(t)}. \tag{4.10}$$

As we can see that, angle and angular velocity are coupled together in this scenario. To estimate the angular velocity, the angle has to be measured beforehand.

To conclude the method of angle and angular velocity measurement using interferometry, we first correlate the signals from two separate receivers and take the average to obtain the interferometric output. We then take the phase of the interferometric output and calculate the target angle per (4.7). Subsequently, time–frequency analysis is applied to the output and extracts the instantaneous interferometric frequency. Finally, we can calculate the angular velocity of the UAV target using either (4.8) or (4.10) based on the specific scenario.

Interferometry is a signal-processing mechanism and can be applied to any type of radar system, such as continuous wave (CW) radar, pulse radar and frequency-modulated continuous wave (FMCW) radar. The main concern with a CW radar is that it only provides velocity information and cannot estimate the relative distance between the UAV target and the observing radar. A pulse radar can be utilized for range estimation; however, its range resolution is relatively low due to the limited bandwidth restricted by sampling frequency. An FMCW radar achieves high-range resolution using small instantaneous transmit power, in turn associated with small size and weight, by continuously transmitting frequency modulated waveforms. For an FMCW radar, the range is calculated by estimating the frequency of the beat signal, which is the difference between the frequencies of received and transmitted signals, thus achieving large bandwidth (corresponds to high range resolution) regardless of sampling frequency. Therefore, the FMCW radar is especially suitable for applications that require high precision and portability, such as automotive radar and airborne radar. In the sequel, we extend the interferometry theory to an FMCW radar for UAV detection and tracking.

4.3 UAV detection

4.3.1 Range-Doppler processing

Detection of UAVs is accomplished based on a range-Doppler profile using an FMCW radar. Assume that the FMCW radar transmits a continuous linear frequency modulated signal with the carrier frequency f_0, bandwidth B and sweep time T. The transmitted signal can be expressed as

$$s_T(t) = \sum_{n=0}^{\infty} \exp\left\{-j2\pi\left[f_0(t-nT) + \frac{K}{2}(t-nT)^2\right]\right\}u(t-nT), \tag{4.11}$$

where $K = B/T$ is the chirp rate, and $u(t - nT)$ is a rectangular pulse with length T. We define the within-period time t_s as

$$t_s = t - nT = t \bmod T, \tag{4.12}$$

which is the time counting from the start of nth sweep period. Let us consider a UAV located at an initial distance of R_0 and moving with a relative radial velocity of v_r along radar LoS. The back-scattered signal from the UAV will have the same form, but with a delay τ. The received signal of the nth chirp can then be described as

$$s_R(t) = \exp\left\{-j2\pi\left[f_0(t_s - \tau(t)) + \frac{K}{2}(t_s - \tau(t))^2\right]\right\}, \tag{4.13}$$

where the time delay $\tau(t)$ is expressed as

$$\tau(t) = \frac{2R(t)}{c} = \frac{2(R_0 - v_r(t)t)}{c}, \tag{4.14}$$

which is related to the range and radial velocity of the UAV. The UAV is usually assumed to locate in the same range gate within each pulse repetition period. Thus, the distance between the UAV and the radar can be approximated by

$$R(t) \approx R[n] = R_0 + nTv_r[n]. \tag{4.15}$$

For a UAV target moving in the FoV of the FMCW radar, compared with the transmit signal, the received signal exhibits a delay induced by the range and a Doppler frequency shift induced by the radial velocity, as illustrated in Figure 4.2.

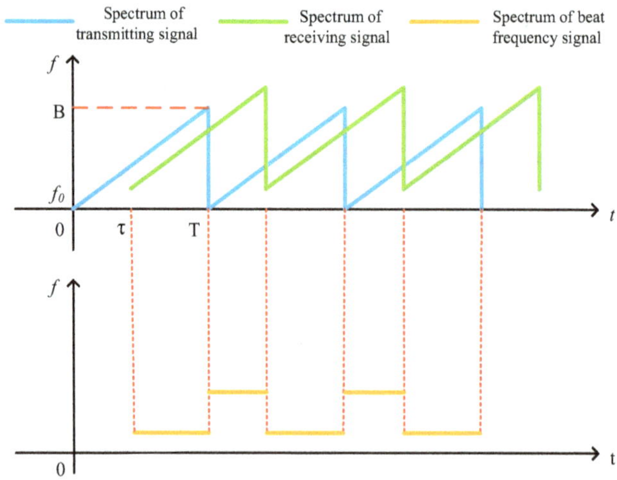

Figure 4.2 Frequency of transmitted signal and received signal (upper plot), the beat frequency signal (lower plot) after stretch processing

According to the FMCW radar principle, mixing the received signal by the first channel with the transmitted signal, we can obtain the beat signal. That is,

$$\begin{aligned} s_{B1}(t) &= s_T(t)s_{R1}^*(t), \\ &= \exp\left\{-j4\pi\left[\frac{K(R_0+nTv_r[n])}{c}t_s+\frac{v_r[n]}{\lambda}nT+\frac{R_0}{\lambda}+\frac{K(R_0+nTv_r[n])^2}{c^2}\right]\right\}, \\ &\approx \exp\left\{-j4\pi\left[\frac{K(R_0+nTv_r[n])}{c}t_s+\frac{v_r[n]}{\lambda}nT+\frac{R_0}{\lambda}\right]\right\}, \end{aligned} \tag{4.16}$$

where $\lambda = c/f_0$ is the wavelength corresponds to carrier frequency. The approximation in (4.16) utilizes the fact that the denominator c^2 is much larger than the numerator $K(R_0 + nTv_r[n])^2$.

We can observe from (4.16) that the range of the target in the nth pulse repetition period, $(R_0 + nTv_r[n])$, is obtained by taking FFT of the beat frequency signal within pulses, also known as fast-time FFT. The Doppler frequency shift, $f_d[n] = 2v_r[n]/\lambda$, is obtained by taking FFT over consecutive pulses, also known as slow-time FFT, supposing the velocity is constant over a CPI. Therefore, the range and radial velocity of the target can be estimated from the range-Doppler map by performing 2-D FFT on the beat signal. Figure 4.3 illustrates the range-Doppler

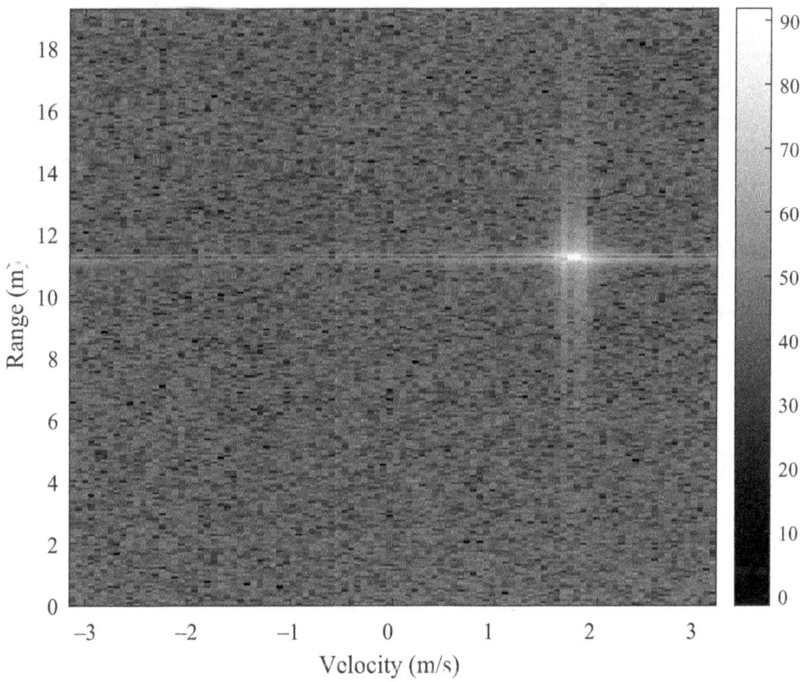

Figure 4.3 The range-Doppler map of the simulated moving target

map of a simulated target at $t = t_0$ with an instantaneous range of $R(t_0) = 11.2$ m and moving with a radial velocity of $v_r(t_0) = 1.8$ m/s. The signal-to-noise ratio (SNR) of the simulated target is 10 dB. The colour bar shows the intensity of each range bin and radial velocity cell in dB.

4.3.2 CFAR detection

In order to complete the detection of UAV targets, a sequence of post-processing steps is also required. Detection of targets is usually accomplished by comparing the output signal power with a predefined threshold [30]. As the noise and clutter power varies with different velocity and range cells and changes over time as well, it is inappropriate to set a constant threshold for UAV detection. A typical constant false-alarm rate (CFAR) algorithm can be employed to the range-Doppler map for UAV detection, which can automatically adjust the threshold to keep the false-alarm rate constant even when the interference and noise level changes dynamically.

Here, we apply cell-averaging CFAR detector in both range and velocity dimensions. For each concerned test cell, we take some neighbouring cells as the training cells and assume that the training cells only consist of background noise and do not contain the signal backscattered from the target. Usually, the target occupies more than one cell, so it is necessary to insert some guard cells between the test cell and training cells to prevent signal power leakage. The relative positions of the test cell, guard cells and training cells are illustrated in Figure 4.4. Taking the average of the intensity of training cells, if the ratio between the intensity of the test cell and the averaged training cell is larger than the threshold, this test cell is preserved. Otherwise, its value is set to zero. If there are non-zero cells after CFAR processing, we consider that a target is detected in the corresponding test cell.

For the range-Doppler map in Figure 4.3, we apply CFAR detector in both range and velocity dimension. We set the size of guard cells as six cells and training cells as ten cells. The ratio threshold is 17 dB. Figure 4.5 shows the range-Doppler map after applying CFAR detector. The colour bar shows the intensity of each range and radial velocity cell in dB, and black colour indicates that the bin is set as zero by CFAR detector. We can observe that the noise is sufficiently suppressed and the cells containing the majority of the target energy are remained. Thus, this algorithm significantly reduces the false-alarm rate.

4.3.3 Clustering

In practical applications, the modelled point scatterer of the UAV target might be perceived as an extended target from the range-Doppler map attributed to a few reasons, such as the sidelobe leakage of antenna beampattern, change of viewpoint, closeness to the observing radar and spectrum spreading induced by rotating rotors,

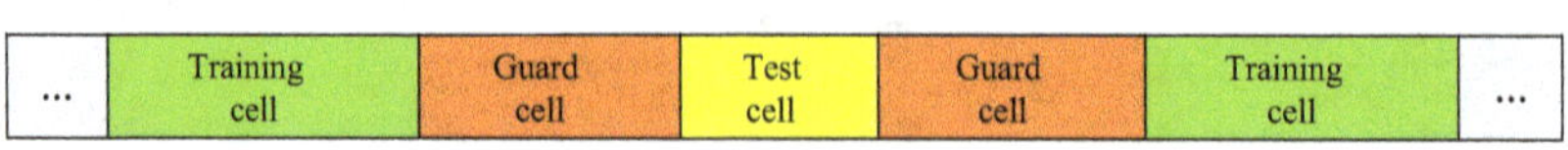

Figure 4.4 The illustration of CFAR

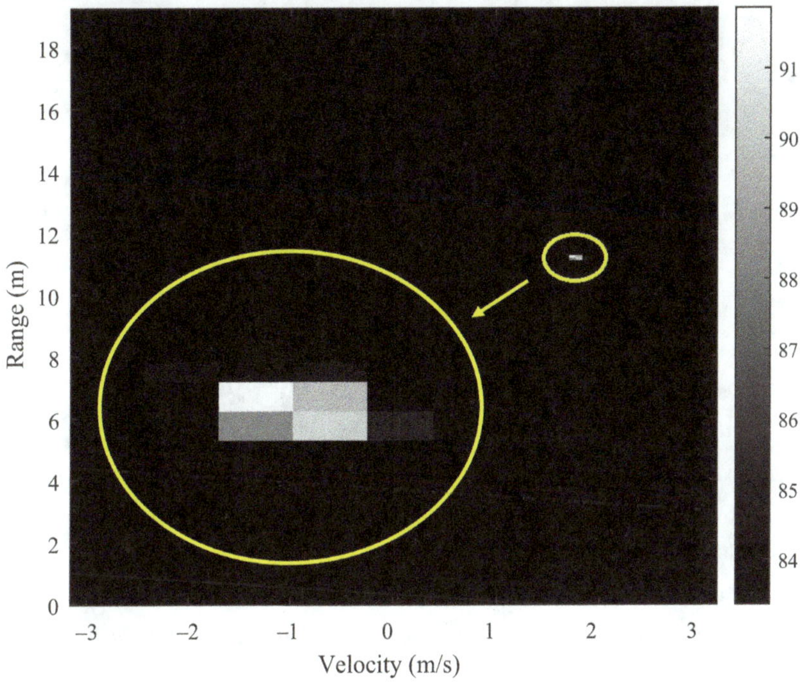

Figure 4.5 The range-Doppler map of the simulated moving target using a CFAR detector

etc. Put it differently, the reflected energy of the target will spread to the adjacent cells, all above the ratio threshold, resulting in the detection of the target in a number of range and velocity cells. In practical applications, we cannot simply treat each detected cell as an independent target. Otherwise, it will not only elevate system computational cost but also increase the false-alarm rate. Therefore, it is necessary to cluster closely spaced cells into one target and estimate the overall range and radial velocity of the target.

Here, we introduce a simple cell clustering algorithm. First, the cell with the highest intensity is located and chosen as the origin. Next, a target region is identified as the one adjacent to the origin cell by using four- or eight-connected criterion. Finally, the centroid is obtained by taking the intensity as the weight and performing a weighted average of the range and radial velocity in the target region. This weighted, averaged range and radial velocity corresponding to the centroid can be regarded as the final range and radial velocity of the target. Figure 4.6 illustrates the connected target region and the centroid of the simulated target using eight-connected criterion in the range-Doppler map after CFAR detection. We can see from Figure 4.6 that the clustered target accurately reflects the overall range and radial velocity of the true target. The procedure of UAV target detection is summarized in Figure 4.7.

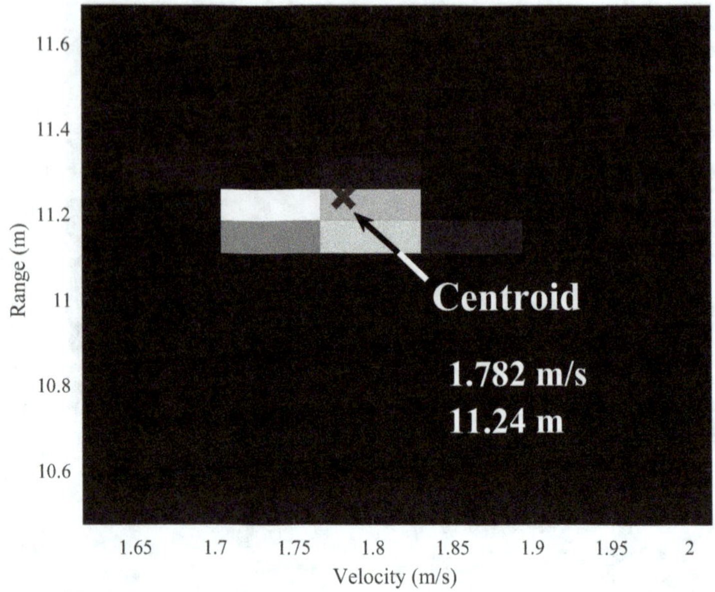

Figure 4.6 Target condensation of the simulated moving target

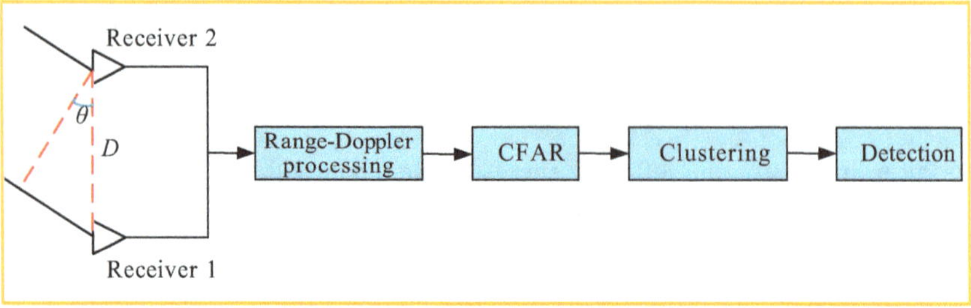

Figure 4.7 Schematic diagram of UAV target detection process

4.4 UAV tracking

Tracking of UAVs is accomplished based on the acquired information of range, angle and 2-D velocity. The principle of UAV tracking is delineated in Figure 4.8. Suppose that there is a target moving along the curve trajectory as shown in Figure 4.8. The initial range and angle of the target are denoted as R_0 and θ_0 at the time instant t_0. The linear velocity $v(t)$ of the target describes the position changing rate and its direction is along the tangential of the trajectory curve, which can be decomposed into two components, one is along the radar LoS, referred to as radial velocity $v_r(t)$, and the other is perpendicular to the radar LoS, referred to as transversal velocity $v_t(t)$. The transversal velocity contributes to the angular

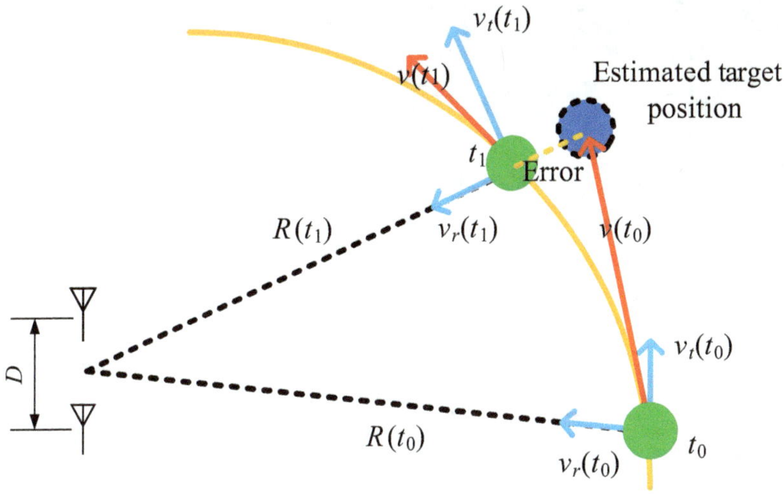

Figure 4.8 The illustration of UAV tracking principle

displacement of the object relative to the antenna boresight, and the angular velocity is $\omega(t)$, linearly proportional to the transversal velocity, that is, $v_t(t) = \omega(t)R(t)$. When the interval between two observations is small, the tangential line starting from the target position at time instant t_0 can be utilized to approximate the curve trajectory with an acceptable error, as illustrated in Figure 4.8. In summary, combining the information of target range, angle and 2-D velocity, namely radial and angular velocities, we are able to estimate the next position of the target, thus fulfilling the mission of UAV tracking.

As explained in Section 4.2, a single-receiver FMCW radar can only measure the range and radial velocity of the target relative to the radar and could not provide the 2-D information. Therefore, we utilize interferometric FMCW radar to simultaneously measure both the radial and transversal parameters, including position and velocity [31], delineated as follows.

4.4.1 Range and angle

It is worth to note that we implement 2-D FFT to obtain the range-Doppler map in the detection step, where a large processing gain can be secured by the long-time accumulation. While, in the tracking step, the timely update of both the range and radial velocity is required, which implies the window length of signal-processing frame needs to be shortened.

If we take the transmitter as a reference, the range of the target relative to the transmitter can be determined by a simple geometrical relationship, as shown in Figure 4.9. That is,

$$\frac{\tilde{D}^2 + R[n]^2 - R_1[n]^2}{2\tilde{D}R[n]} + \frac{\tilde{D}^2 + R[n]^2 - R_2[n]^2}{2\tilde{D}R[n]} = 0, \tag{4.17}$$

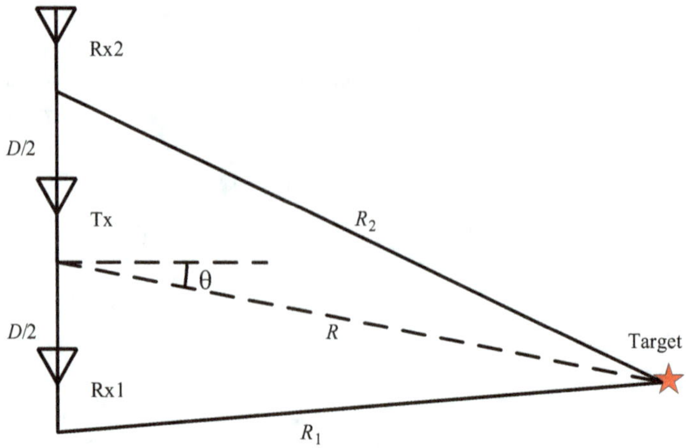

Figure 4.9 The geometry of the interferometric radar

where $R[n]$ is the range of the UAV target relative to the transmitter in nth chirp, the range $R_{Bk}[n]$ is the distance from the transmitter to the target then back to the kth receiver, and thus $R_1[n] = R_{B1}[n] - R[n], R_2[n] = R_{B2}[n] - R[n]$ are the range of the UAV target relative to two receivers. And $\tilde{D} = D/2$ is the distance between transmitter and receiver. We can then solve for $R[n]$, which can be expressed by the following expression:

$$R[n] = \frac{2R_{B1}[n]^2 + 2R_{B2}[n]^2 - D^2}{4(R_{B1}[n] + R_{B2}[n])}. \tag{4.18}$$

We can also derive the angle of the target with respect to the transmitter. That is,

$$\theta[n] = \frac{\pi}{2} - \arccos\frac{\tilde{D}^2 + R[n]^2 - R_1[n]^2}{2\tilde{D}R[n]}, \tag{4.19}$$

where $\theta[n]$ ranges from $-\pi/2$ to $\pi/2$.

If the target satisfies the far-field condition, that is, $R[n] \gg D$, the wave-front of target reflected signal is planar and the propagation path can be considered parallel. In this scenario, the triangular geometrical relationship in Figure 4.9 can be approximated by Figure 4.10.

We can observe that, if the transmitter is placed in the middle of two receivers, the ranges from the target to the transmitter and receivers have the following relationship,:

$$R[n] = \frac{R_1[n] + R_2[n]}{2}. \tag{4.20}$$

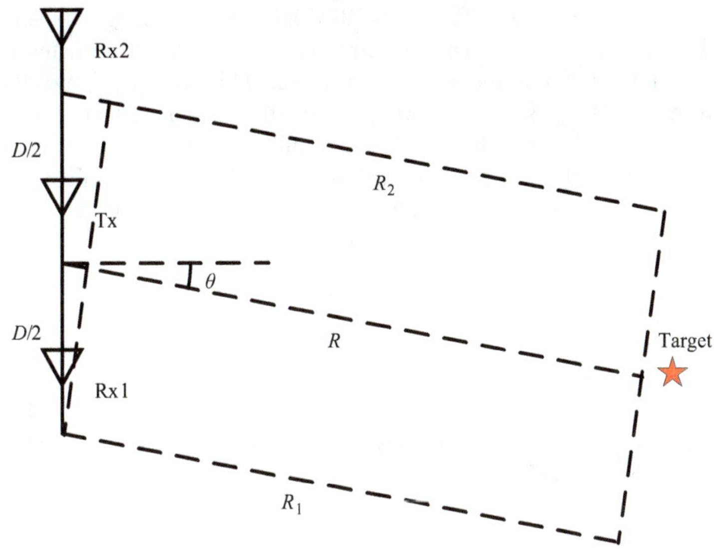

Figure 4.10 The geometry of the interferometric radar under far-field condition

Therefore, the range of the target can be approximated by

$$R[n] = \frac{R_{B1}[n] + R_{B2}[n]}{4}. \tag{4.21}$$

We also notice that angle $\theta[n]$ can be approximately calculated by the following equation:

$$\theta[n] = \arcsin \frac{R_2[n] - R_1[n]}{D} = \arcsin \frac{R_{B2}[n] - R_{B1}[n]}{D}. \tag{4.22}$$

Note that (4.7) and (4.22) are essentially the same as the range difference is linearly proportional to the phase difference between two receiving channels, while (4.22) eliminates the phase ambiguity and argument operation.

In real implementation, we can estimate the return range, $R_{Bk}, k = 1, 2$, by performing fast-time FFT to the beat signals corresponding to two receivers. For receiver $k = 1, 2$, beat signal s_{Bk} is first reshaped into an $M \times N$ matrix $\tilde{s}_{Bk}$, where M is the sampling points in one chirp, and N is the chirp number. That is,

$$\tilde{s}_{Bk}[m, n] = s_{Bk}[m + (n-1)M], \qquad m = 1, \ldots, M; \quad n = 1, \ldots, N. \tag{4.23}$$

Then, for every chirp $n = 1, \ldots, N$, we perform FFT to obtain the frequency $f_{rk}[n]$ with the maximum magnitude. The return range $R_{Bk}[n]$ can then be estimated by

$$R_{Bk}[n] = \frac{f_{rk}[n]c}{K}. \tag{4.24}$$

Substituting (4.24) into (4.21) and (4.22) yields the timely range $R[n]$ and angle $\theta[n]$, both with respect to the transmitter, respectively. We can then determine the 2-D position information of the target and perform UAV target tracking. However, only considering the first-order motion information is not reliable and robust, especially for agile UAVs with high manoeuvrability. As explained in Section 4.2, additional second-order information can be inferred from the reflected signal, including radial and angular velocities, we can add them to the target-tracking process and improve tracking performance.

4.4.2 2-D velocity

Different from the detection stage, where a long-time accumulation is conducted to secure the processing gain, target tracking requires a timely update of the motion information. Therefore, STFT is usually employed instead of FFT. STFT is a conventional method that characterizes the frequency distribution of the signal versus time, which is defined as

$$\mathrm{STFT}(x) = S_p[f] = \sum_{n=1}^{N} x[n]w[n - p\tau]e^{-j2\pi fn}, \tag{4.25}$$

where $x[n]$ is the discrete signal, τ is the unit time delay of the sliding window, p indicates an integer multiple of unit time delay, a rectangular sliding window is defined as

$$w[n] = \begin{cases} 1, & n = 1, \ldots, L, \\ 0, & \text{else}, \end{cases} \tag{4.26}$$

where L is the size of the window. Note that the sliding window length L determines the trade-off between time and frequency resolution, which obeys the classic 'Uncertainty Principle'. Also, when L is large, STFT is equivalent to FFT with a large accumulation gain while sacrificing the information updating rate.

As shown in (4.16), which is copied again as follows:

$$s_{B1}(t) \approx \exp\left\{-j4\pi\left[\frac{K(R_0 + nTv_r[n])}{c}t_s + \frac{v_r[n]}{\lambda}nT + \frac{R_0}{\lambda}\right]\right\}.$$

We can observe that range profile can be obtained by applying fast-time FFT to each pulse, and then STFT is performed to the M_c concerned range bins to attain the Doppler spectrogram. Finally, the amplitude of the spectrograms corresponding to the M_c range bins are added up, which constitutes the final spectrogram. That is,

$$S_p(f, f') = \sum_{m=1}^{M}\sum_{n=1}^{N} \tilde{s}_{Bk}[m, n]e^{-j2\pi fm}w[n - p\tau]e^{-j2\pi f'n}, \tag{4.27}$$

$$S_p(f') = \sum_{m\in M_c} |S_p(f_m, f')|, \tag{4.28}$$

where $|\cdot|$ denotes absolute value. The illustration of this process is shown in Figure 4.11.

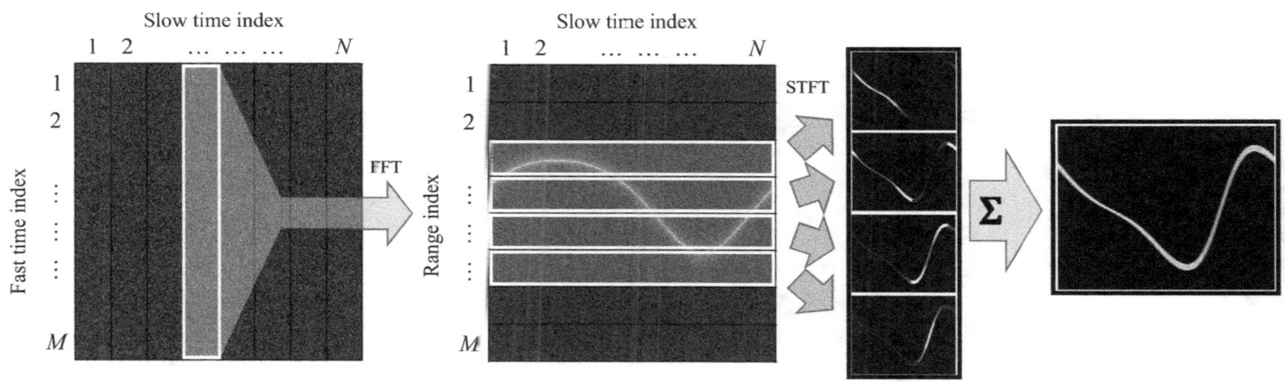

Figure 4.11 The illustration of Doppler spectrogram generation process

The Doppler frequency $f_{dk}[p]$ of the UAV target in the pth sliding window received by the kth antenna can be extracted from the spectrogram $S_p(f')$ by cell averaging algorithm introduced in Section 4.3.3. Then the radial velocity with respect to the kth receiver can be determined by

$$v_{rk}[p] = \frac{f_{dk}[p]\lambda}{2} = \frac{cf_{dk}[p]}{2f_0}. \tag{4.29}$$

In order to reduce the measurement error, we average the radial velocity estimates obtained by the two receivers. The final radial velocity estimate is

$$v_r[p] = \frac{v_{r1}[p] + v_{r2}[p]}{2}. \tag{4.30}$$

Next, we estimate the angular velocity $\omega[p]$ of the UAV target. The interferometry principle introduced in Section 4.2 is still applied here. Proceeding from (4.16), the beat frequency signal of the second antenna can be written as

$$\begin{aligned} s_{B_2}(t) &= s_T(t)s_{R_2}^*(t), \\ &= \exp\left\{-j2\pi\left[\frac{2K(R_0 + v_r nT)}{c}t_s + \frac{2v_r}{\lambda}nT + \frac{2R_0}{\lambda} + f_0\frac{D\sin\varphi}{c}\right.\right. \\ &\quad \left.\left. -\frac{K}{2}\left(\frac{D\sin\phi}{c}\right)^2 + K(t_s - \tau(t))\frac{D\sin\varphi}{c} - \frac{K(2R_0 + 2v_r nT)^2}{2c^2}\right]\right\}, \\ &\approx \exp\left\{-j2\pi\left[\frac{2K(R_0 + v_r nT)}{c}t_s + \frac{2v_r}{\lambda}nT + \frac{2R_0}{\lambda} + (f_0 + Kt_s)\frac{D\sin\varphi}{c}\right]\right\}. \end{aligned} \tag{4.31}$$

The two beat signals in (4.16) and (4.31) are correlated in the interferometer, generating the response,

$$\begin{aligned} s_I(t) &= \left\langle s_{B_1}(t)s_{B_2}^*(t)\right\rangle \\ &= \exp\left\{j2\pi(f_0 + Kt_s)\frac{D\sin\phi}{c}\right\}, \\ &= \exp\left\{j2\pi(f_0 + Kt_s)\frac{D\sin(\varphi_0 + n\omega T)}{c}\right\}. \end{aligned} \tag{4.32}$$

Under the small angle assumption, where the angle of the UAV target is kept near the boresight direction, (4.32) can be rewritten as

$$s_I(t) \approx \exp\left\{j2\pi(f_0 + Kt_s)\frac{D(\varphi_0 + n\omega T)}{c}\right\}. \tag{4.33}$$

We can argue from (4.33) that angle profile can be obtained by applying fast-time FFT to each pulse, and then STFT is performed to the M_{Ic} concerned angle bins to attain the interferometric spectrogram. Finally, the amplitudes of the spectrograms corresponding to the M_{Ic} angle bins are added up, which constitutes the final spectrogram. Specifically, we first reshape the interferometric signal s_I into an $M \times N$ matrix $\tilde{s}_I$, where M is the sampling points in one period, and N is chirp number. That is,

$$\tilde{s}_I[m,n] = s_I[m+(n-1)M], \qquad m = 1,\ldots,M; \quad n = 1,\ldots,N. \tag{4.34}$$

Then, for every period $n = 1,\ldots,N$, we perform fast-time FFT to extract the frequency $f_I[n]$, and the angle can be estimated as

$$\theta[n] = \arcsin \frac{cf_I[n]}{KD}. \tag{4.35}$$

After obtaining the angle profile, we can extract the angular velocity of the UAV target using the similar method for radial velocity. We first select a series of angle bins M_{Ic} which we concern about and perform STFT to these angle bins, respectively. The interested angle bins are usually determined by the monitoring area, or the radar beam width. Then, we perform STFT to these angle bins, respectively, add up the absolute value of the spectrograms corresponding to all M_{Ic} bins and obtain the final interferometric spectrogram. That is,

$$S_{Ip}(f,f') = \sum_{m=1}^{M}\sum_{n=1}^{N} \tilde{s}_I[m,n]e^{-j2\pi fm}w[n-p\tau]e^{-j2\pi f'n}, \tag{4.36}$$

$$S_{Ip}(f) = \sum_{m \in M_{Ic}} |S_{Ip}(f_m,f')|. \tag{4.37}$$

The spectrogram $S_{Ip}[f]$ shows the time-varying interferometric frequency $f_{DI}[p]$ of the UAV target, then the angular velocity can be determined by

$$\omega[p] = f_{DI}[p]\frac{\lambda}{D}, \tag{4.38}$$

which is consistent with (4.9). We refer to (4.35) and (4.38) as the pre-interferometry method.

By comparing the estimation methods of range and radial velocity with those of angle and angular velocity, we conclude that the two pairs are mathematical duals in terms of range corresponding to angle and radial velocity corresponding to angular velocity. This observation inspires another post-interferometry method of estimating angle and angular velocity, specifically,

$$f_I[n] = f_{r1}[n] - f_{r2}[n], \tag{4.39}$$

and

$$f_{DI}[p] = f_{d1}[p] - f_{d2}[p]. \tag{4.40}$$

In practical implementation, we prefer the post-interferometry method of (4.39) and (4.40), due to its simplicity and robustness against the noise.

4.4.3 Kalman filtering

Tracking is responsible for the track maintenance and termination of the detected UAV target. To cope with the noisy output of the detector, we use the well-known Kalman filtering to recursively predict the state variables over time from measurements and assumptions about the dynamics of the tracked object [15,32,33]. The radar measurements are somewhat inaccurate and may not be continuous in every frame. Kalman filtering helps by smoothing the inaccurate measurements and trying to predict the real values before updating the track. In this work, the state vector of the Kalman filter contains the 2-D position and velocity. Consider the state of the UAV target is consecutively measured and updated with a time interval Δt. The kth state of the target can be expressed as

$$\mathbf{x}_k = (\theta_k, R_k, \omega_k, v_{rk})^T. \tag{4.41}$$

An estimate of the state is denoted as $\bar{\mathbf{x}}_k = (\bar{\theta}_k, \bar{R}_k, \bar{\omega}_k, \bar{v}_{rk})^T$. Since there is no known control input, the state at current instant can be expressed as

$$\mathbf{x}_k = \mathbf{F}\mathbf{x}_{k-1} + \mathbf{v}, \tag{4.42}$$

where $\mathbf{F}$ is the state transition matrix defined as

$$\mathbf{F} = \begin{bmatrix} 1 & 0 & \Delta t & 0 \\ 0 & 1 & 0 & \Delta t \\ 0 & 0 & 1 & 0 \\ 0 & 0 & 0 & 1 \end{bmatrix}. \tag{4.43}$$

And $\mathbf{v}$ is the process noise with covariance $\mathbf{Q}$. In the continuous white noise acceleration motion model [34], the covariance $\mathbf{Q}$ obeys

$$\mathbf{Q} = \begin{bmatrix} \frac{1}{3}\Delta t^3 & 0 & \frac{1}{2}\Delta t^2 & 0 \\ 0 & \frac{1}{3}\Delta t^3 & 0 & \frac{1}{2}\Delta t^2 \\ \frac{1}{2}\Delta t^2 & 0 & \Delta t & 0 \\ 0 & \frac{1}{2}\Delta t^2 & 0 & \Delta t \end{bmatrix} q, \tag{4.44}$$

where Δt is the time interval between consecutive measurements, and the process noise q is determined by the motion feature of the target.

An estimation of the state is predicted by the previous state,

$$\bar{\mathbf{x}}_{k|k-1} = \mathbf{F}\bar{\mathbf{x}}_{k-1}; \tag{4.45}$$

and the predict error covariance is given by

$$\mathbf{P}_{k|k-1} = \mathbf{F}\mathbf{P}_{k-1}\mathbf{F}^T + \mathbf{Q}, \tag{4.46}$$

where $\mathbf{P}_{k-1}$ is the estimate covariance.

The state is updated by the measurement via an observation matrix **H**, which represents the 2-D position and velocity in this scenario, that is,

$$\mathbf{H} = \begin{pmatrix} 1 & 0 & 0 & 0 \\ 0 & 1 & 0 & 0 \\ 0 & 0 & 1 & 0 \\ 0 & 0 & 0 & 1 \end{pmatrix}. \tag{4.47}$$

The optimal Kalman gain $\mathbf{K}_k$ is then

$$\mathbf{K}_k = \mathbf{P}_{k|k-1}\mathbf{H}^T\left(\mathbf{H}\mathbf{P}_{k|k-1}\mathbf{H}^T + \mathbf{R}\right)^{-1}, \tag{4.48}$$

where **R** is the covariance of the observation noise.

Since fast-time data is used for the position estimation and slow-time data is used for velocity estimation, the measurements of position and velocity are assumed uncorrelated.

The updated state estimate $\bar{\mathbf{x}}_k$ is determined by

$$\bar{\mathbf{x}}_k = \bar{\mathbf{x}}_{k|k-1} + \mathbf{K}_k\left(\hat{\mathbf{x}}_k - \mathbf{H}\bar{\mathbf{x}}_{k|k-1}\right), \tag{4.49}$$

where $\hat{\mathbf{x}}_k = \left(\hat{\theta}_k, \hat{R}_k, \hat{\omega}_k, \hat{v}_{rk}\right)^T$ is the measurement at time step k. The estimate covariance $\mathbf{P}_k$ is then updated according to

$$\mathbf{P}_k = (\mathbf{I} - \mathbf{K}_k\mathbf{H})\mathbf{P}_{k|k-1}. \tag{4.50}$$

The proposed tracking algorithm based on the 2-D position and velocity information is summarized as follows. First, an interferometric FMCW radar monitors the interested region and obtains the range-Doppler map by implementing 2-D FFT on the received beat signal to obtain the range-Doppler map. Then, a CFAR detector is used to reduce the false-alarm rate and reliably detect the target. Next, cell clustering is performed to obtain the range and radial velocity of the overall target motion, which further reduces the false-alarm rate. Once the target is detected, we perform fast-time FFT and STFT to the received beat signals from two receivers, respectively, and extract the timely information of range, radial velocity, azimuth angle and angular velocity. Finally, we input these measurements into a Kalman filter and predict the trajectory of the UAV target. According to the estimation methods of angle and angular velocity, we summarize two tracking procedures, referred to as pre-interferometry tracking and post-interferometry tracking. The former utilizes (4.35) and (4.38) to estimate the angle and angular velocity, while the latter utilizes (4.39) and (4.40) for estimation. The two tracking procedures are illustrated in Figures 4.12 and 4.13, respectively.

4.5 Simulation results

To better explain and validate the algorithm in Section 4.4, we simulate a target rotating in front of the radar as an example. The radar operates at the carrier frequency of $f_0 = 24$ GHz and transmits a linear frequency modulated wave with a

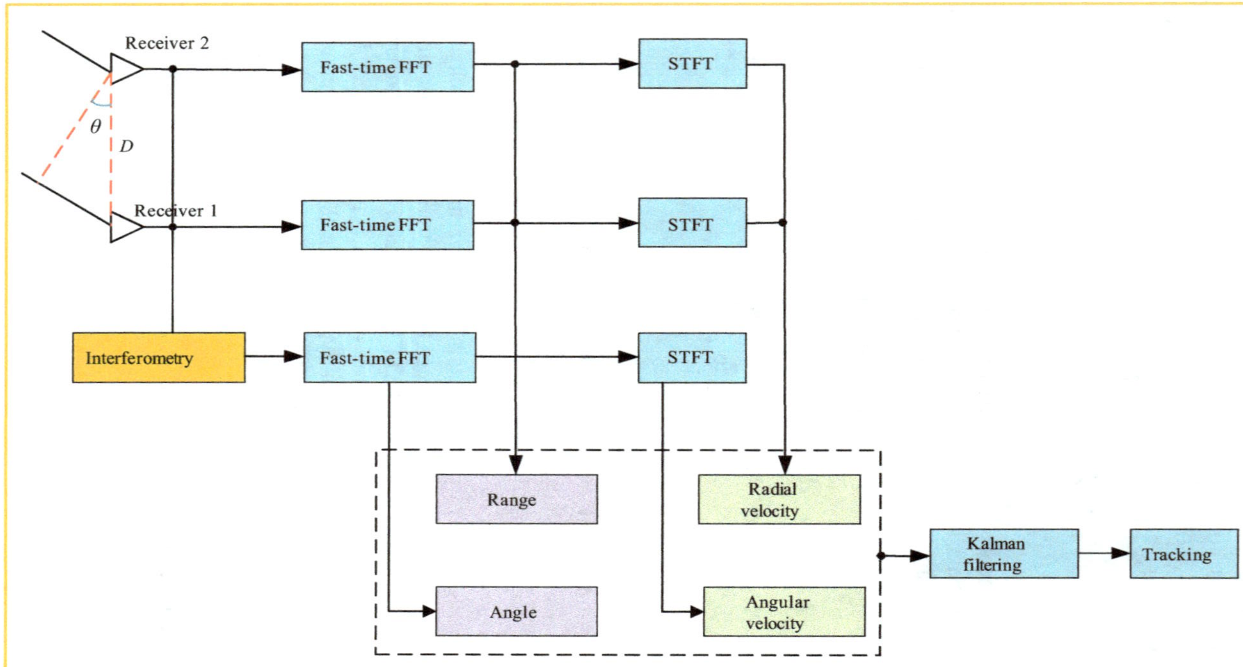

Figure 4.12 Schematic diagram of pre-interferometry tracking method

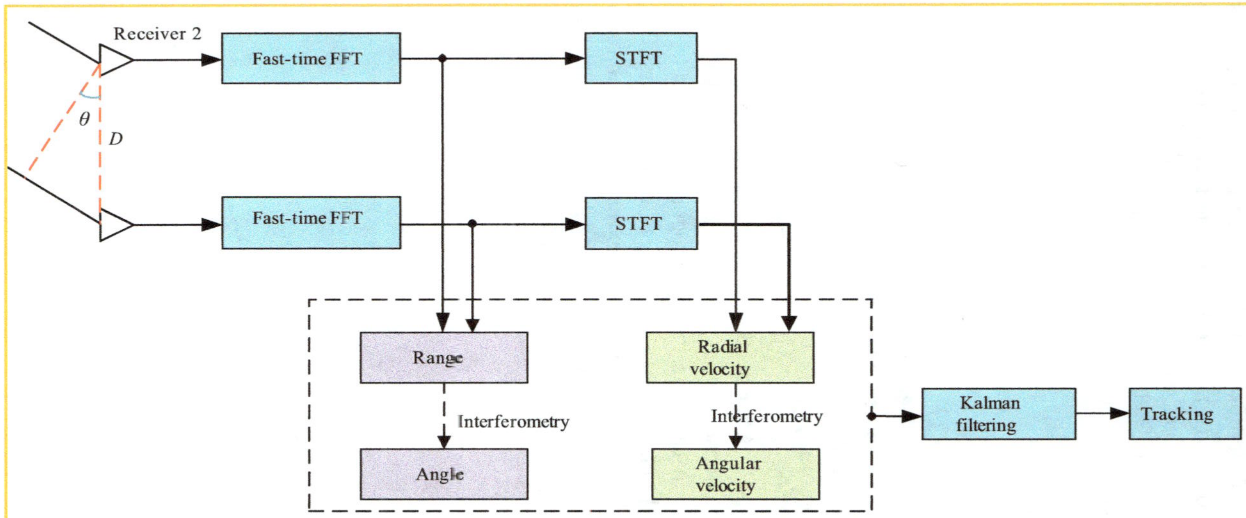

Figure 4.13 Schematic diagram of post-interferometry tracking method

bandwidth of $B = 1$ GHz. The chirp duration $T = 1$ ms, and the sampling rate $f_s = 256$ kHz. The total data length is 10 s, and the SNR is set as 10 dB. The two receivers are placed with a baseline $D = 1$ m. The transmitter, taken as a reference, is located at the origin of coordinate system (0,0) (m), and the receivers are located at $(0, -0.5)$ (m) and $(0, 0.5)$ (m), respectively. The rotation centre of the target is (10,0) (m), and the target moves on a trajectory of an ellipse with angular velocity with respect to the observing radar of $\omega = 0.1$ rad/s. The theoretical trajectory is described by the following equation:

$$\begin{cases} x = 10 + 3\sin(2\pi\omega t) & \text{(m)}, \\ y = 5\cos(2\pi\omega t) & \text{(m)}. \end{cases} \tag{4.51}$$

The simulational set-ups are illustrated in Figure 4.14. The true range, radial velocity, angle and angular velocity of the target are plotted in Figure 4.15.

We first reshape the received signals and perform 256-point fast-time FFT to obtain the range profile of two receivers per (4.24), which are shown in Figure 4.16. The range in Figure 4.16 represents R_{Bk} from the transmitter to the target then back to the kth receiver. All of the colour bars in Figure 4.16 represent the intensity in unit of dB. After obtaining the range profiles, we perform STFT to estimate Doppler frequencies, thus radial velocity. In the simulation, we add up all 256 range bins to calculate the spectrograms. Figure 4.17 presents the spectrograms of radial velocity corresponding to two receivers. From the spectrograms of range and radial velocity, we can extract the overall range and radial velocity of the target by picking up the value of maximum intensity, as shown in Figure 4.19(a). Similarly, we obtain the angle profile and spectrogram of angular

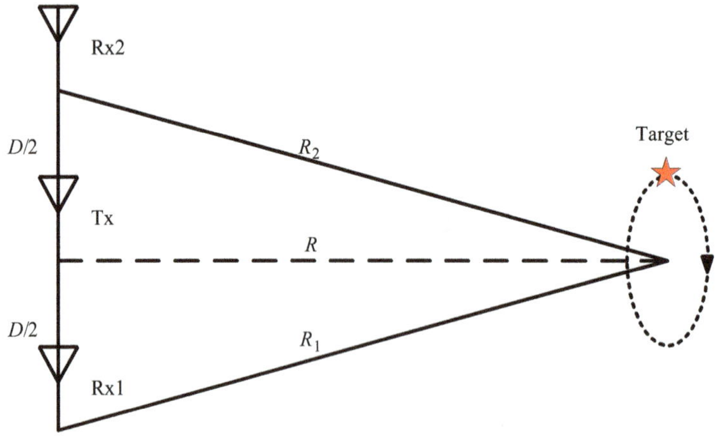

Figure 4.14 The set-ups of simulation

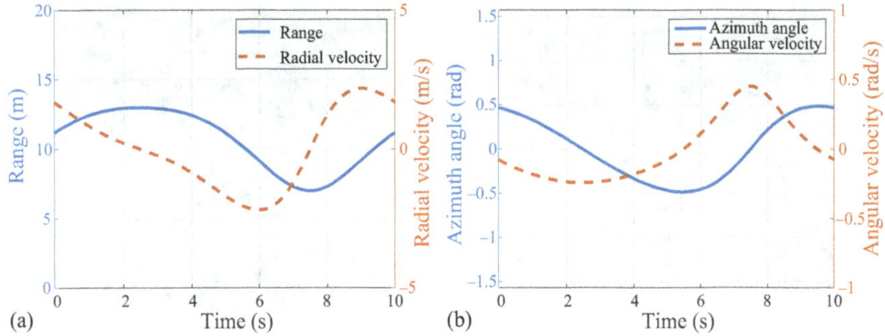

Figure 4.15 The true location and velocity of the simulated target: (a) radial and (b) angular

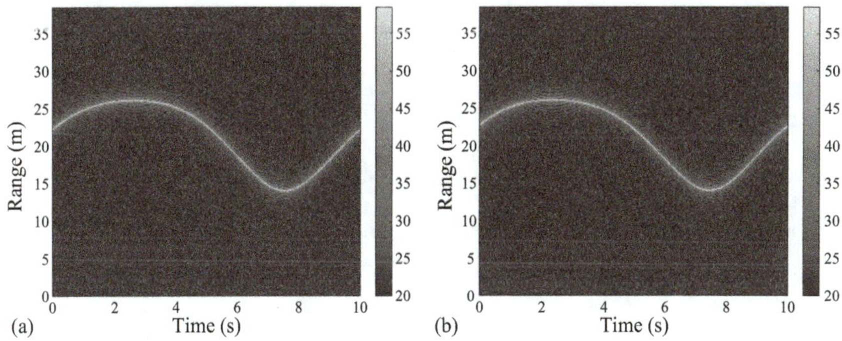

Figure 4.16 The range profiles of the simulated target: (a) receiver 1 and (b) receiver 2

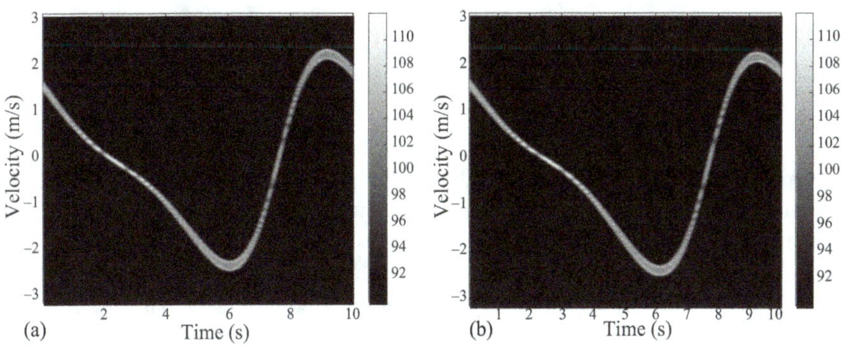

Figure 4.17 The spectrograms of the simulated target: (a) receiver 1 and (b) receiver 2

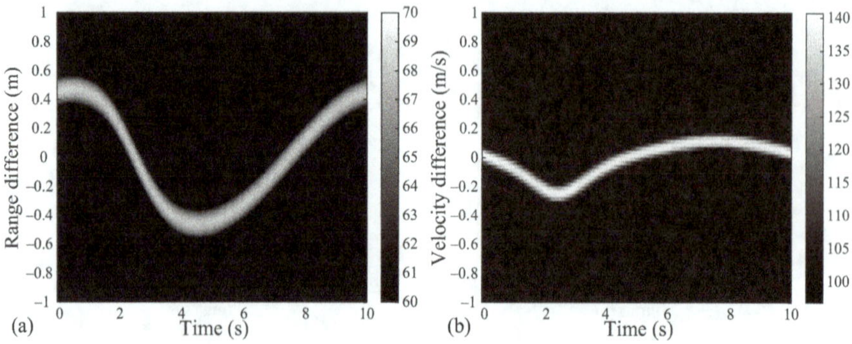

Figure 4.18 (a) The angle profile and (b) spectrogram of angular velocity of the simulated target

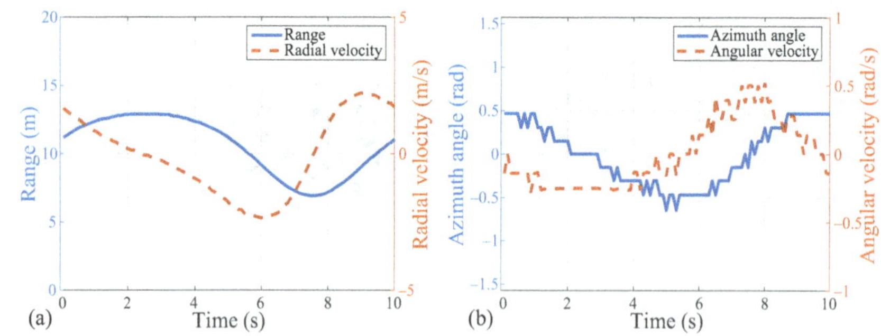

Figure 4.19 The estimated location and velocity of the simulated target: (a) radial and (b) angular

velocity in Figure 4.18, where the overall angle and angular velocity are estimated, as shown in Figure 4.19(b).

With four 2-D parameters, including range, radial velocity, angle and angular velocity ready, we then apply Kalman filtering to the noisy measurement and estimate the trajectory of the target. Parameters of the employed Kalman filter are set as follows: the process noise is $q = 20$ and the variance of the observation noise is a diagonal matrix with $(1, 10, 1, 10)$ populating along the diagonal. The Kalman filtered range, radial velocity, angle and angular velocity are depicted in Figure 4.20. It clearly shows that the Kalman filter can effectively alleviate the effect of noise and quantization error and smooth the measurements. The estimated 2-D trajectory is plotted in Figure 4.21, which approaches the true trajectory of the simulated target.

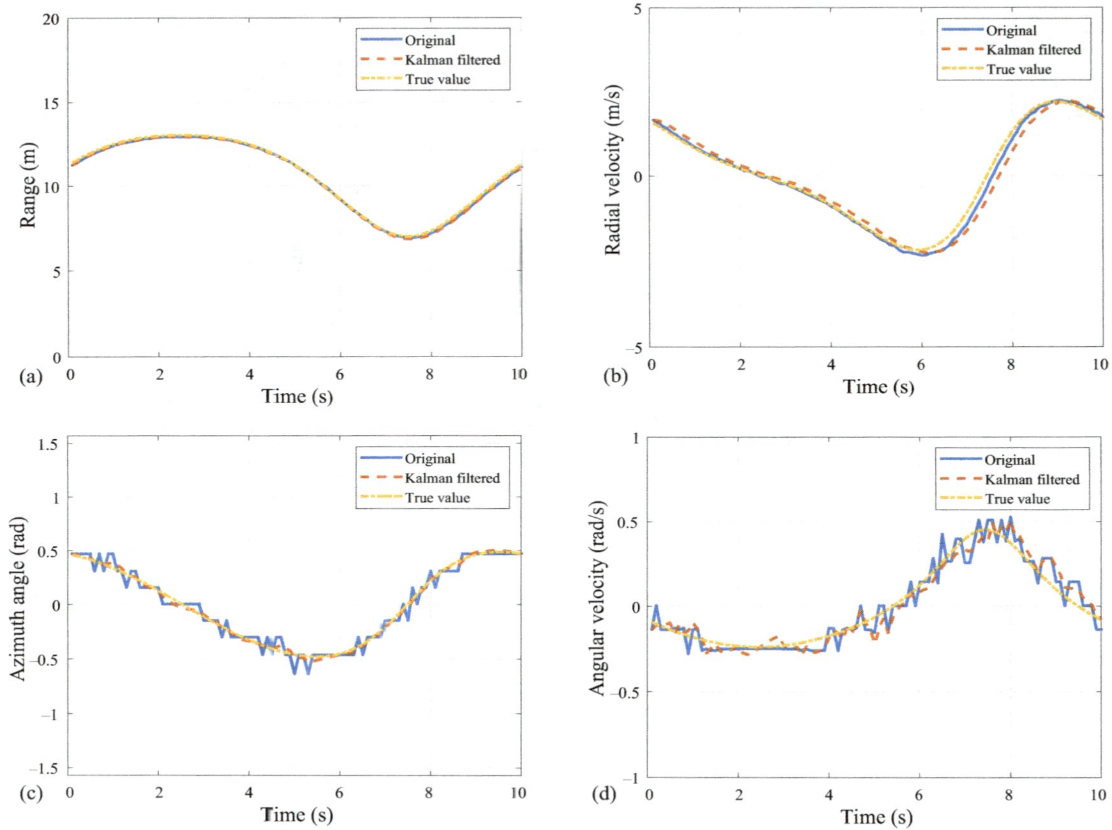

Figure 4.20 *The estimated location and velocity of the simulated target: (a) range, (b) radial velocity, (c) azimuth angle, (d) angular velocity*

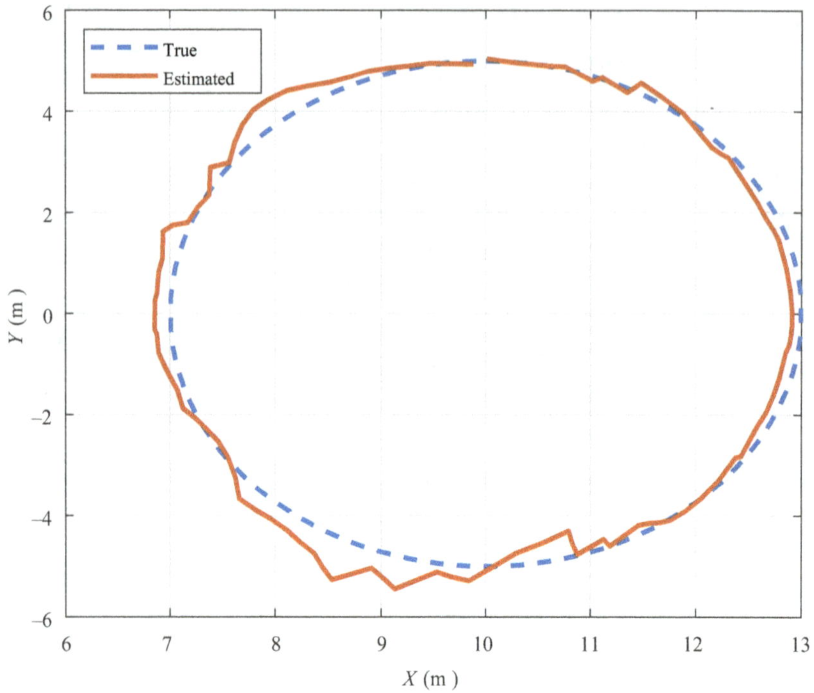

Figure 4.21 Estimated and true trajectory of the simulated target

4.6 Experimental results

We employed a small plastic UAV as the target to test the detection and tracking performance of the proposed algorithm in real experiments. The picture of UAV is shown in Figure 4.22 and the RCS is super small. The maximum flying time is only 2 min, which significantly affects the field experiment. The experiment is conducted indoor with a maximum distance of 30 m due to the UAV control. An FMCW interferometric radar with a carrier frequency of $f_0 = 24$ GHz, the bandwidth of $B = 2$ GHz and sweeping time of $T = 1$ ms is employed in the experiment. The UAV was controlled to move back and forth travelling along a straight line with an azimuth angle of $\theta = -15°$ with respect to radar LoS. It is difficult to stabilize the flying attitude and flying path of the UAV, especially in indoor environment. The antennas are steering towards $\theta = -15°$ by hand such that the UAV can remain staying in the radar main beam for maximum directional gain. The two receivers are placed on the two sides of transmitter by 0.65 m apart, and thus the baseline between the two receivers is $D = 1.3$ m. The experimental set-up is shown in Figure 4.23.

The detection performance of the small UAV is first tested using both simulated and experimental data. The simulation scenario is set as the same as that of the experimental set-up for comparison. The 2-D FFT is implemented on a total of

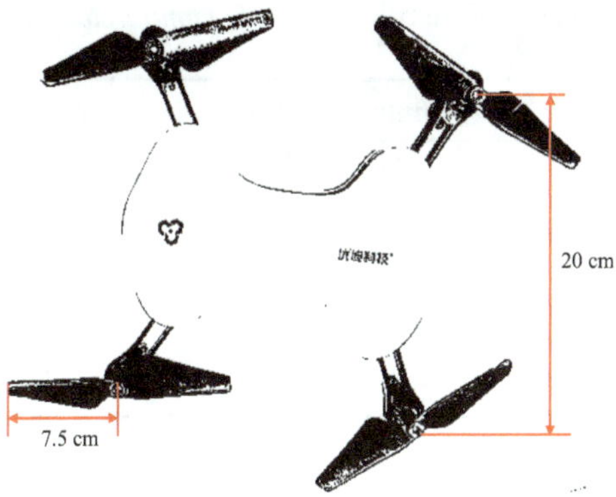

Figure 4.22 The picture of the small plastic UAV

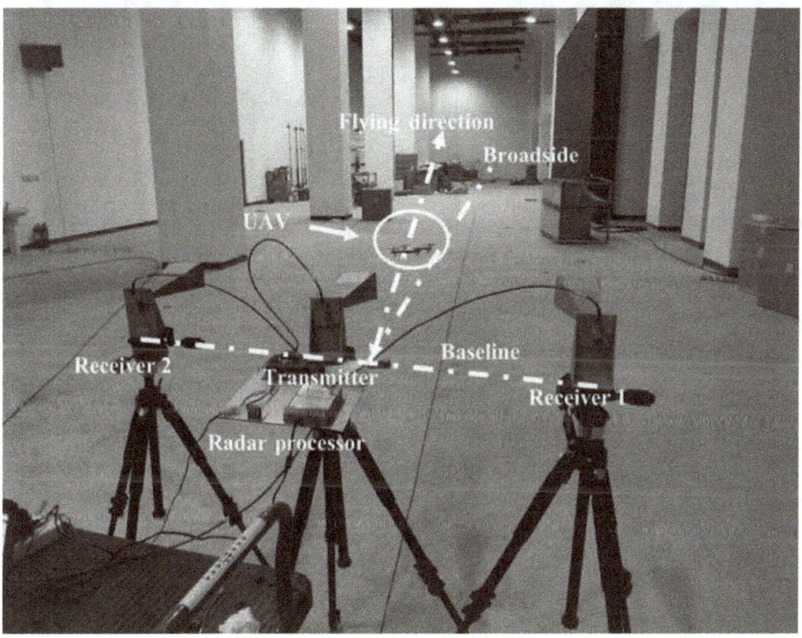

Figure 4.23 The set-up of indoor flying experiment

256 chirps to obtain the range-Doppler map for UAV detection. The distance between the UAV and the radar is varying from 5 to 65 m in varied steps. The detection probabilities of the UAV versus different distances are listed in Table 4.1 using both simulated and experimental data. We can obverse that experimental results agree well with simulation results, and the detection probability decreases dramatically when the distance is larger than 40 m attributed to the super small RCS (around 0.001 m^2) of the employed UAV model.

Table 4.1 Detection probability of UAV using simulated and experimental data

Distance (m)	**Simulation (%)**	**Experiment (%)**
5	100	100
10	100	100
15	100	100
30	99.9	80
40	52.8	50
50	30.1	25
65	0.4	0

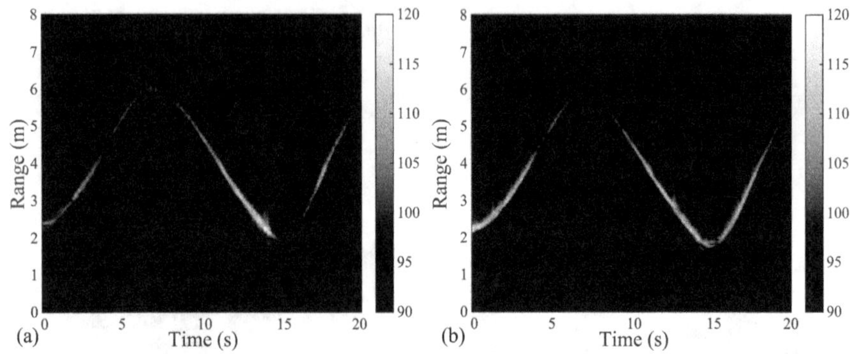

Figure 4.24 The range profiles of the real UAV target: (a) receiver 1 and (b) receiver 2

After collecting the received signals with a length of 20 s and implementing fast-time FFT, we obtained the range profile of the UAV target, which is presented in Figure 4.24. We can observe from the range profile of Figure 4.24 that the UAV target moves back and forth relative to the radar, which agrees with the real situation. Considering the radial-velocity profile, we selected range cells between 0 and 8 m and performed STFT for each range bin. The summation of spectrograms formulated the overall Doppler spectrogram containing the radial velocity of the UAV target, which is shown in Figure 4.25. The positive and negative velocities also indicate that the UAV target moved towards and away from the observing radar.

The range, angle, radial and angular velocity are further extracted, as shown in Figure 4.26(a) and (b) using the proposed post-interferometry method. Finally, we extract 2-D position and velocity information and estimate the trajectory of the UAV target. Parameters of the employed Kalman filter are as follows: the process noise is $q = 10$ and the variance of the observation noise is diagonal with $(20, 20, 20, 20)$ populating along the diagonal. The predicted tracking path is depicted in Figure 4.27, which agrees well with the real flight trajectory. The experiment validates the feasibility and effectiveness of the proposed method.

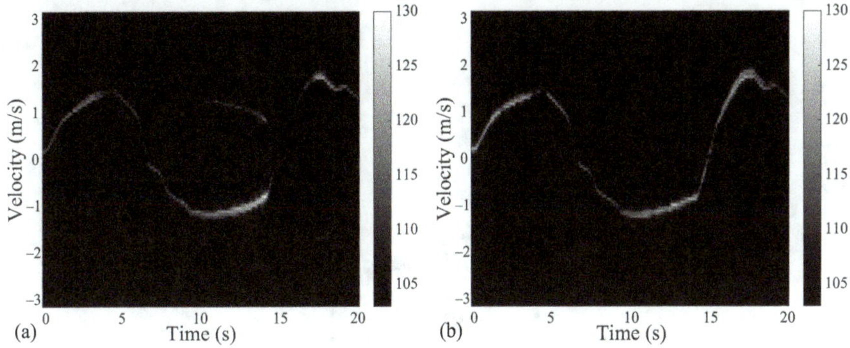

Figure 4.25 The spectrograms of the real UAV target: (a) receiver 1 and (b) receiver 2

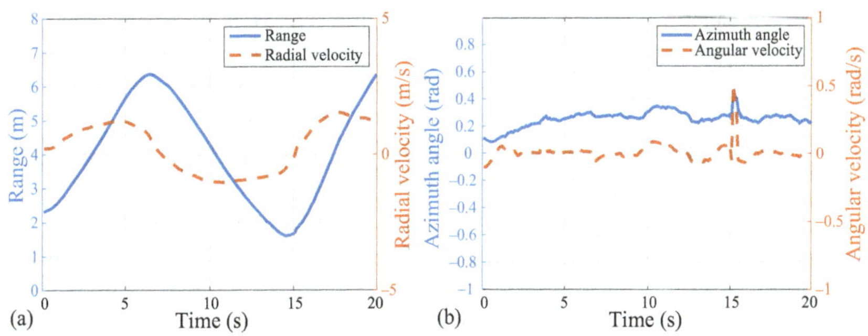

Figure 4.26 The estimated location and velocity of the UAV target: (a) radial and (b) angular

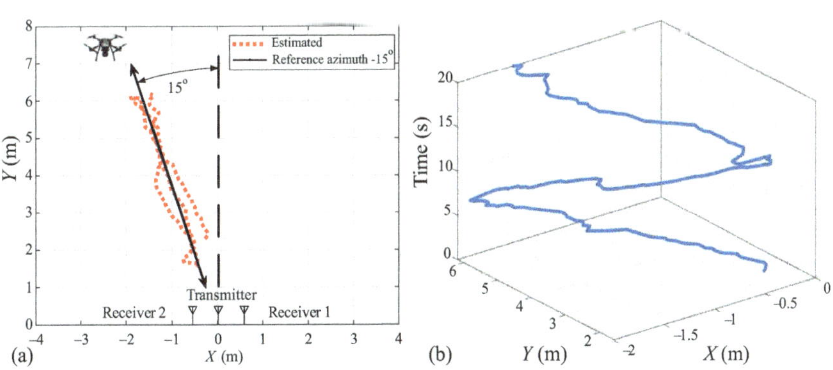

Figure 4.27 The estimated trajectory of the UAV target: (a) trajectory and (b) position versus time

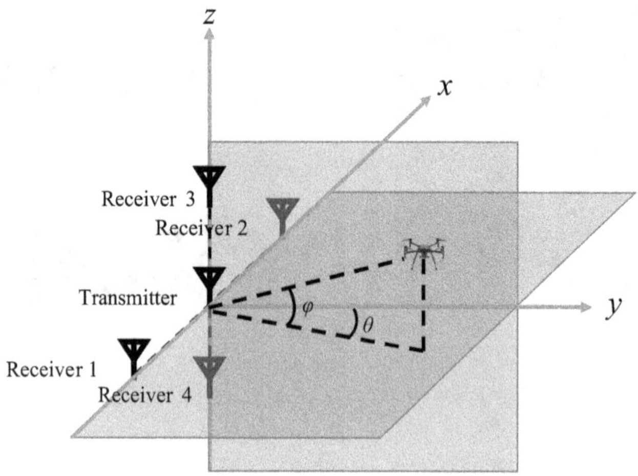

Figure 4.28 Configuration for UAV detection and tracking in 3-D space

4.7 Conclusion and future work

To conclude, interferometric processing of two widely spaced receivers enabled the extraction of 2-D position and velocity of UAV targets, that is, along the radial and transversal direction. Applying Kalman filtering to the estimated time-varying 2-D information, we were able to perform UAV detection and tracking. Simulations and experiments validated the proposed method. However, UAVs usually move in three-dimensional (3-D) space. In order to achieve real-time positioning of UAVs, parameters in azimuth and elevation are required simultaneously. The proposed work can be easily extended to 3-D space tracking by employing another receiver placed vertically above the transmitter (an optional fourth receiver placed vertically under the transmitter), as illustrated in Figure 4.28. The extension of the proposed work to the 3-D UAV detection and tracking will be explored in the near future.

References

[1] Wang X, Liang H, and Wang P. Detection and tracking of UAVs using interferometric radar. In: 2019 International Radar Conference (RADAR). IEEE; 2019. p. 1–6.

[2] Liang H, Wang P, and Wang X. Simultaneous tracking of multiple targets using interferometric FMCW radar. In: 2019 IEEE International Conference on Signal, Information and Data Processing (ICSIDP). IEEE; 2019. p. 1–6.

[3] Fioranelli F, Ritchie M, Griffiths H, *et al.* Classification of loaded/unloaded micro-drones using multistatic radar. Electronics Letters. 2015;51(22):1813–1815.

[4] Ritchie M, Fioranelli F, and Borrion H. Micro UAV crime prevention: can we help Princess Leia? In: Crime Prevention in the 21st Century. Cham: Springer International Publishing AG; 2017. p. 359–376.

[5] Drozdowicz J, Wielgo M, Samczynski P, *et al.* 35 GHz FMCW drone detection system. In: 2016 17th International Radar Symposium (IRS). IEEE; 2016. p. 1–4.
[6] Chen VC, Li F, Ho SS, *et al.* Micro-Doppler effect in radar: phenomenon, model, and simulation study. IEEE Transactions on Aerospace and Electronic Systems. 2006;42(1):2–21.
[7] de Wit JM, Harmanny R, and Premel-Cabic G. Micro-Doppler analysis of small UAVs. In: 2012 9th European Radar Conference. IEEE; 2012. p. 210–213.
[8] Tahmoush D. Detection of small UAV helicopters using micro-Doppler. In: Radar Sensor Technology XVIII. vol. 9077. International Society for Optics and Photonics; 2014. p. 907717.
[9] Björklund S. Target detection and classification of small drones by boosting on radar micro-Doppler. In: 2018 15th European Radar Conference (EuRAD). IEEE; 2018. p. 182–185.
[10] Harmanny R, De Wit J, and Cabic GP. Radar micro-Doppler feature extraction using the spectrogram and the cepstrogram. In: 2014 11th European Radar Conference. IEEE; 2014. p. 165–168.
[11] Ritchie M, Fioranelli F, Griffiths H, *et al.* Micro-drone RCS analysis. In: 2015 IEEE Radar Conference. IEEE; 2015. p. 452–456.
[12] Ritchie M, Fioranelli F, Griffiths H, *et al.* Monostatic and bistatic radar measurements of birds and micro-drone. In: 2016 IEEE Radar Conference (RadarConf). IEEE; 2016.
[13] Wang X, Wang P, and Chen VC. Simultaneous measurement of radial and transversal velocities using interferometric radar. IEEE Transactions on Aerospace and Electronic Systems. 2019.
[14] Wang X, Wang P, Cao X, *et al.* Interferometric angular velocity measurement of rotating blades: theoretical analysis, modeling and simulation study. IET Radar, Sonar & Navigation. 2018;13(3):438–444.
[15] Hoffmann F, Ritchie M, Fioranelli F, *et al.* Micro-Doppler based detection and tracking of UAVs with multistatic radar. In: 2016 IEEE Radar Conference (RadarConf). IEEE; 2016. p. 1–6.
[16] Jian M, Lu Z, and Chen VC. Drone detection and tracking based on phase-interferometric Doppler radar. In: 2018 IEEE Radar Conference (RadarConf18). IEEE; 2018. p. 1146–1149.
[17] Klare J, Biallawons O, and Cerutti-Maori D. Detection of UAVs using the MIMO radar MIRA-CLE Ka. In: Proceedings of EUSAR 2016: 11th European Conference on Synthetic Aperture Radar. VDE; 2016. p. 1–4.
[18] Rankin G, Tirkel A, and Leukhin A. Millimeter wave array for UAV imaging MIMO radar. In: 2015 16th International Radar Symposium (IRS). IEEE; 2015. p. 499–504.
[19] Nanzer JA. Micro-motion signatures in radar angular velocity measurements. In: 2016 IEEE Radar Conference (RadarConf). IEEE; 2016. p. 1–4.
[20] Wang X, Wang P, and Chen VC. Simultaneous measurement of radial and transversal velocities using a dual-frequency interferometric radar. In: 2019 IEEE Radar Conference (RadarConf). IEEE; 2019. p. 1–6.

[21] Holder J. Angle-of-Arrival Estimation Using Radar Interferometry: Methods and Applications. vol. 2. Edison, NJ: SciTech Publishing; 2013.

[22] Mollai S and Farzaneh F. Compact cross form antenna arrays intended for wideband two dimensional interferometric direction finding including the channel phase tracking error. AEU-International Journal of Electronics and Communications. 2018;83:558–565.

[23] Van Doan S, Vesely J, Janu P, *et al.* Optimized algorithm for solving phase interferometer ambiguity. In: 2016 17th International Radar Symposium (IRS). IEEE; 2016. p. 1–6.

[24] Ly PQ, Elton SD, Li J, *et al.* Computationally fast AOA estimation using sparse large aperture arrays for electronic surveillance. In: 2013 International Conference on Radar. IEEE; 2013. p. 526–531.

[25] Ly PQ, Elton SD, Gray DA, *et al.* Unambiguous AOA estimation using SODA interferometry for electronic surveillance. In: 2012 IEEE 7th Sensor Array and Multichannel Signal Processing Workshop (SAM). IEEE; 2012. p. 277–280.

[26] Lin E and Monte L. Joint frequency and angle of arrival estimation using the Chinese Remainder Theorem. In: 2017 IEEE Radar Conference (RadarConf). IEEE; 2017. p. 1547–1551.

[27] Jacobs E and Ralston EW. Ambiguity resolution in interferometry. IEEE Transactions on Aerospace and Electronic Systems. 1981; AES-17(6):766–780.

[28] Searle S. An examination of bias in SODA interferometry. IEEE Signal Processing Letters. 2017;24(4):470–474.

[29] Liang H, Wang X, and Farina A. Performance analysis and optimal placement for unambiguous DOA estimation using interferometry. In: 2020 IEEE Radar Conference (RadarConf). Florence, Italy; 2020. p. 1–6.

[30] Chen WK. The VLSI Handbook. New York, NY: CRC Press; 2018.

[31] Nanzer JA. Interferometric detection of the angular velocity of moving objects. In: 2010 IEEE MTT-S International Microwave Symposium. IEEE; 2010. p. 1628–1631.

[32] Fuhrmann L, Biallawons O, Klare J, *et al.* Micro-Doppler analysis and classification of UAVs at Ka band2. IEEE; 2017. p. 1–9.

[33] Wang X and Lu X. Three-dimensional impact angle constrained distributed guidance law design for cooperative attacks. ISA Transactions. 2018;73:79–90.

[34] Bar-Shalom Y, Li XR, and Kirubarajan T. Estimation With Applications to Tracking and Navigation: Theory Algorithms and Software. New York, NY: John Wiley & Sons; 2004.

Chapter 5

Passive radar detection of drones with staring illuminators of opportunity

Nasyitah Ghazalli[1], *Alessio Balleri*[1], *Mohammed Jahangir*[2], *Fabiola Colone*[3] *and Chris Baker*[2]

Passive Bistatic Radars (PBRs) use non-cooperative illuminators of opportunity to detect, localise and track targets. They have attracted considerable research interest in recent years because they can be operated and deployed at a relatively low cost, are difficult to detect and hence allow covert operations in hostile environments and do not require the allocation of an increasingly more congested frequency spectrum. Various analogue and digital communication systems (such as television (TV) and radio broadcast systems) have been studied and exploited as illuminators of opportunity. However, despite the extensive work carried out on PBRs that exploit random communication signals, there has been limited research investigating the use of existing non-cooperative radar systems as illuminators of opportunity. The exploitation of radar signals to achieve passive bistatic detection is attracting attention as it may offer significant advantages. Because common radar waveforms are deterministic, a reference channel is essentially not required to detect a target. Prior intelligence or live estimations of the deterministic waveform design parameters allow the passive receiver to be matched with the illuminator of opportunity and thus generate a range–Doppler map. Radar signals are also designed for detection and provide Doppler tolerance, large bandwidths (which provide good range resolutions) and good compression ratios.

This chapter presents a PBR solution that exploits non-random signals transmitted by a non-cooperative staring radar system to detect drones. Staring radars offer a constant illumination of the volume under surveillance and, unlike radar systems that deploy a rotating antenna, offers a continuous signal of opportunity. They are very attractive illuminators in particular for short-range applications to detect low Radar Cross Section (RCS) and slow moving targets.

[1]Centre for Electronic Warfare, Information and Cyber, Cranfield University, Defence Academy of the UK, Shrivenham, United Kingdom
[2]School of Engineering, University of Birmingham, Birmingham, United Kingdom
[3]Department of Information Engineering, Electronics and Telecommunications (DIET), University of Rome "La Sapienza", Rome, Italy

5.1 Introduction

PBRs can detect, localise and track targets passively by processing signals of opportunity transmitted by non-cooperative sources of illumination. They are also often referred to as passive coherent location radars, passive covert radars, non-cooperative radars and opportunistic radars [1].

In recent years, PBRs have attracted significant interest among the radar community due to some attractive characteristics that offer key operational advantages. Because PBRs are completely passive, a dedicated transmitter is not required and, as a result, they are smaller and relatively lower cost than active radars as well as easy to install and operate. Because they operate silently, PBRs are covert and difficult to detect and localise in hostile environments [1]. In addition to all these, PBRs exploit signals already available in the environment and do not require the allocation of a dedicated portion of the frequency spectrum which is becoming increasingly congested [2].

The general shifting of illuminators of opportunity from analogue to digital, together with larger available processing power at the receiver, has brought greater bandwidth, better range resolution and significant performance improvement. Typical sources of non-cooperative illumination include signals from broadcast systems (including TV and radio signal), cellular base stations and local area networks. The choice of different types of illuminators is driven by their strengths and weaknesses for each specific application. These include the signal parameters, the experimental configuration and power budget characteristics [3].

Although PBRs have matured rapidly in the past few years, they are continuing to attract significant research interest. Ongoing research efforts are, for example, looking at improving passive receiver systems and processing techniques and at identifying optimal deployment and configuration solutions that exploit a wider range of transmitters of opportunity with better waveform properties [4]. Despite all these efforts, to date, there has still been very little work to develop PBRs that exploit existing non-cooperative radar systems of opportunity.

The exploitation of radar signals rather than communication signals to achieve passive bistatic detection is attractive as it may offer significant advantages that are worth investigating. Because common radar waveforms are deterministic, the bistatic configuration does not essentially require a reference channel to detect a target, as is required when non-deterministic communications signals are used. The "knowledge" of the deterministic waveform allows the passive receiver to be matched with the illuminator of opportunity and thus to generate a range–Doppler map. Radar signals are often characterised by large bandwidth and a good compression ratio and hence they provide enhanced detection and range resolution. Despite these advantages, using radar platforms as illuminators of opportunity also introduces some technical challenges, especially when the transmitter is non-stationary. Airborne illuminators, for example, are continuously moving and can provide coverage only for a limited amount of time. PBRs need to rely on some prior knowledge of the presence of the signal and, if the platform is moving, there will be Doppler effects induced by the platform dynamics that have to be taken into account.

To date, research attempting to use radar signals for PBRs has been mostly based on the exploitation of Synthetic Aperture Radar (SAR) systems with the use of a reference channel. In the literature, there have been few studies of PBRs that do not use a reference channel but the vast majority of these exploit random signals of opportunity, such as Digital Video Broadcasting-Terrestrial (DVB-T) [5] and Digital Television (DTV) [6] signals.

The Thales-Aveillant ground based staring array radar technology can be used for Air Traffic Control (ATC), including the monitoring of drones. The radar, with a wide beam transmitter antenna, offers a constant illumination of the entire volume under surveillance. Unlike classic ATC radar systems that deploy a rotating antenna, the Thales-Aveillant Gamekeeper provides continuous target illumination and offers a constant signal of opportunity. This makes it a very attractive illuminator, in particular for short-range applications detecting low-RCS and slow-moving targets, such as drones.

Drones are small targets that have increasingly become a significant danger to manned air traffic and pose a security [7] and privacy threat. Passive solutions to detect and track drones represent a very attractive option, especially in the vicinity of airports, where the risk of interfering with other existing radar and communication systems is high. A passive receiver network sited in the vicinity of staring array radar technology provides continuously available signals of opportunity and can support long integration times to provide fine Doppler resolution for detection, tracking and classification of small targets.

This chapter investigates and demonstrates the feasibility of PBR target detection exploiting non-random radar signals of opportunity transmitted by existing non-cooperative staring radar illuminators. A passive radar prototype was developed and performance investigated on data collected in a set of short-range experimental trials with the Thales-Aveillant Gamekeeper radar. Results show drones can be successfully detected with and without the use of the reference channel at the receiver.

5.2 Overview of a passive bistatic radar

A comprehensive reference on a bistatic radar and its applications is available in [8,9]. In this section, some examples of the main applications are reviewed to emphasise the limited work available on PBR that exploit radar signals.

5.2.1 PBR exploiting analogue signals

PBR systems using an analogue terrestrial TV as an illuminator of opportunity were investigated by Griffiths and Long [10] in 1986. A Phase Alternating Line TV signal, carrying the colour-encoding system for analogue TV signals, was exploited, which found the presence of high sidelobes, strong ambiguities and dynamic range limitations [10]. Similar results were corroborated by Ringer *et al.* [11] who showed that the echo received from a single target was composed of a number of multiple returns. It was shown that using TV signals as an illuminator of opportunity introduced high clutter returns which, however, could be controlled by the use of large bandwidth to integration time products. The work on PBR with TV signals was continued by Howland in [12] to demonstrate that such systems could provide a target detection and tracking of aircraft experimentally. Target tracks were estimated from the measurements of the target Doppler shift and angle to successfully show that aircraft detection and tracking could be achieved, despite limitations in measurement accuracies, at ranges up to 260 km.

In 1997, Sahr and Lind [13] investigated commercial FM broadcasts to measure ionospheric turbulence. The advantage of FM transmitters is that they cover the wider 88–108-MHz frequency range and transmit relatively high power. FM signals containing speech and music were investigated to show that the FM signal properties resulted in better range resolution and Doppler shifts. The ambiguity function (AF) of FM radio signals was investigated in [14]. This study showed that the range and Doppler resolution is a function of the type of content of the FM signal. Signals that presented a high spectral content, such as rock music, provided better performance than those with a lower spectral content, such as speech, especially within pauses between words. A further analysis of performance predictions for FM radio signals was presented in [15]. This study investigated performance for a PBR exploiting two FM transmitters transmitting different powers and showed that the detection range could be improved (to over 30 km) when the two transmitters were exploited together. High transmit power and wide coverage made FM radio signals suitable for long-range detection. One of the problems of PBR is that the receiver needs to detect low-power target echoes in the presence of a continuous signal from the illuminator of opportunity. Because most broadcast transmitters are continuous, PBR does not have the privilege of a pulsed transmission and thus no periods of silence exist in which to listen for echoes. This is called Direct Signal Interference (DSI). An example of DSI mitigation for FM signals was studied in [16] and the limiting effects of DSI at the receiver were investigated in [17].

Range resolution performance of PBR exploiting multiple FM channels from the same co-located transmitter was analysed in [18]. It was theoretically shown that the range resolution could be improved from 6–7 to 1 km. A multi-frequency approach for FM PBR was also investigated by Bongioanni *et al.* who exploited high-frequency (HF) diversity of FM radio signals [19]. An alternative method to improve range resolution by exploiting multiple adjoining channels coming from a single transmitter was presented in [20]. The study focused on the matched filter improvement for better range resolution via correlation accounting for differences in Doppler shifts amongst channels and channel synchronisation.

A recent study in FM radio signals investigated the collection of the reference signal using beamforming [21]. Beamforming was also used to perform target angle estimation.

5.2.2 PBR exploiting digital signals

PBRs exploiting digital transmissions of opportunity have attracted significant interest, due to the increased available bandwidth offered by digital signals of opportunity, and recent efforts are leading to better range resolution and AF characteristics.

Global System for Mobile (GSM) communication is one of the digital signals of opportunity that have been used in PBRs. The performance of PBRs using the transmissions of a mobile phone base station, operating at 1,800 MHz with a bandwidth of 55 kHz, was investigated in [15]. A prototype PBR exploiting GSM signals was developed in [22] to demonstrate a successful detection of ground moving targets. The use of adaptive digital beamforming algorithms and a compact antenna array was also proposed to reduce direct path signal interference. A method to estimate the GSM reference signal from the direction of arrival using a two-element antenna array was proposed in [23]. The optimum beamforming was employed to extract the reference signal and to attenuate DSI. Results showed that a high speed train with a velocity of 150 km/h could be successfully detected. In [24], the GSM-Railway radio communication infrastructure was exploited to monitor trains and estimate their positions and velocities. A fast-moving train and a slow moving train were successfully detected. Passive bistatic Forward Scattering Radar (FSR) geometries have also attracted interest [25] and the use of GSM as an illuminator of opportunity in an FSR geometry was shown in [26].

Another mobile network that has been exploited for PBR is the Long-Term Evolution (LTE) network which provides high-speed wireless communication and data for mobile devices. An experimental study on PBR with LTE was carried out in [27] to detect a moving car with a minimum range resolution of 15 m and a minimum velocity resolution of 0.284 m/s. Further studies within LTE have shown the capability of LTE-based passive radars to detect ground moving targets, including cars, motorbikes and humans [28,29]. LTE-based passive radars were also investigated for air traffic surveillance [30]. A very recent study has also investigated 5G signals for PBR applications [31].

In 2006, BAE Systems developed a wideband PBR to demonstrate the use of DAB and DVB-T signals [32]. A further study implementing a more advanced testbed with analogue cancellations was implemented in [33] to show a civil passenger aircraft could be detected at a bistatic range of over 80 km using cross ambiguity processing. Palmer *et al.* studied DVB-T to detect automotive and airborne targets at various ranges [34]. Five targets from automotive traffic were detected with velocities ranging from 30 to 60 km/h at ranges of 200–1,050 m. In 2014, the use of DVB-T was further explored for maritime surveillance capabilities by Langellotti *et al.* [35]. This study investigated the detection and localisation of vessels beyond territorial waters and showed that target ranges of 20–100 km could be successfully achieved.

DVB-T signals present undesired peaks in the AF which are generated by guard intervals and pilot subcarriers. This limitation means that there are some restrictions to target detection. Most of the techniques used to enhance the AF of

DVB-T are based on modifying the reference signal before it is cross-correlated with the echo signal. Undesired peaks reduction, for example, was investigated in [36,37]. In 2012, Colone *et al.* [38] proposed a technique to further overcome these limitations by cascading the linear filters responsible for power equalisation and the pilot carriers.

Wireless networking transmissions are another potential source to enable PBRs. The most common WiFi standards used in previous research are the IEEE 802.11a, 802.11b and 802.11g operating in the 2.4 and 5-GHz spectrum bands. The detection performance of a PBR exploiting IEEE 802.11 beacon transmissions, against two moving human targets separated by of 12 and 35 m, was studied by Guo *et al.* in [39]. A feasibility study to demonstrate stand-off through-the-wall detection using WiFi signals was presented in [40]. Results showed that a moving person behind a wall with a multilayer structure could be detected. PBRs based on WiFi were further investigated by Falcone *et al.* in [41] to detect targets in outdoor environments. A moving vehicle was detected at a range of 95 m and a human target was detected at a range of 85 m. In [42], WiFi signals were used to detect signs of life from breathing with PBR and another recent study investigated the use of WiFi in passive FSR [43].

With the development of space systems and the increased availability of transmitting satellites, space-based illuminators of opportunity have also attracted the interest of the PBR community. Satellites have been exploited that reside in Geostationary Orbits (GEO), Medium Earth Orbits and Low Earth Orbits (LEO). In 1992, Griffiths *et al.* investigated broadcast signals by TV GEO satellites as an illuminator of opportunity [44]. In 1993, Tsui and Shaw investigated a method to determine the distance of a target utilising four Global Positioning System (GPS) satellites [45]. They first measured the angle of arrival of echoes from the target so that the distance between the target and the receiver could be calculated. In 1995, Koch and Westphal further studied the application of Global Navigation Satellite Systems in PBR for the detection of airborne targets [46]. The US NAVSTAR-GPS and the Russian GLONASS systems were used to form a passive multistatic space-borne radar that exploited 48 satellites transmitting at L-band (1–2 GHz) to successfully detect military and civil aircraft. However, because these satellites are not designed for remote sensing applications, there was a drawback in terms of resolution capability [47]. This was also investigated by Glennon *et al.* [48] where it was shown that performance was limited by the limited transmitted power level of the GPS signals (about 25–30 W). The study of communication satellites for PBR was expanded with the launch of Iridium, a satellite system which resides in an LEO orbit. The Iridium satellites were used to detect an air target at a distance of 30 km from a passive receiver on the ground [49]. It was observed that a challenge of exploiting satellite on LEO orbits is the relative motion of the satellite with respect to the ground. GPS signals have been used to detect an airplane in forward scatter geometry in [50].

Solutions exploiting both analogue and digital illuminators have also been investigated. In [51], for example, a prototype was developed to show that ground and aerial targets could successfully be detected and tracked by combining FM,

DAB and DVB-T. In [52], a hybrid passive radar sensor was proposed to deliver air guidance and surveillance in three dimensions under the ARGUS 3D project. The project aim was to improve ATC radar for civilian applications and showed that the exploitation of FM radio, DVB-T and DAB signals could offer high target position performance and good range estimates.

5.2.3 PBR exploiting radar signals

To date, studies exploiting radar systems as illuminators of opportunity have been limited to space-borne SAR and over the horizon radar transmitters [53]. SAR satellites mainly transmit chirp signals with wide bandwidth and hence offer very good range resolution. There are some drawbacks in using SAR satellites in LEOs. First of all, the transmitter needs to be tracked so that the passive receiver is always pointing towards the transmitter to collect the reference signal. Furthermore, such satellites do not have a fixed geographical coverage and move with a high velocity. This motion results in displacements or shifts in target detection [54].

A study on commercial-based SAR by Griffiths *et al.* in 2002 investigated SAR illuminators using a satellite in an LEO orbit [53]. The Envisat satellite was selected, carrying the Advanced ASAR operating at C-band with a carrier frequency of 5.331 GHz. SAR transmissions were exploited to detect air targets. That study only involved theoretical analysis, introducing the concept of implementation but no experiments were carried out.

In 2003, the work was continued for moving target detection by Whitewood *et al.* [55]. A stationary ground based receiver was employed to receive the echo from a flying target. An electronic Displaced Phase Centre Antenna pulse canceller was implemented to improve the detection performance of platform motions that are subject to clutter. A later study by Whitewood *et al.* [56] on Envisat focussed on bistatic imaging experiments. However, the technique presented could be also used for the detection of targets such as aircraft. The study with Envisat stopped with the end of the mission following the unexpected loss of contact in 2012 [57].

Another study on PBRs using non-cooperative radar signals of opportunity was carried out in [58]. In this study, the pulse radar signal transmitted by the mechanically scanning transmitter of an ATC radar was exploited to detect an aircraft. Results showed that the aircraft could be successfully detected with a passive receiver prototype using a reference channel, demonstrating the capability of ATC radars as potential illuminators of opportunity.

In [59], the exploitation of an Airport Surveillance Radar was further investigated. A replica of the pulse signal was extracted from the reference signal and this was used for pulse compression.

HF signals have high Effective Radiated Powers compared to other illuminators of opportunity [15]. Exploiting HF signals in PBR provides very long range detection because these signals can propagate beyond the radio horizon. The exploitation of the digital HF broadcast band (3–30 MHz) in PBR applications was discussed in [60]. The study exploited Digital Radio Mondiale (DRM) HF transmissions. It was found that HF radar gives course range resolution given that it has low bandwidth due to either ionospheric or interference constraints. The use of HF radar also required a phased array for sufficient coverage in long-range propagation in PBR applications [60].

The issue of exploiting HF PBR based on DRM as the illuminator of opportunity was studied by Xianrong *et al.* [61]. They addressed several problems arising in an HF passive radar: the presence of various secondary peaks in the AF of the DRM signal, direct path rejection, clutter from multipath in the surveillance channel and extraction of the reference signal. A suppression method was applied to reduce the secondary AF peaks by reducing pilot symbol amplitudes, imposing guard intervals and setting unwanted parts to zeros. Multipath clutter rejection was achieved by applying temporal and spatial filters. A further study by Zhao *et al.* presented an experimental study on HF PBR exploiting DRM signals based on hybrid sky–surface waves [62]. The preliminary results showed that DRM-based HF can be exploited in PBR applications for sea state remote sensing. However, future studies and improvements are needed before this system can be applied in practice. HF Surface-Wave Radars are an example of beyond the horizon radar that propagates by utilising the surface-wave propagation mode that diffracts along the surface of the earth. In the monostatic case, this radar has potential to detect targets beyond 300 km [1].

5.2.4 Drone detection in PBR

Using a traditional radar for drone detection, tracking and classification is a challenging task because drones have a small RCS and fly slowly and at low altitudes compared to conventional aircraft. Passive solutions to detect and track drones represent a very attractive option, especially in the vicinity of airports, where the risk of interfering with other existing radar and communication systems is high. A passive solution does not require a dedicated transmitter and therefore provides a low-cost non-interfering solution whilst allowing covert operations which may be relevant to military and anti-terrorism applications.

The first experimental study to exploit GSM signals to detect a small drone was presented in [63]. Results from the experiment showed that the drone was successfully detected from the Doppler shift obtained in the time-Doppler map. However, the GSM signal used was locally generated, rather than using actual GSM network signals.

A WiFi-based passive radar was also exploited for the detection of ultra-light aircrafts and drones in [64]. The aim of that study was to improve the security of small airfields and outdoor areas. Results showed the capability to detect and localise ultra-light aircraft and drones in 3D in short-range PBR by exploiting WiFi signals.

A study on PBR exploiting Universal Mobile Telecommunications System (UMTS) 3G signals was presented by Chadwick in [65] for micro-drone detections. Results showed the potential of UMTS transmitters as illuminators of opportunity for drone detection.

The feasibility of using DTV signals for drone detection was shown in [66]. In a very recent study by Fang *et al.*, the feasibility of multistatic single-antenna PBR based on DTV was presented [6]. The multistatic configuration consisted of one transmitter and three receivers. The reference signal was reconstructed from the surveillance signal. A drone was successfully detected and tracked with the suggestion to perform more experiments in different environments.

5.2.5 PBR with single-channel detection

In the literature, there are a few studies of PBR without the use of a reference channel, and most of them exploit random signals of opportunity. Single-receiver processing was studied in [67] by exploiting DTV signals. The reference channel was recovered by remodulation and demodulation of the surveillance channel. The target was detected at ranges between 1.4 and 3.4 km.

The reconstruction of the reference signal may result in poor cancellation performance as was shown in [5] with DVB-T signals. Zhang *et al.* [68] used an antenna array as a surveillance antenna in contrast to previous studies that used omnidirectional antennas to study single-channel detection. The reference signal was recovered from the covariance matrix and it was shown that cancellation performance could be improved with a maximisation cancellation ratio criterion. The authors claimed that their method is not restricted to DTV or DVB-T signals.

5.3 The staring radar signals

The Thales-Aveillant Gamekeeper is a counter-drone staring radar that transmits a continuous train of pulses with a low-gain, wide-beam antenna. The radar continuously covers a 90° azimuth sector centred around the antenna boresight and to an elevation of 30° from the horizontal. It operates in L-band with a bandwidth of 1 MHz and a PRI of 136 μs. On receive, the radar uses a separate digital array that allows a generation of multiple simultaneous digital beams. Because the radar illuminates targets continuously, long integration times (up to roughly 0.25 s) and high Doppler resolution (down to 4 Hz) are used to detect small airborne targets with RCS values of 0.01 m^2 up to ranges of 5 km. References [69,70] report on results with the monostatic staring Gamekeeper radar from large field trials conducted with control targets illustrating good detection sensitivity against small drones. The radar is particularly well suited for exploitation in a PBR system because it provides constant coverage and no need for pulse-chasing.

5.4 Experimental set-up

PBR experiments were carried out at Deenethorpe Airfield, a former Royal Air Force station, located about 3 km east of Corby in England. The experimental configuration is shown in Figure 5.1. The Gamekeeper radar transmitter location is shown at (A), the passive bistatic receiver reference antenna at (B) and the co-located bistatic surveillance antenna to receive echoes from targets is at (C). The flightpath of the drone is drawn in yellow and the beam widths and directions of the

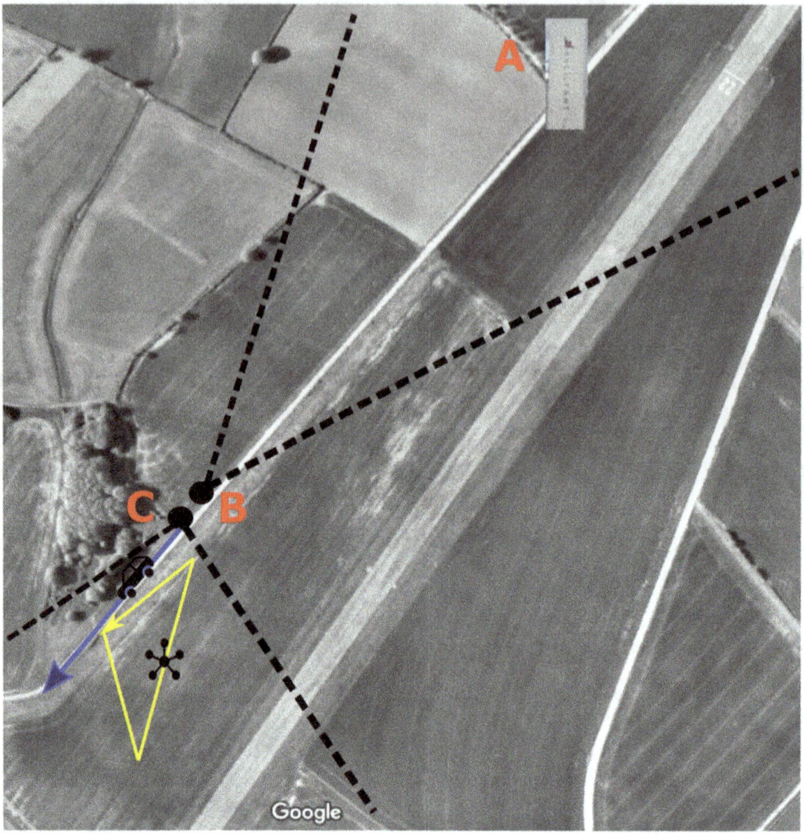

Figure 5.1 *Geometrical configuration: part (A) is the Gamekeeper radar, (B) indicates the position of the reference antenna and (C) indicates the surveillance antenna. The yellow line represents the trajectory of the drone*

reference and surveillance antennas of the PBR are indicated by the dotted black lines. The distance between the Gamekeeper transmitter and the passive receiver (the baseline distance) was approximately 635 m and the surveillance antenna was vertically polarised to match the polarisation of the transmitted signal. The trajectory of the drone was recorded with a GPS receiver providing the 'air truth' information to help with the analysis of the experimental results. The drone used in this experiment was a DJI Inspire 1 drone which has a diameter of 0.5 m and weighs 3 kg. The drone was flown at a height of 10 m during the measurements. Continuous measurements were taken and several passes of the drone were recorded.

The passive receiver used for the experiments was a National Instrument USRP-2943R, providing a 2×2 MIMO transceiver. Two of the available channels were used in the experiments to acquire the reference and surveillance signals. The received signals in the reference channel and the surveillance channel were down-converted to baseband using a common built-in receiver Local Oscillator (LO). Data was stored in a laptop for off-line processing in MATLAB®. During signal processing, the reference and surveillance channel were treated separately. The block diagram of the system is shown in Figure 5.2.

5.5 Detection results with reference channels

The I/Q signals from each channel were matched-filtered to form a slow-time fast-time map. The signal was received in long blocks of data and arranged in a vector form. Each pulse in the surveillance channel was cross-correlated with the corresponding pulse measured in the reference channel assuming no range ambiguities occurred given the relative short target range investigated. Finally, a Fast Fourier Transform (FFT) in the slow-time domain was calculated to produce the range–Doppler maps. In the processing, a Hamming window was used to suppress side-lobes in both range and Doppler. The range–Doppler map enables discrimination between stationary and moving returns, to separate the moving target from the stationary clutter.

5.5.1 Experimental results

This section presents the results of the experiment obtained with the use of the reference channel. The integration time was 0.1 s corresponding to integrating 734 pulses. The received signals were down-converted and sampled at 14.704 MHz by the passive receiver.

Due to the lower RCS and speed of the drone, a suitably long integration time was selected to improve the detection ability of the receiver and to increase Doppler resolution. The drone was successfully detected within the beamwidth of the surveillance antenna when the drone flew away and towards the receiver. Figure 5.3 shows the range–Doppler map for the drone flying away from the receiver and shows the peak response with a negative Doppler shift of −85.48 Hz. Figure 5.4 shows the corresponding Doppler cut at the target range. Figure 5.5 shows the range–Doppler map for the drone flying towards the receiver. The peak response can be seen with a Doppler shift of 85.48 Hz. The drone was also detected in other frames from the few passes it made within the proximity of the passive receiver and similar Doppler shifts were measured. Figure 5.6 shows the corresponding Doppler cut at the target range. Results also show a weak response at −10 dB around zero-Doppler. This is most likely coming from the relatively strong non-stationary vegetation clutter returns and due to a drop in clutter suppression gain because of the use of a longer integration time. The resulting long integration time made it more difficult to remove the clutter with a standard low-pass filter (i.e., clutter coherency was partially lost during the dwell time).

The GPS truth data was recorded and this gives the Cartesian coordinates for the target as a function of time. Together, with the known coordinates of the

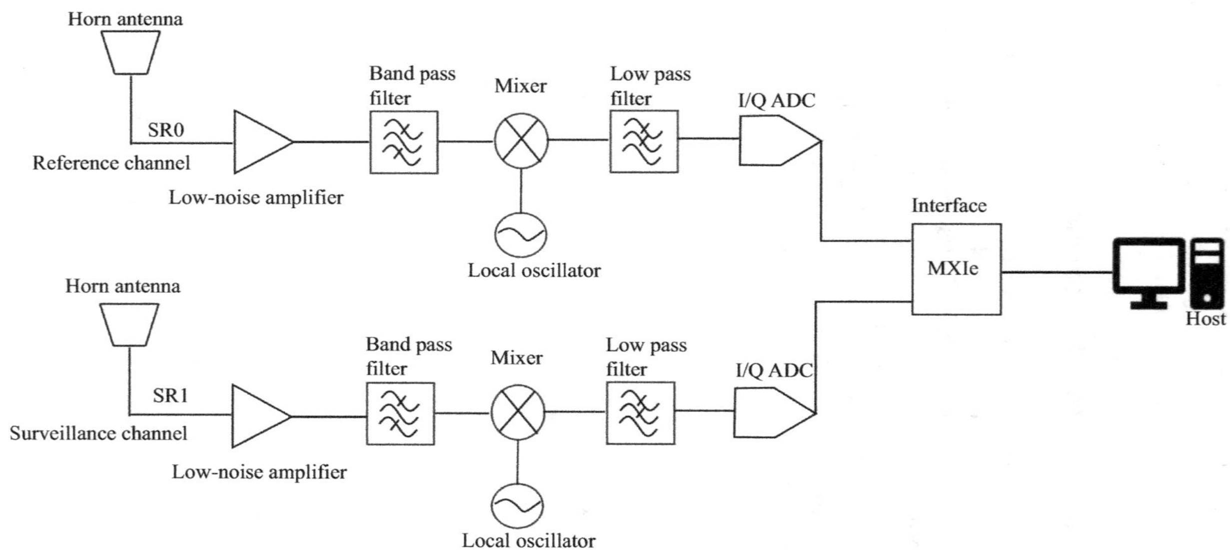

Figure 5.2 Block diagram for staring radar target detection

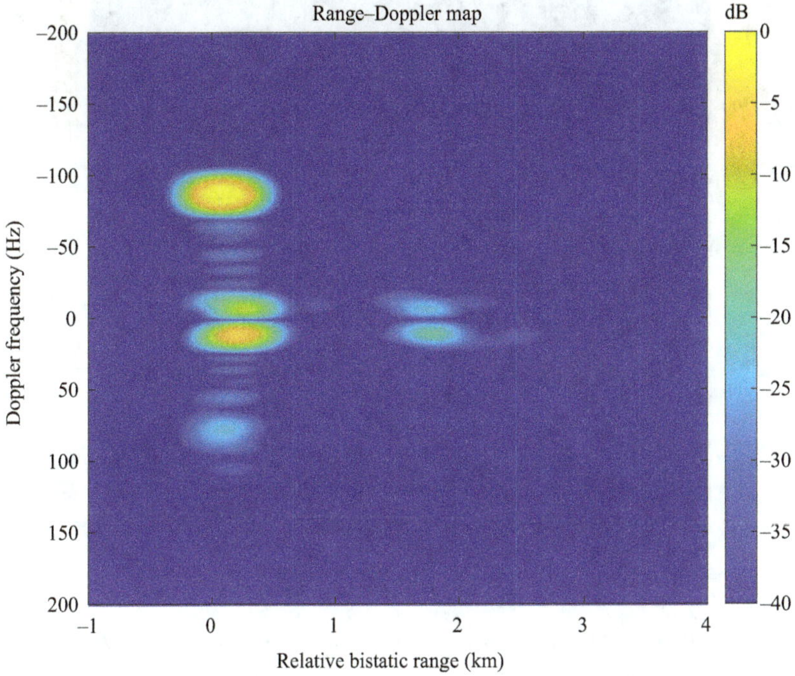

Figure 5.3 Range–Doppler map after zero-Doppler suppression for a drone flying away from the receiver with 0.1-s integration time

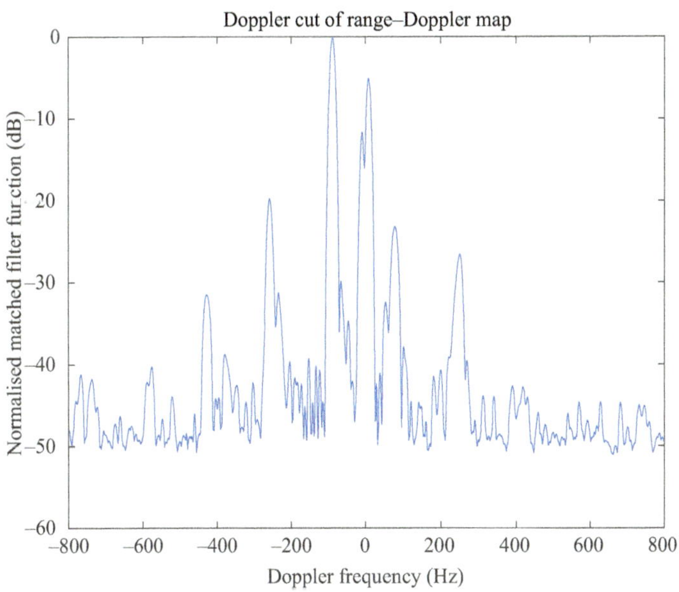

Figure 5.4 Doppler cut of the response at the target range for the drone flying away from the receiver obtained with 0.1-s integration time

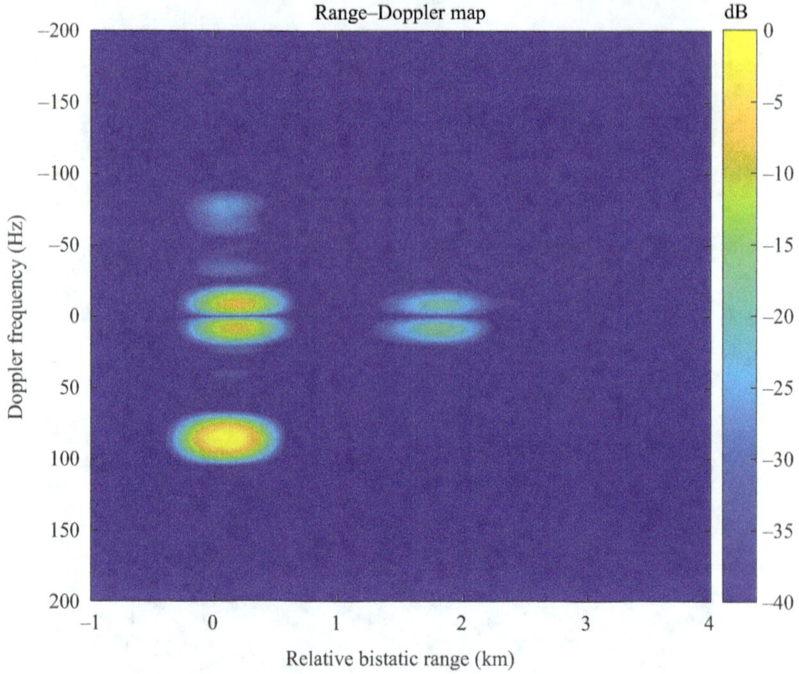

Figure 5.5 Range–Doppler map after zero-Doppler suppression for a drone flying towards the receiver with 0.1-s integration time

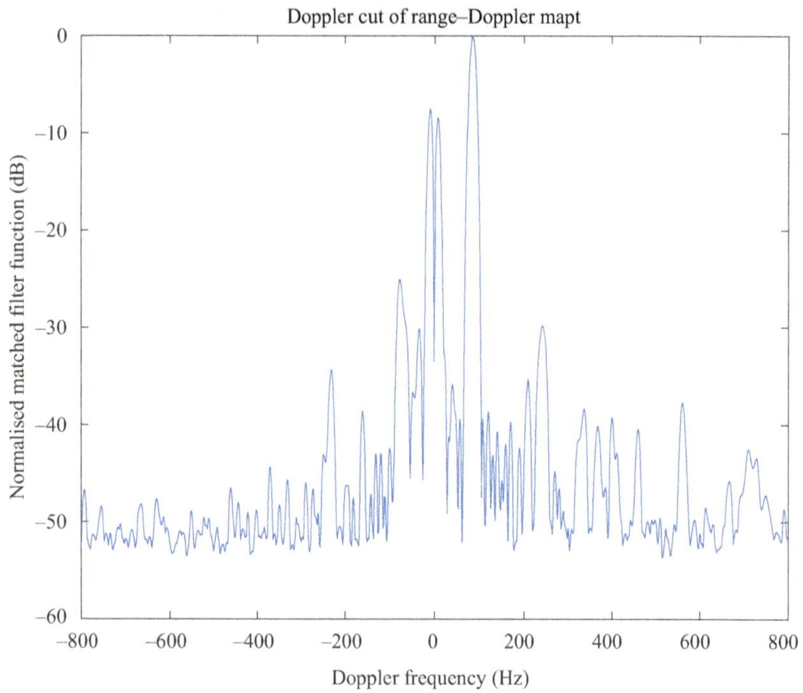

Figure 5.6 Doppler cut of the response at the target range for the drone flying towards the receiver obtained with 0.1-s integration time

transmitter and passive receiver, the range rate of the targets with respect to the transmitter and the passive receiver can be accurately estimated. The bistatic Doppler frequency derived from the GPS is given by the following equation:

$$f_D = -\frac{f_0}{c}\left[\frac{dR_{Tx}}{dt} + \frac{dR_{Rx}}{dt}\right] \tag{5.1}$$

where f_0 is the carrier frequency, c is the speed of light, dR_{Tx}/dt is the time derivative of the changing range between the transmitter and the targets, and dR_{Rx}/dt is the time derivative of the range between the drone and passive receiver. The velocity and Doppler frequency calculated from the GPS data is shown in Figure 5.7. The zero time in the GPS data plotted in these figures is with reference to the sample start time.

The radar data could not be fully synchronised with the GPS ground truth. However, results relative to the flying drone are within the Doppler interval obtained with the GPS ground truth data. For example, Figure 5.7 shows at 48 s, the Doppler frequency is −86.2 Hz with a velocity of 11.3 m/s (25.3 mph) (the drone flying away from the surveillance antenna) and at 63 s (the drone flying towards the surveillance antenna), the Doppler frequency is 85.3 Hz with a velocity of 10.9 m/s (24.4 mph).

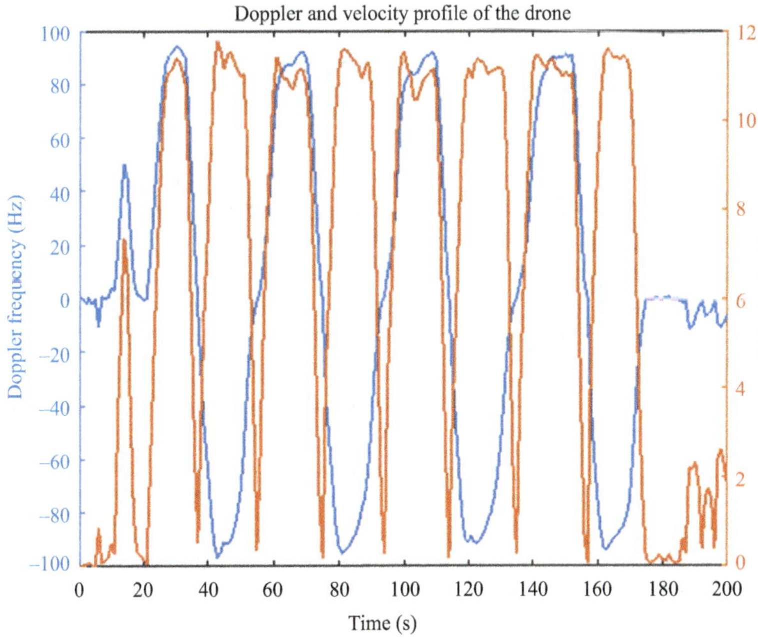

Figure 5.7 Velocity profile of the drone and relative expected Doppler shift frequency calculated with the GPS ground-truth data

5.6 PBR without a reference channel

This section presents the results of passive target detection using the same experimental set-up as in Section 5.5 but obtained without the use of the reference channel. The aim of this section is to enable comparisons to be made between the solution with and without a reference channel.

The total echo signal at the surveillance channel, containing the contributions of the signal from the transmitter–target–receiver and the direct signal captured through the antenna sidelobes, was split into slow- and fast-time and placed in a matrix for matched-filter processing. This was done by reshaping the total echo signal to the length of F_s/PRF, with F_s being the sampling frequency, in every row of PRI (under the assumption of a known PRF). The matched-filter was applied, on each row of the received signal matrix, by implementing a cross-correlation with a representative software copy of a transmitted pulse. The output of the matched-filter gives a range profile map. However, in this case, the range information is only an apparent range of the target allowing the measurement of relative distances only. Nonetheless, by exploiting deterministic signals, the passive receiver becomes less computationally expensive, smaller and even lower cost at the expense of the measurement of the bistatic range. The Doppler frequency of targets was found by applying an FFT to the output of the matched-filter results to generate a range–Doppler map.

Two key processing challenges had to be addressed to allow single-channel detection:

1. Time misalignments between consecutive PRIs.
2. Undesired Doppler frequency offsets due to independent LOs in the transmitter and receiver.

The first challenge is in the cross-correlation stage. Because F_s/PRF is not a perfect integer, a shift in slow time (in every pulse in the data matrix) is observed. An example of a slow-time diagram before and after alignment corrections is shown in Figures 5.8 and 5.9. Because no reference channel is used, the apparent bistatic range for this case is a random value that depends only on the randomly selected starting point of a PRF. However, one could use the reflection from a reference target or the direct interfering signal from the transmitter (when available) and provide the relative bistatic range to these.

Undesired and drifting Doppler shifts were observed due to the inevitable noncoherency between the receiver and transmitter LOs. This was evident by observing the direct signal from the transmitter entering through the receiving antenna sidelobes away from zero-Doppler. This undesired effect was compensated by using the peak response of the range–Doppler map which, for this short-range application, was that of the direct signal from the transmitter. After this Doppler offset compensation, zero-Doppler removal was applied to the data to mitigate stationary clutter.

Finally, a Hamming window was applied to the final processed data to reduce sidelobes.

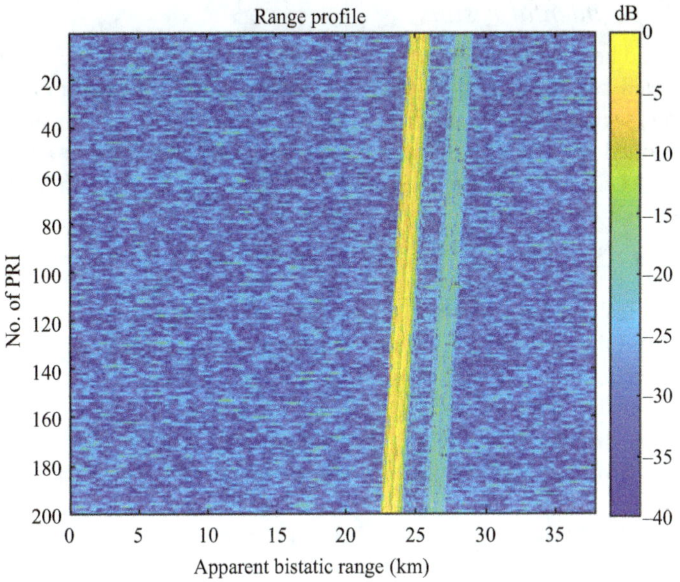

Figure 5.8 Example of a slow-time diagram before alignment corrections

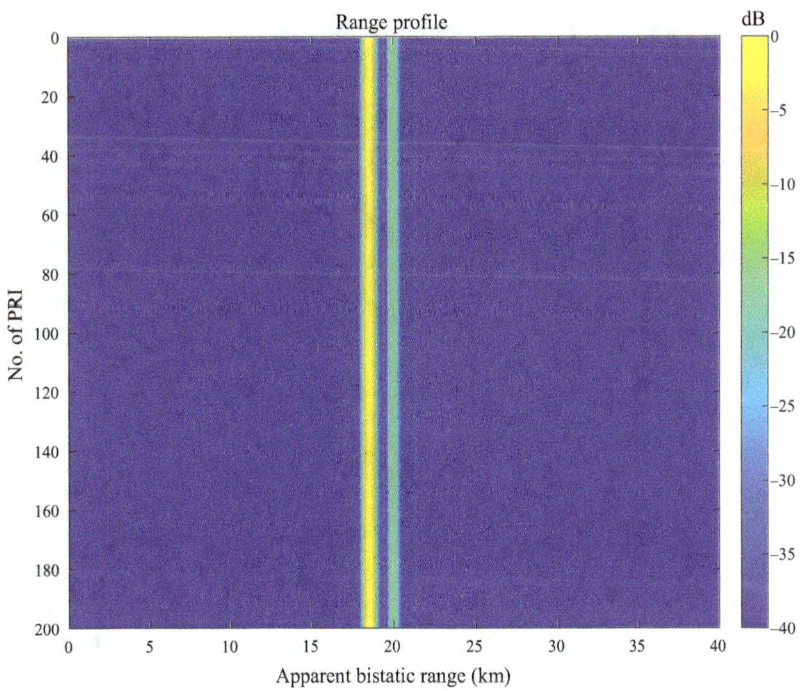

Figure 5.9 Example of a slow-time diagram after alignment corrections

5.6.1 Experimental results

The data used here is the same as that in Section 5.5 but the reference channel data was discarded and target echoes were processed using the algorithm presented in the previous section. Results obtained in this experiment will be compared with the results obtained with the use of the reference channel and therefore all processing parameters (i.e., FFT points, sampling frequency and integration time) are the same.

Figures 5.10 and 5.11 show the range–Doppler maps for a drone flying away and towards the receiver after alignment corrections and Doppler offset compensations. Both figures show that there is a strong response around zero Doppler, resulting from the contributions of the DSI and clutter, demonstrating satisfactory Doppler offset compensations. Figure 5.10 shows that there is a weak response at negative Doppler which is the target. A similar result was found in Figure 5.11 which shows that there is a weak response at positive Doppler. The effects of clutter and DSI are then mitigated by applying mean value removal to both results.

Figure 5.12 shows the result for a drone flying away from the receiver. Results show a peak response at −88.24 Hz corresponding to 10.6 m/s (23.7 mph). This is also evident in the corresponding Doppler-cut in Figure 5.13. The Doppler-cut also

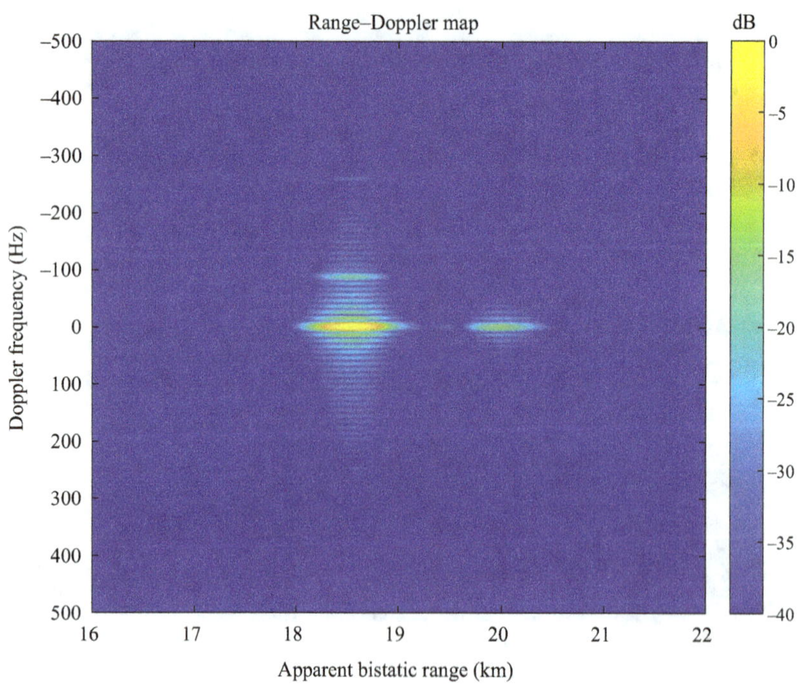

Figure 5.10 Range–Doppler map for a drone flying away from the receiver before mean value removal

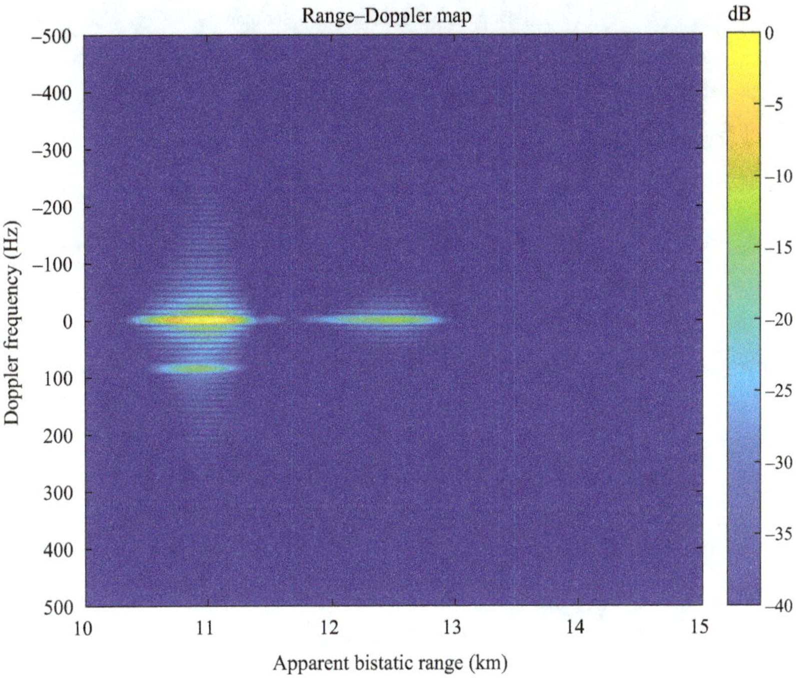

Figure 5.11 Range–Doppler map for a drone flying towards the receiver before mean value removal

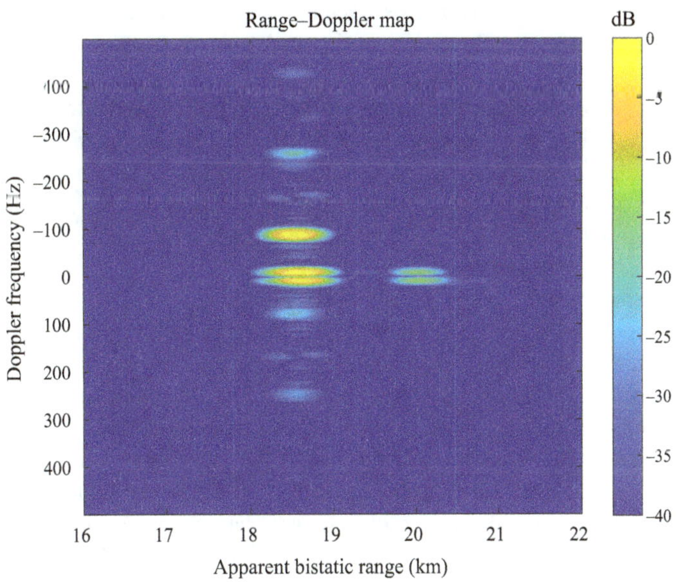

Figure 5.12 Range–Doppler map for a drone flying away from the receiver after mean value removal and windowing

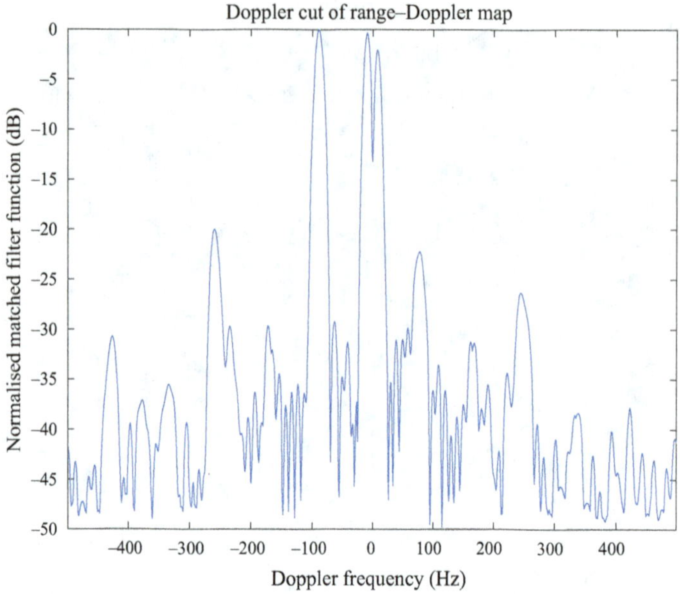

Figure 5.13 Doppler-cut for a drone flying away from the receiver after mean value removal and windowing

shows a higher clutter response with respect to that obtained with the use of the reference channel with a difference between the target and the clutter of only 0.3 dB. However, from the range–Doppler map, it can be clearly seen that the target was separated in frequency from the clutter.

Figure 5.14 shows the result for the drone flying towards the receiver. Results show the peak Doppler response is at 84.56 Hz corresponding to 10.1 m/s (22.6 mph). Results are also corroborated by the Doppler-cut shown in Figure 5.15. In this case, clutter isolation comparable to that obtained with the use of the reference channel is achieved. The difference between the highest peak response and the clutter response is 8 dB.

Measurements were taken continuously along the drone trajectory with the drone passing in front of the receiver a few times (with 0.1 s integration time). Variations in clutter between different detection snapshots might be due sudden gusts of wind that contributed to clutter non-stationarity and to antenna tripod vibrations. Higher clutter levels without the use of a reference channel might instead be due to the use of a software reference signal which clearly does not provide any live information of the non-idealities of the system which can result in self-compensation. Nevertheless, results showed that by increasing the integration time, targets could be successfully detected despite the small RCS. This was clearly possible because of the nature of the staring transmitter of opportunity which was exploited.

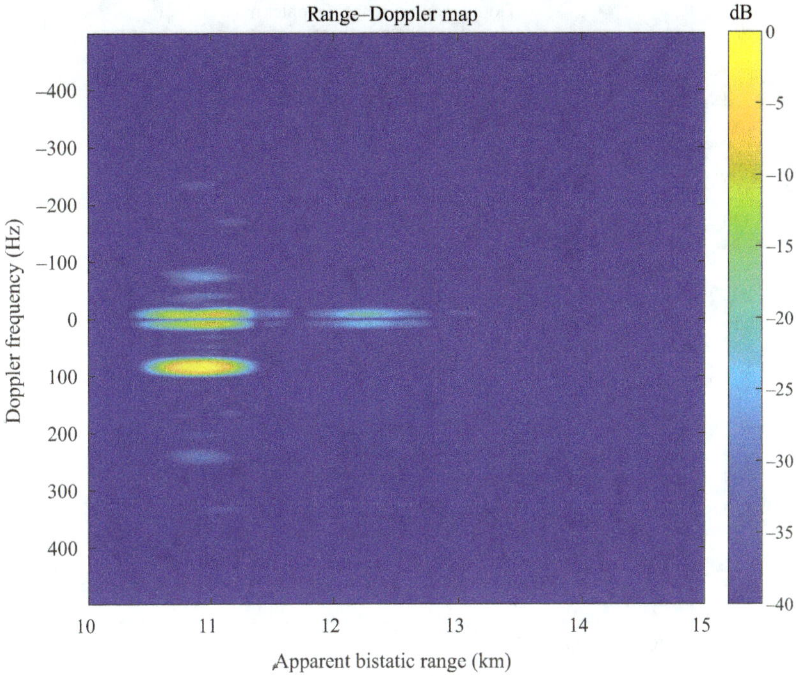

Figure 5.14 Range–Doppler map for a drone flying towards the receiver after mean value removal and windowing

5.7 Comparisons

To make comparisons between the two processing solutions, Doppler frequencies obtained from the two experiments are given in Table 5.1.

The table shows the peak Doppler for the detected target, both with and without the use of a reference channel, with a percentage difference calculated with respect to the case with the reference channel. Results show that, for these two measurements, the percentage error between two cases was less than 5%.

5.8 Conclusions

In this chapter, we have presented a review of the literature on PBRs which covered the use of commonly used communication transmitters of opportunity and the less researched use of radar transmitters of opportunity. A survey of the small amount of work reported to date on using PBR to detect drones has also been covered.

The exploitation of radar signals to achieve passive bistatic detection, with and without the use of a reference channel, was investigated using the Thales-Aveillant counter-drone Gamekeeper radar as an illuminator of opportunity. This radar was

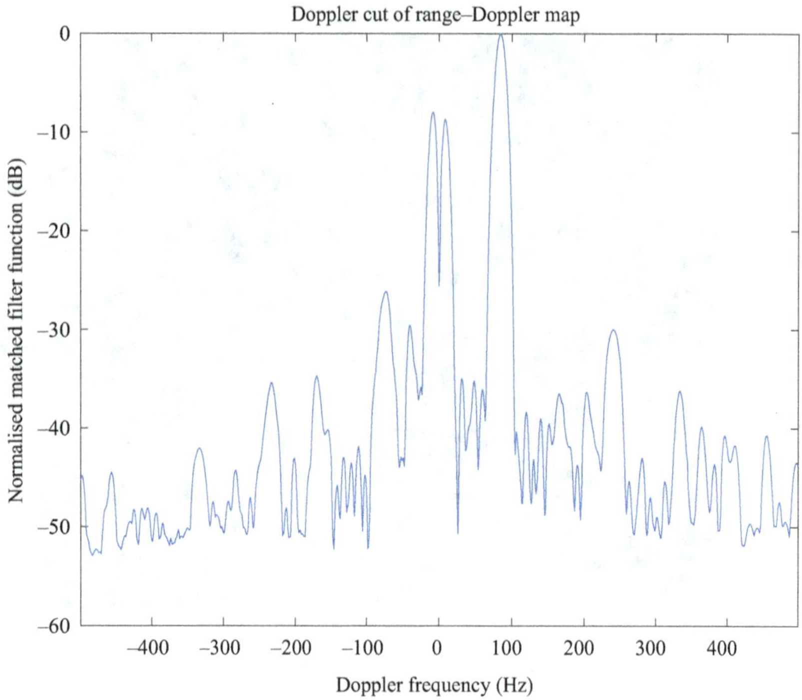

Figure 5.15 Doppler-cut for a drone flying towards the receiver after mean value removal and windowing

Table 5.1 Peak measured Doppler frequency for the detected moving target with the reference channel and without the reference channel

Target	Trajectory	Reference channel (Hz)	No reference channel (Hz)	Percentage of difference
Drone	Away from the receiver	−85.48	−88.24	3.22
	Towards the receiver	85.48	84.56	1.07

selected as a staring one that provides continuous target illumination and offers a constant signal of opportunity.

The results from an experimental trial detecting small drones using the staring radar illuminator of opportunity were presented. The results showed that a drone flying towards and away from a low-cost passive receiver could provide a magnitude response that was straightforwardly detected. The estimates of the drone velocities were in good agreement with independent air-truth GPS data. It was also shown that target detection without the use of a reference channel is feasible provided that knowledge of the signal transmit is available a priori or has been estimated. These results demonstrate that low-cost passive solutions could potentially be successfully employed as part of a more comprehensive air traffic surveillance system that is able to detect and track emerging targets such as drones in the vicinity of airports and possibly beyond with the added attraction of not posing interference risks to existing sensors.

Results presented in this chapter investigated the Doppler return from the targets. Because the results in this chapter are based on very limited experimental data of flying drone, a statistical analysis to characterise performance with Receiver Operating Characteristic (ROC) curves and Monte Carlo simulations was not possible and this represents an opportunity for future work when a larger dataset becomes available. For further work, the study can also be extended to micro-Doppler signature analysis. Data was also processed offline but the technique could be extended for real-time operation given suitable hardware and software.

References

[1] Willis NJ and Griffiths HD. Advances in Bistatic Radar. Willis NJ and Griffiths HD, editors. Raleigh: SciTech Publishing, Inc; 2007.

[2] Bayat S, Nayebi MM, and Norouzi Y. Target detection by passive coherent FM based bistatic radar. In: Proceedings of the 2008 International Conference on Radar, Radar 2008. IEEE; 2008. p. 412–415.

[3] Howland P. Editorial: Passive radar systems. IEE Proceedings – Radar, Sonar and Navigation. 2005;152(3):105.

[4] Farina A and Kuschel H. Guest editorial special issue on passive radar (Part I). IEEE Aerospace and Electronic Systems Magazine. 2012;27(10):5.

[5] Searle S, Davis L, and Palmer J. Signal processing considerations for passive radar with a single receiver. In: 2015 IEEE International Conference on Acoustics, Speech and Signal Processing (ICASSP). IEEE; 2015. p. 5560–5564.

[6] Fang G, Yi J, Wan X, *et al.* Experimental research of multistatic passive radar with a single antenna for drone detection. IEEE Access. 2018;6:33542–33551.

[7] Katy W. Venezuela President Maduro survives 'drone assassination attempt' – BBC News. Venezuela; 2018.

[8] Griffiths HD and Baker CJ. An Introduction to Passive Radar. Norwood, MA: Artech House; 2017.

[9] Baker CJ and Griffiths HD. Bistatic and multistatic radar sensors for homeland security. In: Advances in Sensing with Security Applications. Dordrecht: Kluwer Academic Publishers; 2006. p. 1–22.

[10] Griffiths HD and Long NRW. Television-based bistatic radar. IEE Proceedings F – Communications, Radar and Signal Processing. 1986; 133(7):649–657.

[11] Ringer MA, Frazer GJ, and Anderson SJ. Waveform Analysis of Transmitters of Opportunity for Passive Radar. Defence Science and Technology Organisation; 1999.

[12] Howland PE. Target tracking using television-based bistatic radar. IEE Proceedings – Radar, Sonar and Navigation. 1999;146(3):166–174.

[13] Sahr JD and Lind FD. The Manastash Ridge radar: A passive bistatic radar for upper atmospheric radio science. Radio Science. 1997;32(6):2345–2358.

[14] Griffiths HD and Baker CJ. Measurement and analysis of ambiguity functions of passive radar transmissions. In: IEEE International Radar Conference, 2005. Arlington, VA: IEEE; 2005. p. 321–325.

[15] Griffiths HD and Baker CJ. Passive coherent location radar systems. Part 1: Performance prediction. IEE Proceedings – Radar, Sonar and Navigation. 2005;152(3):153.

[16] Jiabing Z, Liang T, and Yi H. A direct path interference cancellation approach to passive radar based on FM radio transmitter. In: 2006 IEEE International Conference on Electro/Information Technology. East Lansing, MI: IEEE; 2006. p. 55–59.

[17] Howland PE, Maksimiuk D, and Reitsma G. FM radio based bistatic radar. IEE Proceedings – Radar, Sonar and Navigation. 2005;152(3):107.

[18] Tasdelen AS and Koymen H. Range resolution improvement in passive coherent location radar systems using multiple FM radio channels. In: The Institution of Engineering and Technology Forum on Waveform Diversity and Design in Communications, Radar and Sonar. London, UK: Institution of Engineering and Technology; 2006. p. 23–31.

[19] Bongioanni C, Colone F, and Lombardo P. Performance analysis of a multi-frequency FM based passive bistatic radar. In: 2008 IEEE Radar Conference. Rome: IEEE; 2008. p. 1–6.

[20] Olsen KE and Woodbridge K. FM based passive bistatic radar range resolution improvement. In: 2011 12th International Radar Symposium (IRS). Leipzig: IEEE; 2011. p. 327–332.

[21] Wan L, Wang X, and Bi G. A range-Doppler-angle estimation method for passive bistatic radar. In: 2018 International Conference on Signals and Systems (ICSigSys); 2018. p. 213–218.

[22] Tan DKP, Sun H, Lu Y, *et al.* Passive radar using Global System for Mobile communication signal: Theory, implementation and measurements. IEE Proceedings – Radar, Sonar and Navigation. 2005;152(3):116.

[23] Kubica M, Kubica V, Neyt X, *et al.* Optimum target detection using illuminators of opportunity. In: 2006 IEEE Conference on Radar. Verona, NY: IEEE; 2006. p. 417–424.

[24] Chetty K, Chen Q, and Woodbridge K. Train monitoring using GSM-R based passive radar. In: 2016 IEEE Radar Conference (RadarConf). Philadelphia, PA: IEEE; 2016. p. 1–4.

[25] Gashinova M, Daniel L, Hoare E, *et al.* Signal characterisation and processing in the forward scatter mode of bistatic passive coherent location systems. EURASIP Journal on Advances in Signal Processing. 2013; 2013(1):36.

[26] Krysik P, Kulpa K, and Samczyński P. GSM based passive receiver using forward scatter radar geometry. In: 2013 14th International Radar Symposium (IRS). Dresden: IEEE; 2013. p. 637–642.

[27] Salah AA, Abdullah RR, Ismail A, *et al.* Experimental study of LTE signals as illuminators of opportunity for passive bistatic radar applications. Electronics Letters. 2014;50(7):545–547.

[28] Raja Abdullah RSA, Hashim FH, Salah AA, *et al.* Experimental investigation on target detection and tracking in passive radar using long-term evolution signal. IET Radar, Sonar & Navigation. 2016;10(3):577–585.

[29] Raja Abdullah RSA, Salah AA, Ismail A, *et al.* LTE-based passive bistatic radar system for detection of ground-moving targets. ETRI Journal. 2016; 38(2):302–313.

[30] Klock C, Winkler V, and Edrich M. LTE-signal processing for passive radar air traffic surveillance. In: 2017 18th International Radar Symposium (IRS). IEEE; 2017. p. 1–9.

[31] Thomä RS, Andrich C, Galdo GD, *et al.* Cooperative passive coherent location: A promising service for future mobile radio networks. IEEE Communications Magazine. 2018.

[32] Gould D, Pollard R, Sarno C, *et al.* A multiband passive radar demonstrator. In: 2006 International Radar Symposium. Krakow: IEEE; 2006. p. 1–4.

[33] Gould D, Pollard R, Sarno C, *et al.* Developments to a multiband passive radar demonstrator system. In: 2007 IET International Conference on Radar Systems. Edinburgh, UK: IET; 2007. p. 1–5.

[34] Palmer J, Merrett D, Palumbo S, *et al.* Illuminator of opportunity bistatic radar research at DSTO. In: 2008 International Conference on Radar. IEEE; 2008. p. 701–705.

[35] Langellotti D, Colone F, Lombardo P, *et al.* Over the horizon maritime surveillance capability of DVB-T based passive radar. In: European Microwave Week 2014: "Connecting the Future", EuMW 2014 – Conference Proceedings; EuRAD 2014: 11th European Radar Conference. Rome, Italy: IEEE; 2014. p. 509–512.

[36] Gao Z, Tao R, Ma Y, *et al.* DVB-T signal cross-ambiguity functions improvement for passive radar. 2006 CIE International Conference on Radar. Shanghai: IEEE; 2006. p. 1–4.

[37] Barcena-Humanes JL, Martin-de Nicolas J, Solis-Carpintero C, *et al.* DVB-T ambiguity peaks reduction in passive radar applications based on signal reconstruction. In: 2014 44th European Microwave Conference. IEEE; 2014. p. 1900–1903.

[38] Colone F, Langellotti D, and Lombardo P. DVB-T signal ambiguity function control for passive radars. IEEE Transactions on Aerospace and Electronic Systems. 2014;50(1):329–347.

[39] Guo H, Woodbridge K, and Baker CJ. Evaluation of WiFi beacon transmissions for wireless based passive radar. In: 2008 IEEE Radar Conference, RADAR 2008. 1. Rome; 2008. p. 1–6.

[40] Chetty K, Smith GE, and Woodbridge K. Through-the-wall sensing of personnel using passive bistatic WiFi radar at standoff distances. IEEE Transactions on Geoscience and Remote Sensing. 2012;50(4): 1218–1226.

[41] Falcone P, Colone F, and Lombardo P. Potentialities and challenges of WiFi-based passive radar. IEEE Aerospace and Electronic Systems Magazine. 2012;27:15–26.

[42] Chen Q, Chetty K, Woodbridge K, *et al.* Signs of life detection using wireless passive radar. In: 2016 IEEE Radar Conference (RadarConf). IEEE; 2016. p. 1–5.

[43] Martelli T, Colone F, and Lombardo P. First experimental results for a WiFi-based passive forward scatter radar. In: 2016 IEEE Radar Conference (RadarConf). IEEE; 2016. p. 1–6.

[44] Griffiths H and Garnett A. Bistatic Radar using satellite-borne illuminators of opportunity. In: 92 International Conference on Radar. vol. 6; 1992. p. 276–279.

[45] Tsui JB and Shaw RL. Passive ranging through global positioning system. Google Patents; 1993.

[46] Koch V and Westphal R. New approach to a multistatic passive radar sensor for air/space defense. IEEE Aerospace and Electronic Systems Magazine. 1995;10(11):24–32.

[47] Antoniou M and Cherniakov M. Passive GNSS-based SAR resolution improvement using joint Galileo E5 signals. IEEE Geoscience and Remote Sensing Letters. 2015;12(8):1640–1644.

[48] Glennon EP, Dempster AG, Rizos C, *et al.* Feasibility of air target detection using GPS as a bistatic radar. Journal of Global Positioning Systems. 2006; 5(1–2):119–126.

[49] Cherniakov M, Nezlin D, and Kubik K. Air target detection via bistatic radar based on LEOS communication signals. IEE Proceedings-Radar, Sonar and Navigation. 2002;149(1):33–38.

[50] Suberviola I, Mayordomo I, and Mendizabal J. Experimental results of air target detection with a GPS forward-scattering radar. IEEE Geoscience and Remote Sensing Letters. 2012;9(1):47–51.

[51] Edrich M, Meyer F, and Schroeder A. Design and performance evaluation of a mature FM/DAB/DVB-T multi-illuminator passive radar system. IET Radar, Sonar & Navigation. 2014;8(2013):114–122.

[52] Kuschel H, Ummenhofer M, Lombardo P, *et al.* Passive radar components of ARGUS 3D. IEEE Aerospace and Electronic Systems Magazine. 2014;29:15–25.

[53] Griffiths H, Baker C, and Baubert J. Bistatic radar using satellite-borne illuminators. In: RADAR 2002. Edinburgh, UK: IET; 2002. p. 1–5.

[54] Raney RK. Synthetic aperture imaging radar and moving targets. IEEE Transactions on Aerospace and Electronic Systems. 1971;(3):499–505.

[55] Whitewood AP, Muller BR, Griffiths HD, *et al.* Bistatic synthetic aperture radar with application to moving target detection. In: 2003 Proceedings of the International Conference on Radar (IEEE Cat. No. 03EX695). Adelaide: IEEE; 2003. p. 529–534.

[56] Whitewood AP, Baker CJ, and Griffiths HD. Bistatic radar using a space-borne illuminator. In: 2007 IET International Conference on Radar Systems. Edinburgh, UK; 2007. p. 1–5.

[57] ESA. ESA Declares End of Mission for Envisat/Envisat/Observing the Earth/Our Activities/ESA; 2012.

[58] Samczyński P, Wilkowski M, and Kulpa K. Trial results on bistatic passive radar using non-cooperative pulse radar as illuminator of opportunity. International Journal of Electronics and Telecommunications. 2012; 58(3):291–294.

[59] Ito T, Takahashi R, Morita S, *et al.* Experimental result of passive bistatic radar with unknown transmitting radar pulse. In: 2013 European Radar Conference. Nuremberg: IEEE; 2013. p. 455–458.

[60] Thomas JM, Baker CJ, and Griffiths HD. DRM signals for HF passive bistatic radar. In: IET International Conference on Radar Systems 2007. IEE; 2007. p. 166–166.

[61] Wan X, Zhao Z, Zhang D, *et al.* HF passive bistatic radar based on DRM illuminators. In: Proceedings of 2011 IEEE CIE International Conference on Radar. IEEE; 2011. p. 157–160.

[62] Zhao Z, Wan X, Zhang D, *et al.* An experimental study of HF passive bistatic radar via hybrid sky-surface wave mode. IEEE Transactions on Antennas and Propagation. 2013;61(1):415–424.

[63] Aldowesh A, Shoaib M, Jamil K, *et al.* A passive bistatic radar experiment for very low radar cross-section target detection. In: 2015 IEEE Radar Conference. IEEE; 2015. p. 406–410.

[64] Martelli T, Murgia F, Colone F, *et al.* Detection and 3D localization of ultralight aircrafts and drones with a WiFi-based Passive Radar. In: International Conference on Radar Systems (Radar 2017). Belfast: Institution of Engineering and Technology; 2017. p. 1–6.

[65] Chadwick AD. Micro-drone detection using software-defined 3G passive radar. In: International Conference on Radar Systems (Radar 2017). Belfast: Institution of Engineering and Technology; 2017. p. 1–6.

[66] Liu Y, Wan X, Tang H, *et al.* Digital television based passive bistatic radar system for drone detection. In: 2017 IEEE Radar Conference (RadarConf). IEEE; 2017. p. 1493–1497.

[67] Barott WC and Engle J. Single-antenna ATSC passive radar observations with remodulation and keystone formatting. In: 2014 IEEE Radar Conference. IEEE; 2014. p. 0159–0163.

[68] Zhang Y, Li Y, Li W, *et al.* Target detection with single surveillance channel for PBR. In: 2016 IEEE International Geoscience and Remote Sensing Symposium (IGARSS). IEEE; 2016. p. 6409–6412.

[69] Jahangir M and Baker CJ. CLASS U-space drone test flight results for non-cooperative surveillance using an L-band 3-D staring radar. In: 2019 20th International Radar Symposium (IRS); 2019. p. 1–11.

[70] Jahangir M, Ahmad BI, and Baker CJ. Robust drone classification using two-stage decision trees and results from SESAR SAFIR Trials. In: 2020 IEEE International Radar Conference (RADAR); 2020. p. 636–641.

Chapter 6

DVB-T-based passive radar for silent surveillance of drones

María-Pilar Jarabo-Amores[1], David Mata-Moya[1], Nerea del-Rey-Maestre[1], Tatiana Martelli[2] and Fabiola Colone[2]

6.1 Introduction

Nowadays, Unmanned Aerial Vehicles (UAVs) and drones are used as threat's vectors that create personal and public security issues [1–4]. The unpredictable and complex motion along with the small Radar Cross Section (RCS) and low velocity makes the drone detection a challenging task for any radar system [5].

In the context outlined earlier, the security level enabled by conventional active radar systems could be augmented by the cost-effective, non-intrusive and eco-friendly Passive Radar (PR) technology. As a PR system does not have its own transmitter, this allows reduced costs, intrinsic covert operation capability and the lack of additional electromagnetic pollution [6,7]. To guarantee complete and continuous coverage, PR can effectively be integrated within conventional active radars not only to extend the surveillance coverage, acting as 'gap-filler', but also to reduce the probability of out of service of the surveillance system. Moreover, aiming at the monitoring of airport terminal areas or harbours, where the installation of additional sensors might be limited by regulations related to public safety and risk of interference with pre-existing systems, a network of PRs could easily be deployed to provide continuous and complete coverage.

The stationary nature and the isotropic characteristic of many of the employable Illuminators of Opportunity (IoO) provide a persistent illumination of the targets of interest to generate Coherent Processing Intervals (CPIs) of long integration times (T_{int}) on receive to counteract the limited power density offered by the emitter. This certainly applies to many ground-based transmitters for analogue or digital radio/TV broadcasting. Among them, the emitters of the Digital Video

[1]Signal Theory and Communications Department, Superior Polytechnic School, University of Alcalá, Alcalá de Henares, Spain
[2]Department of Information Engineering, Electronics and Telecommunications (DIET), Sapienza University of Rome, Rome, Italy

Broadcasting-Terrestrial (DVB-T) are particularly attractive for counter-drone applications. Specifically, the high radiated power of these transmitters and the excellent coverage make them suitable for the detection of these small RCS and low altitude targets. In addition, the continuous emissions and the fine range resolution of about 20 m (equivalent monostatic range resolution yielded by a signal bandwidth of approximately 8 MHz) make them potentially able to continuously detect and discriminate closely spaced targets. Aiming at the detection of the low Signal-to-Noise Ratio (SNR) targets and at widening the DVB-T-based PR coverage area, very long integration times (up to few seconds) can be exploited if the migration effects are properly compensated [8,9]. It is worth noticing that the use of long integration time allows also to improve the Doppler resolution as well as to discriminate between slowly moving targets and clutter contributions, which is of particular interest in a scenario with a high density of targets. By employing an Orthogonal Frequency-Division Multiplexing (OFDM) modulation, DVB-T signals are noise-like waveforms; thus, they provide ambiguity function with attractive properties that are nearly independent of the signal content and almost time-invariant [10,11]. Eventually, since a DVB-T transmitter typically broadcasts multiple channels at different carrier frequencies, this provides the desired diversity of information that could be successfully exploited for both target detection and its localization [7,12,13].

Recently, different authors have investigated the use of such sensor for counter-drone operations [14–21] proving the capability of such technology to detect and localize small and medium drones up to a few kilometres from the PR receiver. Moreover, the capability of such sensor in simultaneous detection of drones flying near the airport area along with the conventional civil air traffic at farther ranges has been proved in [19,20].

This chapter reports the latest results of DVB-T-based PR for counter-drone operations obtained by the research groups of the University of Alcalá and Sapienza University of Rome. First, Section 6.2 reports a power budget analysis to provide a preliminary evaluation of the expected coverage of PR against drones. Subsequently, Section 6.3 describes the adopted processing scheme, properly tailored to be effective in the considered application. In detail, a significant effort has been devoted to the disturbance cancellation stage that represents one of the key stages within a conventional PR processing scheme. The Neyman–Pearson (NP) detector approximation and clutter modelling are investigated in Section 6.4. Multichannel strategies, based on frequency and spatial diversity, are presented in Section 6.5 to improve the detection and localization performance of PRs. Finally, Section 6.6 draws our conclusion.

6.2 DVB-T-based PR coverage study

System coverage studies are carried out to analyse the capabilities of PRs in real scenarios. This analysis reflects how well an area can be monitored by the radar and is used to estimate the expected performance. System coverage is defined for specific target and interference models, as the maximum range where a target is detected fulfilling the Probability of Detection (P_D) and Probability of False Alarm (P_{FA}) requirements. In this section, a general methodology is presented and evaluated for a rural radar scenario and a target of interest.

6.2.1 Coverage estimation methodology

The key tool for estimating the PR coverage is the bistatic radar equation [22]:

$$(R_R R_T)^2 = \frac{p_T \cdot g_T \cdot g_R \cdot \lambda^2 \cdot \sigma_{bis}}{(4\pi)^3 \cdot p_R \cdot l_{IoO\text{-}target} \cdot l_{target\text{-}PR}} \tag{6.1}$$

where R_R and R_T are target-to-receiver and target-to-transmitter distances, respectively, p_T is transmitter power, p_R is available power at the receiver, λ is signal wavelength, g_T and g_R are the gains of the transmitting and receiving antennas, respectively, σ_{bis} is target's Bistatic RCS (BRCS), and $l_{IoO\text{-}target}$ and $l_{target\text{-}PR}$ are path losses associated with both IoO-target and target-PR paths, respectively. Path losses include free-space losses and diffraction and reflection effects influenced by terrain relief.

When the path losses are negligible and p_R is set to the minimum signal required at the input of the RF-front end of the PR, (6.1) will allow to estimate the Cassini oval, which is widely used as a first approach for coverage calculations. Nonetheless, there are scenarios, especially those oriented to small targets at low altitudes, where path losses due to terrain relief will not be negligible, making the Cassini oval estimation very inaccurate. More precise coverage estimations can be achieved using electromagnetic simulators to compute the path losses, including relief information.

Taking into account this assumption, the radar coverage area is estimated as the area in which the presence of a target produces an acquired echo power level greater than the system sensitivity, S_{min}. The power collected by the PR due to the presence of a hypothetical target in a specific position of the Area of Interest (AoI) identified by the Cartesian coordinates (x,y) can be obtained from the following equation:

$$\begin{aligned} \mathbf{p}_R(x,y) &= \mathbf{p}_{T\text{-}target}(x,y) \cdot \sigma_{bis} \cdot \mathbf{f}_{target}(x,y); \\ \mathbf{p}_{T\text{-}target}(x,y) &= \frac{p_T \cdot g_T(x,y)}{(4\pi) \cdot R_T^2 \cdot l_{IoO\text{-}target}}; \\ \mathbf{f}_{target}(x,y) &= \frac{g_R(x,y) \cdot \lambda^2}{(4\pi)^2 \cdot R_R^2 \cdot l_{target\text{-}PR}}; \end{aligned} \tag{6.2}$$

where $p_{T\text{-}target}(x,y)$ is the power density impinging on the target, and $f_{target}(x,y)$ is the term grouping the propagation factors and receiver antenna gain for the target and PR location. Both terms can be estimated through electromagnetic simulations. The evaluation of (6.2) at each point of the AoI produces $\mathbf{p}_{T\text{-}target}$, $\mathbf{f}_{target}$ and $\mathbf{p}_R$ matrices for the selected geometry, path simulation model and target characteristics. A high-accuracy-estimated coverage map is obtained after applying a threshold value determined from the system sensitivity level.

6.2.1.1 PR sensitivity computation

The sensitivity of a PR system, $S_{\min}$, is defined as the minimum received power level required at the input of the RF-front end to fulfil detection requirements. $S_{\min}$ can be estimated from (6.3) considering the antenna temperature, T_{ant}, the receiver noise factor, NF_R, the integration gain due to the coherent processing, $G_{int} = 10 \cdot \log_{10}(B \cdot T_{int})$ and the required SNR at the detector input, $(S/N)_{DET}$.

$$S_{\min} = 10 \cdot \log_{10}(kT_{ant}B) + NF_R - G_{int} + (S/N)_{DET} \tag{6.3}$$

k is Boltzmann's constant, and B is acquired signal bandwidth. $(S/N)_{DET}$ depends on the target and interference models, the applied detection rule and the desired P_D and P_{FA}. To obtain $S_{\min}$, a margin of 5 dB is usually considered due to the presence of interference sources and clutter returns [23].

6.2.2 Coverage analysis

For a preliminary analysis and to show an example of a coverage study in as realistic conditions as possible, the capabilities of DVB-T-based PRs for detecting small-size drones are analysed in a rural scenario located in Spain. A commercial DJI Phantom 4 drone is selected as target of interest due to its reduced size and its high popularity in the civil area. The drone will be flying at low altitudes, selecting 10 m and 50 m with respect to the local topographic altitude as representative values for taking off/landing manoeuvres and usual flight altitudes.

To perform a complete coverage study, the dependencies on the operating frequency (f_c), acquisition bandwidth (B) and integration time (T_{int}) are analysed assuming two different values for each parameter. DVB-T operating frequencies of 570 MHz and 700 MHz are selected as representative values of the low and high bands, respectively. The use of DVB-T consecutive channels has a direct impact on PR range resolution, which is exploited for improving system capabilities for detection and localization of low reflectivity targets. Because of that $B = 8$ MHz (one channel) and $B = 24$ MHz (three channels) are analysed.

In a classical PR processing scheme, T_{int} is limited, by the time a target remains in the same resolution cell. A target can overpass the resolution cell due to two different effects: the range walk, the movement of the target exceeds the range resolution; and the Doppler walk, the acceleration of the target (a_t) produces a Doppler cell migration. To avoid integration losses due to both effects, equation (6.4) should be fulfilled [24], where v_t is the target speed.

$$T_{int} < \frac{c}{2 \cdot B \cdot v_t}; a_t < \frac{c}{2 \cdot f_c \cdot T_{int}^2} \tag{6.4}$$

In Table 6.1, maximum T_{int} and a_t values to avoid range and Doppler walks are summarized, considering the maximum flight speed of the Phantom 4 drone, $v_t = 20$ m/s. If longer integration times are selected, range and Doppler compensation techniques must be applied to maximize the coherent integration gain. Taking into account these results, $T_{int} \in \{250, 500\}$ ms are selected for the coverage study.

Table 6.1 T_{int} maximum values to avoid range and Doppler migration for $v_t = 20$ m/s

	T_{int_max} (s)	a_{t_max} at $f_c = 570$ MHz (m/s^2)	a_{t_max} at $f_c = 700$ MHz (m/s^2)
$B = 8$ MHz	0.937	0.3	0.24
$B = 24$ MHz	0.313	2.69	2.19

Table 6.2 Experimental parameters for the coverage study

f_c (MHz)	T_{int} (ms)	B (MHz)	Target altitudes with respect to topographic altitude (m)
570 and 700	250 and 500	8 and 24	10 and 50

In Table 6.2, the experimental parameters selected for the coverage analysis are summarized. The coverage estimation is carried out applying the two methodologies detailed in Section 6.2.1: (1) Cassini's ovals are obtained using (6.1) considering only free-space propagation losses, (2) coverage maps are estimated with (6.2) taking into account the path losses, including the effects of the terrain relief and the radiation antenna pattern. Path losses are calculated using WinProp (Altair Feko) electromagnetic simulator [25]. This software allows the selection of different propagation models and the integration of geographic information systems data to model the relief of the PR scenario. The accuracy of the electromagnetic estimations has a high dependence on the cell size and the precision of the relief database. Topographic data were obtained from NASA's Shuttle Radar Topography Mission version 4.1 [26]. A configuration of 5° longitude ×5° latitude area with approximately 58×92 m^2 cell size was chosen. WinProp tool was configured to compute $\boldsymbol{p}_{T\text{-}target}$ and $\boldsymbol{f}_{target}$ matrices using the Dominant Path Prediction Model as full 3D electromagnetic simulation method. The BRCS study of the Phantom 4 drone is presented in the following subsection.

6.2.2.1 Target BRCS estimation

Monostatic RCS studies of commercial small-size drones have been reported in the literature using numerical methods and/or experimental measurements at different operating frequencies. In [27], the RCS of a quadcopter IRIS drone is analysed in the 1–4 GHz frequency range, obtaining measurement values ranging from −18 dBsm to −1 dBsm for vertical polarization. The characterization of the RCS of a DJI Phantom 3 drone at 10 GHz is presented in [28,29]. Simulated RCS results varied between −20 dBsm and 4 dBsm for horizontal polarization and between −12 dBsm and 4 dBsm for vertical polarization. However, few studies of BRCS for small-size drones are available in the literature and they are mainly made using frequencies higher than DVB-T ones. Because of that the study of the BRCS of a Phantom 4 drone at UHF frequencies is performed in this section.

BRCS modelling is carried out using Ansys HFSS electromagnetic simulator with a full-wave simulation method [30]. The simulation geometry is defined by the incidence and observation directions, which are defined by pairs (θ, ϕ), where θ is the elevation angle measured with respect to the Earth's surface normal, and ϕ is the azimuth one measured in the plane tangent to the Earth's surface, with respect to the North. A bistatic angle-dependent approach is performed taking into account the radar scenario geometry: β is set to a constant value and the target model is rotated in the horizontal plane to vary the incident and associated observation directions.

The 3D model of the Phantom 4 was customized with the materials included in the datasheet for their main parts: Kevlar was selected for the main body and the blades, copper for the rotors and lithium for the battery. The following conditions were assumed for the BRCS simulations:

- Simulations were set up with an incident plane wave impinging on the centre of the target 3D model.
- The geometrical characteristics of the radar scenario determined the incidence (θ_i) and observation (θ_o) elevation angles. Due to the low flight altitudes and high distances assumed for the targets, θ_i and θ_o were set to 90°.
- Four bistatic angles were selected as representative values of the real scenarios and targets' trajectories used in this chapter: $\beta \in \{0°, 30°, 60°, 90°\}$. The azimuth incidence angle (ϕ_i) was set to 0°, and four azimuth observation angles (ϕ_o) were selected to fulfil the bistatic angle equation $\beta = \phi_o - \phi_i$.
- The target model was rotated in steps of 2°.
- Horizontal polarization was used (the one selected in Spain).

In Figure 6.1, simulated BRCSs for the different β values at 570 MHz and 700 MHz are depicted in polar coordinates as a function of the rotation angle in the target reference system. BRCS mean values for each case study are summarized in Table 6.3. For f_c = 570 MHz, the BRCS mean value differs approximately 1.5 dBsm for the different β values under study, whereas a difference of approximately 5 dBsm is obtained for f_c = 700 MHz. These results show the dependence on scenario geometry and frequency, obtaining higher BRCS values for lower DVB-T frequencies.

In [31], three RCS frequency regions are defined in terms of target physical dimensions (dim_{target}) and radar wavelength (λ): Rayleigh region ($dim_{target} \ll \lambda$), resonance region ($dim_{target} \approx \lambda$) and optical region ($dim_{target} \gg \lambda$). For f_c = 570 MHz, results show that the Phantom 4 drone BRCS varies smoothly with the azimuth incident angle or the bistatic angle (Figure 6.1(a)), a behaviour similar to that observed for a point source, which is expected in the Rayleigh region of the BRCS. At f_c = 700 MHz, the BRCS approaches the resonance region and higher fluctuations are observed (Figure 6.1(b)). For coverage studies, the lower estimated BRCSs for each f_c are selected (in red in Table 6.3).

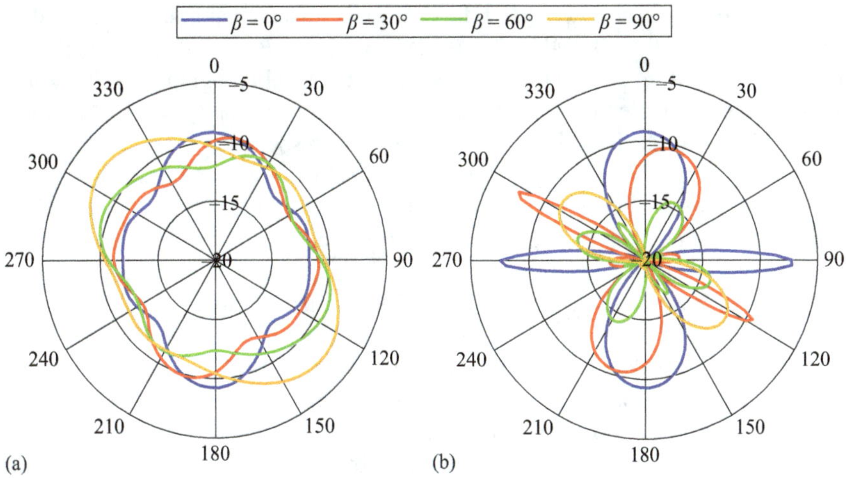

Figure 6.1 *Phantom 4 BRCS (expressed in dBsm) for 570 MHz (a) and 700 MHz (b)*

Table 6.3 *Phantom 4 mean BRCS estimated values*

	$\beta = 0°$ (dBsm)	**$\beta = 30°$ (dBsm)**	**$\beta = 60°$ (dBsm)**	**$\beta = 90°$ (dBsm)**
570 MHz	−10.55	−11.03	−11.17	−9.64
700 MHz	−12.84	−13.99	−17.32	−16.24

6.2.2.2 Coverage estimations

The combination of the path propagation losses estimated, including the terrain relief, the target BRCS estimation and the system sensitivity calculation, allows the generation of the estimated coverage maps using (6.2). To analyse the effects of path losses and evaluate the methodology proposed based on electromagnetic simulations, the Cassini oval is also calculated using (6.1) considering only free space propagation losses.

As a general case, system sensitivity was estimated assuming a Swerling I target model, desired $P_D = 80\%$ and $P_{FA} = 10^{-5}$, and the acquisition components of a DVB-T-based PR system such as the IDEPAR demonstrator developed by the University of Alcalá [32]. $S_{\min} = -139.55$ dBm and $S_{\min} = -142.56$ dBm were calculated for $T_{int} = 250$ ms and 500 ms, respectively.

To perform the coverage estimations in as realistic conditions as possible, a similar scenario to those real ones used in the following sections is considered: a rural scenario located in the South-East of Madrid, Spain, where the PR is deployed in an inland valley with small terrain obstacles. A DVB-T transmitter located at 16.6 km far from the PR and with an Equivalent Radiated Power (ERP) of 20 kW per DVB-T channel is assumed as IoO.

Coverage maps are calculated for the experimental parameters detailed in Table 6.2: $f_c \in \{570, 700\}$ MHz, $T_{int} \in \{250, 500\}$ ms, $B \in \{8, 24\}$ MHz and two target altitudes, 10 m and 50 m, with respect to the local topographic altitude. For saving space, examples of the estimated coverage maps for $f_c = 570$ MHz and $f_c = 700$ MHz are depicted in Figures 6.2 and 6.3, respectively, for $T_{int} = 250$ ms and $B = 8$ MHz. The coverages for the two methodologies and target altitudes overlap: the largest coverage area includes the smaller ones. The use of the relief information in the estimation process produces irregular coverage areas, with the presence of distant isolated covered regions. Taking into account these coverage maps, range coverages are estimated as the distance between the PR and the furthest coverage point along the broadside direction of the surveillance antenna (150°N in the considered scenario), guaranteeing a continuous coverage without isolated areas. Examples of the range coverage estimation are presented in the zoomed area on the bottom left corner of Figures 6.2 and 6.3.

In Table 6.4, range coverage estimations obtained by the Cassini oval and the methodology based on electromagnetic simulations are summarized for the cases of Table 6.2. In the Cassini oval approach, range coverages decrease as f_c increases, and largest values are obtained for higher T_{int} and B due to the improvement of the system sensibility and total IoO ERP, respectively.

On the other hand, the methodology based on electromagnetic simulations provides conservative coverage estimations and results close to the expected ones in a real scenario, due to it takes into account the path losses, including the terrain relief information. In this case, range coverage estimations are specific of the considered radar scenario, varying when a different geographic profile is used. As it

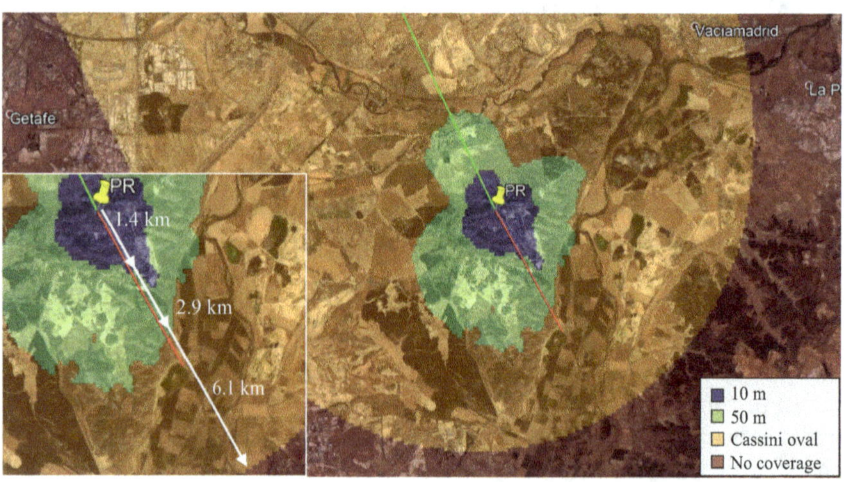

Figure 6.2 Coverage results for a Phantom 4 drone. $T_{int} = 250$ ms, $B = 8$ MHz, $f_c = 570$ MHz. Red line: broadside direction. Green line: baseline

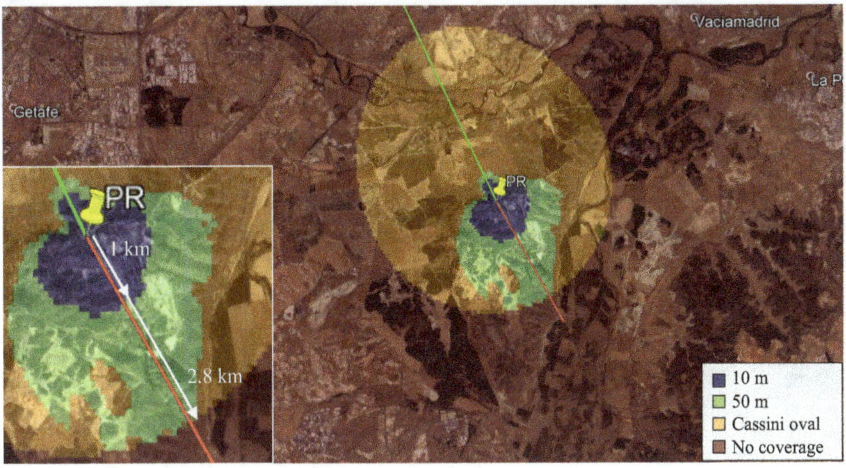

Figure 6.3 Coverage results for a Phantom 4 drone. $T_{int} = 250$ ms, $B = 8$ MHz, $f_c = 700$ MHz. Red line: broadside direction. Green line: baseline

Table 6.4 Maximum continuous estimated coverage at the PR broadside direction

	570 MHz			700 MHz		
	Cassini	**50 m**	**10 m**	**Cassini**	**50 m**	**10 m**
$T_{int} = 250$ ms; $B = 8$ MHz	6.1 km	2.9 km	1.4 km	2.8 km	2.8 km	1.0 km
$T_{int} = 250$ ms; $B = 24$ MHz	9.4 km	3.3 km	2.2 km	4.6 km	2.9 km	1.3 km
$T_{int} = 500$ ms; $B = 8$ MHz	8.0 km	3.1 km	2.1 km	3.8 km	2.9 km	1.2 km
$T_{int} = 500$ ms; $B = 24$ MHz	12 km	3.8 km	2.3 km	6.1 km	3.0 km	1.5 km

is observed in Table 6.4, for both target flight altitudes, range coverages do not drop drastically as f_c increases, and they do not improve significantly for higher T_{int} and B values. This fact could be related to an obstruction in the line of sight between the IoO and the target, which will reduce the received signal level. Figure 6.4 shows the difference between the received power estimated considering only free-space losses and the received power obtained taking into account the simulated path losses for a target located at 50 m altitude with respect to the local topographic altitude and $f_c = 700$ MHz.

In Figure 6.4, a valley area placed at 2.8 km far from the PR is identified, which presents higher path propagation losses and limits the improvement in the range coverage estimation. Figure 6.5 shows the non-line of sight conditions between the IoO and a target located at 2.9 km far from the PR with a flight altitude

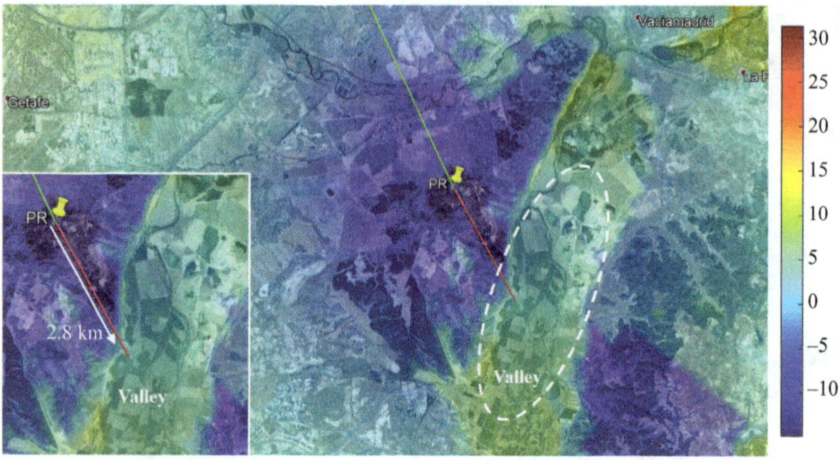

Figure 6.4 Difference (dB) between the received power estimated considering only free-space losses and the received power obtained considering simulated path losses at 700 MHz

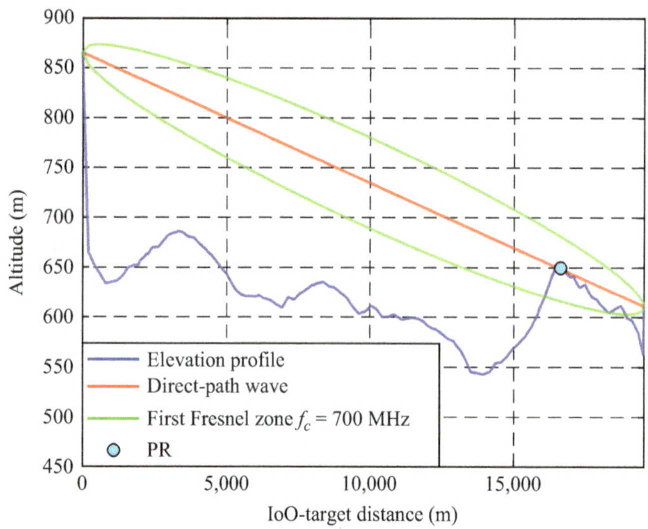

Figure 6.5 Analysis of the terrain profile and the line-of-sight conditions in the considered radar scenario. The target is located at 2.9 km from the IoO with an altitude of 50 m with respect to the topographic altitude

of 50 m. A high terrain obstacle close to the PR location intercepts the direct ray, causing a decrease of the received signal level due to higher diffraction losses. This fact makes insignificant the improvement expected in the range coverages for higher B and/or T_{int} values and lower f_c (Table 6.4, 50 m column). A similar performance is observed for a target with a flight altitude of 10 m with respect to the local topographic one (Table 6.4, 10 m column). In this case, the direct ray is obstructed by the terrain obstacle at a lower distance from the PR.

The results presented in this section show the dependence of the coverage maps and range coverage estimations on the radar scenario, signal processing parameters and selected target. These results also confirm the impact of the relief information in the coverage estimation process, validating the proposed methodology based on electromagnetic simulators combined with BRCS estimations and system sensitivity calculations.

6.3 DVB-T-based PR processing scheme and the disturbance cancellation stage

6.3.1 DVB-T-based PR processing scheme

The potential of a DVB-T-based PR has been widely demonstrated for both air traffic control and coastal/maritime surveillance applications [33–41]. However, differently from aircraft and vessels, drones are characterized by very small RCS values, being mostly made of non-metallic materials and by unpredictable and complex kinematics. Moreover, since they usually fly at low altitudes, their echoes might be masked by natural or man-made elements embedded in the scenarios of interest (e.g. top of trees and buildings). All these characteristics make drone detection a challenging task for any radar system. Therefore, each stage of the DVB-T-based processing scheme should be properly adapted to be effective in the considered drone surveillance application.

Figure 6.6 sketches the employed DVB-T-based processing scheme for target detection and localization, while the main required processing stages are briefly illustrated in the following.

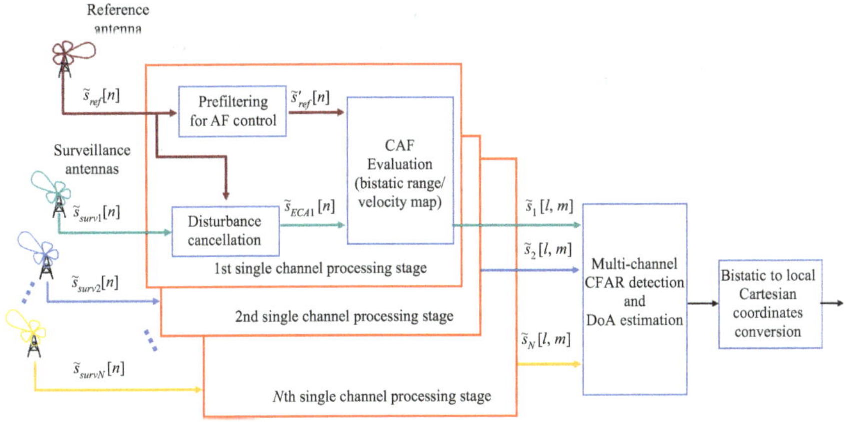

Figure 6.6 Block diagram of the DVB-T-based PR processing scheme

For the sake of generality, we refer to the case of a PR equipped with multiple surveillance channels, which collect the signals $\tilde{s}_{surv\ n}$ ($n = 1,\dots,N$) from an array of N antennas. The passive receiver also features a reference channel to acquire the signal $\tilde{s}_{ref}$ transmitted by the IoO; alternative implementations might be based on the reference signal reconstruction from the surveillance signals according to a demodulation/remodulation approach.

We assume that the whole acquisition time (T_{adq}) is fragmented into CPIs of duration T_{int} seconds. The total number of CPIs and the distance between consecutive CPIs are defined by a selected update time (T_{upd}). Both T_{upd} and T_{int} are carefully selected for a specific PR application taking into account the desired resolutions, expected targets' dynamics and the processing techniques to be applied.

According to a basic PR processing scheme, at each CPI, the signal collected at the reference channel is first used to remove undesired contributions in the signals collected at the surveillance channels (i.e. direct signal, clutter and multipath echoes). As it is well known, the disturbance cancellation represents one of the key stages within a conventional PR processing scheme since the strong stationary contributions could mask the weak targets and, consequently, degrade the target detection performance of the system. It should be noted that such undesired contributions cannot be removed by applying conventional Moving Target Indicator (MTI) techniques. A standard canceller is based on subtracting two or more consecutive echoes, with the idea that the echo from stationary targets in two consecutive echoes would be exactly the same and the difference will cancel the undesired contributions from the stationary background. In PR, two successive batches of the received signals are different even if they have been reflected by the same stationary background, since the transmitted waveform changes continuously. Hence, proper signal-dependent disturbance cancellation approaches have been designed to guarantee an effective clutter removal. Among them, we resort here to the sliding version of the Extensive Cancellation Algorithm (ECA-S) [42,43]. In particular, as demonstrated in [43,44], the ECA-S makes the system more robust to the time-varying characteristics of the environment with respect to the conventional ECA. In addition, it allows a better trade-off between disturbance cancellation and the capability to preserve low-Doppler target echoes, thus improving the detection performance of the resulting PR system.

The ECA operates by subtracting from the generic surveillance signal $\tilde{s}_{surv\ q}(t)$, delayed replicas of the reference signal $\tilde{s}_{ref}(t)$ properly weighted according to adaptively estimated coefficients [35]. Assuming that the multipath and clutter echoes are backscattered from the first K range bins, and the received signals are sampled with sampling frequency f_s, for each receiving channel the output of ECA is evaluated as

$$\begin{aligned}&\tilde{s}_{ECA\ q}[n] = \tilde{s}_{surv\ q}[n] - \textstyle\sum_{k=0}^{K-1} a_{qk} \cdot \tilde{s}_{ref}[n-k];\\ &n = 0,\dots,N_{int}-1, q = 1,\dots,N\end{aligned} \tag{6.5}$$

where $N_{int} = T_{int} \cdot f_s$ is the number of samples within the CPI, being f_s (Hz) the sampling frequency. The filter coefficients $\alpha_q = [\alpha_{q0}\ \alpha_{q1} \cdots \alpha_{q(K-1)}]^T$ are evaluated by resorting to a least square approach that minimizes the power of the signal at the output of the filter:

$$\boldsymbol{a}_q = \left(\boldsymbol{S}_{ref}^H \boldsymbol{S}_{ref}\right)^{-1} \cdot \boldsymbol{S}_{ref}^H \boldsymbol{s}_{surv\ q} \tag{6.6}$$

where $\boldsymbol{s}_{surv\ q}$ is an $N_{int}\times 1$ vector containing N_{int} samples of the qth surveillance signal and $\boldsymbol{S}_{ref}$ is an $N_{int}\times K$ matrix, columns of which are the delayed versions of the reference signal.

In its original version, the ECA requires the filter weights to be estimated by averaging over the whole CPI. In contrast, the batch version (ECA-B) estimates and applies the filter weights over smaller portions (batches) of the integration time. Thus, ECA-B makes the system more robust to the time-varying characteristics of the environment. However, as reported in [43], it yields some limitations in the presence of slowly moving targets or targets moving mainly along the cross-range direction. To overcome the previous limitations, we adopt the ECA-S approach which operates over partially overlapped signal fragments [43]. Specifically, the ECA-S output at the pth fragment of duration T_S is written as

$$\begin{aligned} &\tilde{s}_{ECA-S\ q}[n] = \tilde{s}_{surv\ q}[n] - \textstyle\sum_{k=0}^{K-1} a_{qk}^{(p,T_A)} \cdot \tilde{s}_{ref}[n-k]; \\ &n = pN_S, \ldots, (p+1)N_S - 1; p = 0, \ldots, B_S - 1; q = 1, \ldots, N \end{aligned} \tag{6.7}$$

where N_S is the dimension of each fragment (i.e. $N_S = T_S \cdot f_s$), $B_S = [N_{int}/N_S]$ is the number of consecutive fragments in the CPI, and $\boldsymbol{a}_q^{(p,T_A)} = \left[a_{q0}^{(p,T_A)} \quad \cdots \quad a_{q(K-1)}^{(p,T_A)}\right]^T$ are the current filter coefficients. The latter are adaptively estimated on a longer signal fragment of duration $T_A = N_A/f_s$, symmetrically taken around the current signal fragment to be processed. Basically, $\boldsymbol{a}_q^{(p,T_A)}$ are evaluated as

$$\boldsymbol{a}_q^{(p,T_A)} = \left[\boldsymbol{S}_{ref}^{(p,T_A)^H} \boldsymbol{S}_{ref}^{(p,T_A)}\right]^{-1} \cdot \boldsymbol{S}_{ref}^{(p,T_A)^H} \boldsymbol{s}_{surv\ q}^{(p,T_A)} \tag{6.8}$$

where $\boldsymbol{s}_{surv\ q}^{(p,T_A)}$ is the $(N_A\times 1)$ surveillance vector at the qth channel and $\boldsymbol{S}_{ref}^{(p,T_A)}$ is an $N_A\times K$ matrix collecting the delayed copies of the corresponding reference signal fragment of duration T_A.

As illustrated in [43], the batch duration T_A is selected to obtain remarkable cancellation performance while preserving targets moving at the minimum Doppler of interest (notch Doppler extension $\Delta f_n = 1/T_A$). In contrast, T_S is selected in order to move the undesired structures that arise from the batch processing of the received signals ($T_S<(f_{Dmax}+1/(2T_A))^{-1}$), out of the Doppler range of interest (f_{Dmax}).

Based on the previous considerations, aiming at the detection of slowly moving objects with low RCS, longer batches are to be preferred to yield a narrower cancellation notch, thus preserving the target echo.

After the cancellation stage, the output signals from ECA-S filter are processed with a properly mismatched reference signal, $\tilde{s}'_{ref}$ [45], to evaluate the bistatic range/velocity Cross Ambiguity Function (CAF):

$$\tilde{s}_q[l,m] = \sum_{n=0}^{N_{int}} \tilde{s}_{ECA-S\,q}[n] \cdot \tilde{s}'^{*}_{ref}[n-l]^{-j2\pi(nm/N_{int})} \tag{6.9}$$

where l is the time bin associated with a delay $\tau_l = l/f_s$ and a relative bistatic range $r_1 = cl/f_s$, and m is the Doppler shift corresponding to $f_m = f_s(m/N_{int})$ and a bistatic velocity $v_m = (mf_s/N_{int})(c/f_c)$ being c the speed of light and f_c the carrier frequency. Specifically, to evaluate (6.9), the correlation-FFT technique is employed as it represents the most efficient solution for the considered application [7].

As previously mentioned, the reference signal is properly filtered before it is employed for CAF evaluation. Such filtering stage is intended to remove the high side lobes and spurious peaks appearing in the DVB-T signal ambiguity function. To this purpose, we resort to the approach presented in [45] which is based on the cascade of two filters, namely the pilot signals equalization and the Residual Peaks Removal to remove the zero-Doppler peaks. It is worth noticing that both filters are not strictly necessary in the short-range case since the spurious peaks appear outside the observed range/velocity region. Nevertheless, their application guarantees that the surveillance range/velocity area observed in short-range applications is not affected by undesired side lobe structures arising from the presence of strong targets outside that area. To reduce further the computational load, suboptimum techniques can be used to evaluate the CAF with limited SNR loss [7,46,47].

Once the CAF has been evaluated at all the available surveillance channels, the output maps are jointly exploited for both target detection and its Direction of Arrival (DoA) estimation. Specifically, target detection can be sought according to different, centralized or decentralized, approaches. As per a centralized strategy, target detection is performed in the beam space after an appropriate, potentially adaptive, beamforming based on the available spatial channels. In contrast, a simple decentralized strategy can be adopted by separately applying a conventional Cell Average–Constant False Alarm Rate (CA-CFAR) threshold to each intensity map and then combining the binary outputs according to an M-out-of-N criterion. Obviously, in that case, the detection threshold at each channel should be properly adjusted to guarantee the desired global P_{FA}. Once the target has been detected and its bistatic range and DoA have been estimated, its coordinates are converted into local Cartesian coordinates.

6.3.2 Experimental results and impact of the cancellation stage on the drone detection performance

To show the capability of a DVB-T-based PR in counter-drone operations, we report the results of experimental tests carried on in cooperation with the colleagues of Leonardo SpA.

The acquisition geometry is sketched in Figure 6.7. The DVB-T transmitter of Monte Cavo (22.5 km away) was exploited as IoO, while the DVB-T-based PR receiver was installed close to the runway of the military airport of Pratica di Mare (Italy) (see the enlarged view of the local geometry on the bottom left corner of Figure 6.7). Two or three Yagi–Uda surveillance antennas were employed during

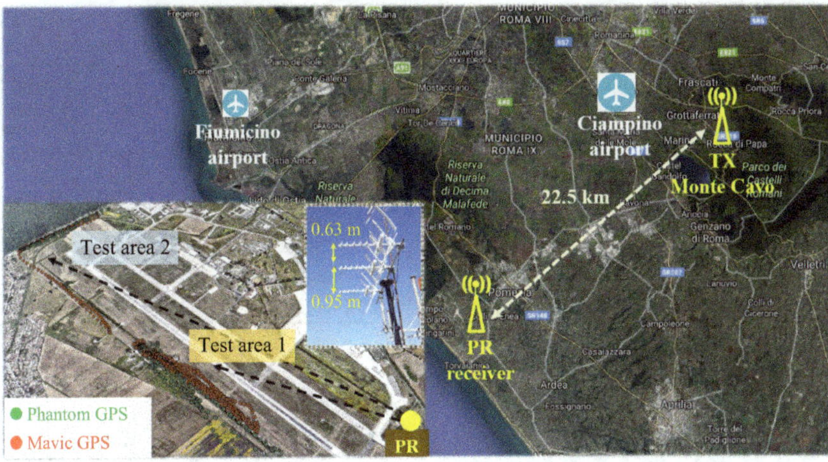

Figure 6.7 Sketch of the acquisition geometry for the test campaigns in the military airport of Pratica di Mare Airport. © 2020 IET. After [20]

the performed tests, each one with a main beamwidth of 36°. They were displaced in the horizontal plane according to different layouts. Specifically, the tests with two surveillance channels exploited two receiving antennas mounted at distance $d = 0.63$ m, whereas when three surveillance channels were deployed, they were connected to a three-elements Non-Uniform Linear Array (NULA) with inter-element distances between the antennas equal to 0.63 and 0.95 m, respectively.

We focus on three data sets, each composed of sequential data files (namely CPIs). The number of processed CPIs, the temporal duration of each CPI (T_{int}) and the total acquisition time (T_{acq}) of each data set are synthesized in Table 6.5. One or two DVB-T frequency channels have been simultaneously collected at carrier frequency F0 = 570 MHz and F1 = 754 MHz, both emitted by the same transmitter. Table 6.5 reports the selected DVB-T channels for each data set.

During the test campaigns, one or two very small drones were employed as cooperative targets (see Figure 6.8) flying in the surrounding area of the airport. They were a DJI Phantom 4 Pro of size approx. 25 cm×25 cm (propellers excluded) and a DJI Mavic Pro of size approx. 20-cm long and 8.3-cm wide (propellers excluded). In the test campaigns, both drones were equipped with a GPS receiver to record their position. In Figure 6.7, the green and red plots represent the GPS trajectories of the Phantom and the Mavic during the collected data sets, respectively.

Two different test areas have been considered for the counter-drone application. The centre of the first area is ideally located at a distance of about 1.7 km away from the RX (Test area 1 in Figure 6.7), and the two drones flew along various paths reaching a minimum distance of 0.85 km up to a maximum distance of 1.93 km from the RX site. In the second case, the small flying objects flew up to the allowed boundaries of the airport area, yielding a maximum distance from the RX of approximately 3 km (Test area 2 in Figure 6.7). In all cases, the surveillance

Table 6.5 Details of the data sets collected during the performed test campaigns

Data set	No. of drones	No. of CPIs	CPI duration, T_{int} (s)	Update time, T_{upd} (s)	Acquisition duration, T_{adq} (min)	Carrier Freq. (MHz)	No. of surv antennas	Test area [min–max] distance
1	2	107	0.5	4.25	≈8	F0 = 570 F1 = 754	3	1 [0.85–1.79] km
2	2	195	1	5.3	≈17	F0 = 570 F1 = 754	3	2 [2.15–3.18] km
3	1 (Phant.)	147	0.5	2.75	≈7	F0 = 570	2	1 [1.45–1.93] km

(a) (b)

Figure 6.8 Picture of the drones used as cooperative targets: DJI Phantom 4 Pro (a); DJI Mavic Pro (b)

antennas were appropriately pointed to include the small flying objects in the main beam. Table 6.5 details the considered test area of each data set as well as the employed drones and the number of surveillance antennas exploited.

All the available data files have been first processed according to the processing scheme of Figure 6.6. Due to the low velocity and accelerations of the targets of interest, the adopted T_{int} values (i.e. $T_{int} = 0.5$ s for Data sets 1 and 3 and $T_{int} = 1$ s for Data set 2) allowed the desired coverage without experiencing severe migration in both range and Doppler dimension.

First, we show the impact of the disturbance cancellation stage on drone targets characterized by small RCS and low-Doppler values. To this purpose, all the available data files have been first processed according to different ECA versions over a range of 6 km. The conventional ECA requires the filter weights to be estimated by averaging over the whole CPI, namely T_{int}. Instead, the ECA-S operates with $T_B = 0.2$ s (i.e. $\Delta f_n = 5$ Hz) and, since the maximum observed velocity can be limited at 50 m/s, we obtain that $T_S = 5.2$ and $T_S = 3.9$ ms for F0 = 570 MHz and F1 = 754 MHz, respectively.

As an example, Figure 6.9 shows the range–velocity maps obtained for a single CPI from Data set 2 when different cancellation algorithms are applied. Specifically,

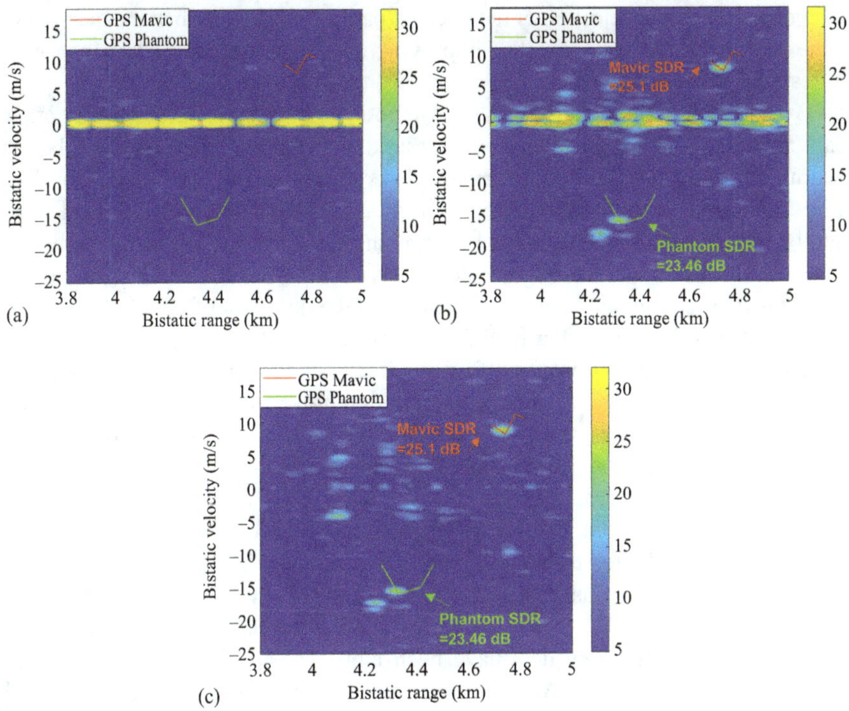

Figure 6.9 Data set 2 – short-range case: enlarged view of the bistatic range–velocity maps obtained at a single CPI using different cancellation techniques: (a) no disturbance cancellation; (b) ECA; (c) ECA-S

based on the availability of GPS information, we report an enlarged view of the maps around the targets of interest. We observe that at the time of the selected CPI, both drones were flying in the monitored area with a bistatic range greater than 4 km and opposite bistatic velocities (see the red and green GPS trajectories). All the reported maps have been scaled for the average background level (both clutter residuals and thermal noise) so that each value represents the estimated Signal-to-Disturbance Ratio (SDR). Figure 6.9(a) shows the range–velocity maps obtained in the absence of disturbance cancellation. As it is evident, a strong contribution corresponding to the direct signal and multipath/clutter echoes is present along the zero velocity. Consequently, both drones are completely masked. Using the ECA for the disturbance removal (Figure 6.9(b)), two high peaks with SDR values of 25.1 and 23.46 dB are clearly visible around the Mavic and Phantom GPS tracks, respectively. However, disturbance residuals are observed around the zero velocity since the stationary contribution is not perfectly removed. Resorting to the tailored ECA-S approach (Figure 6.9(c)), both drone peaks are visible and meanwhile, all the disturbance echoes have been effectively removed, thus showing a better cancellation capability compared to ECA. As expected, the choice of the ECA-S approach together with a proper selection of the relevant parameters yields a better cancellation of the stationary contributions, thus improving the drone detection performance of the resulting PR system.

By observing Figure 6.9(b) and (c), we notice that in both figures, many other peaks appear. These can be caused by undesired contributions, such as side lobes of the ambiguity function, disturbance residuals as well as other noise sources, or they can correspond to other targets moving in the considered area. All the peaks are equally treated but, as these maps undergo the CFAR detection stage, only the highest peaks will exceed the threshold, thus yielding a raw detection. Due to the high-density scenario and aiming at reducing the number of false alarms, we operate in the following with a final $P_{FA} = 10^{-7}$.

The detection results obtained for the three tests are shown in Figures 6.10–6.12. In all cases, for a direct comparison, we employed just two of the available surveillance antennas. In each figure, sub-figure (a) illustrates the detection results in the bistatic range–Doppler domain. They are obtained by exploiting a decentralized integration approach among the two surveillance channels: specifically, a CFAR detection threshold is separately applied at each channel, and a 2-out-of-2 strategy is then adopted to integrate the binary outputs, being all the parameters set in order to obtain a final $P_{FA} = 10^{-7}$. Sub-figure (b) reports the results of the targets' localization in Cartesian coordinates obtained using the two employed surveillance antennas. Grey dots have been employed to plot the raw detection results from the PR obtained across 50 consecutive CPIs, while we report in red and green the Mavic and Phantom GPS trajectories, respectively. Eventually, red and green plots are employed to plot the correct target detections, namely the detections associated with the Mavic and Phantom GPS, respectively.

First of all, we observe that both drones are continuously detected along their trajectories from a minimum distance of 1.6 km up to a maximum distance of 5 km bistatic range (which corresponds to $R_R \simeq 3.2$ and $R_T \simeq 25.1$ km) across the different

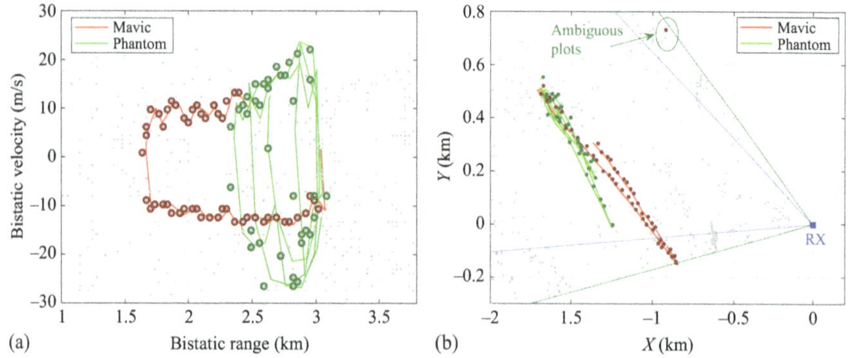

Figure 6.10 Results for Data set 1: (a) detections in the range–Doppler plane; (b) localization in Cartesian coordinates. © 2020 IET. After [20]

tests performed (see Figures 6.10(a)–6.12(a)). This is indeed a remarkable result and successfully demonstrates the sensitivity of the employed system and the effectiveness of the adopted signal processing strategies.

When the drones flew at a shorter distance from the receiver, both targets were detected with a notable continuity along the acquisition time (see Figure 6.10(a)). The corresponding localization results in Figure 6.10(b) show that the system is also able to localize both targets with high accuracy. However, an ambiguous plot is observed for the Mavic drone as highlighted by the green circle. For the considered DVB-T channel at carrier frequency 570 MHz, when two surveillance antennas at distance $d = 0.63$ m are employed, the target DoA is estimated unambiguously only in an angular sector equal to 49.34° (green dotted lines in Figure 6.10(b)). Notice that the obtained unambiguous angular sector is slightly larger than the antenna main beamwidth of approx. 36° (blue dotted lines in Figure 6.10(b)). Consequently, the target flying just outside the unambiguous sector is likely to be detected but the corresponding plots are miss-localized in the Cartesian domain.

As the drone's distances increase (Data set 2), we observe some missed detections (see Figure 6.11(a)) and a less accurate localization capability is obtained (Figure 6.11(b)). This is because the target SNR decreases, thus yielding a degraded DoA estimation accuracy. This, in turn, determines higher positioning errors as the small objects move away from the receiver position. In addition, a large number of ambiguous plots are observed for the considered cases. This effect gets worse when Data set 3 is considered (see Figure 6.12). In the reported test, the Phantom flew totally outside of the surveillance antenna main beam. This demonstrates that the small target can still be detected with a good continuity, although its reflections impinge on the side lobes of the surveillance antennas. However, as the drone location falls outside the unambiguous angular sector, the passive plots are totally miss-localized (see the green plots in Figure 6.12(b)). For illustrative purpose only, by properly compensating for the a priori knew target

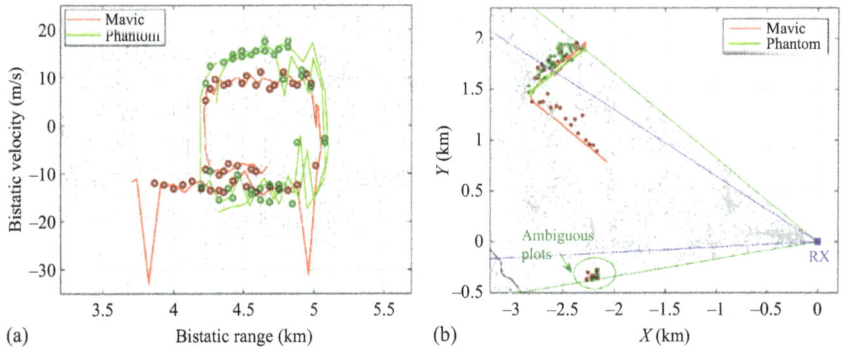

Figure 6.11 Results for Data set 2: (a) detections in the range–Doppler plane; (b) localization in Cartesian coordinates. © 2020 IET. After [20]

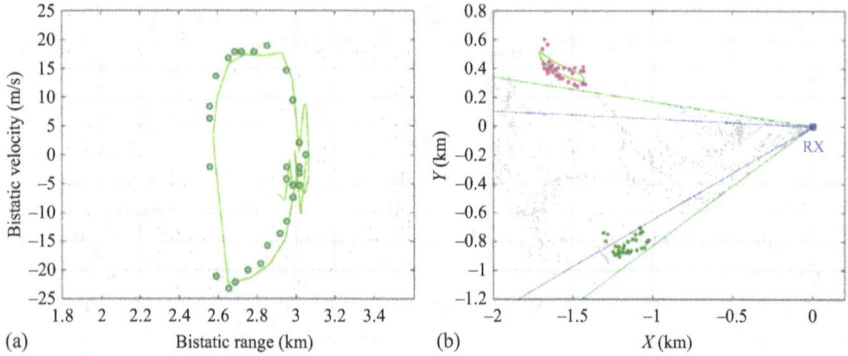

Figure 6.12 Results for Data set 3: (a) detections in the range–Doppler plane; (b) localization in Cartesian coordinates. © 2020 IET. After [20]

DoA ambiguity error, the drone is correctly localized, as shown by the magenta plots in Figure 6.12(b).

Apparently, the effect of the angular ambiguity potentially applies to other plots provided by the PR sensor, i.e. to all the detections resulting from targets laying outside the unambiguous sector. Also, all the false alarms are projected into the Cartesian domain within the same region. This results in a dense distribution of plots within a small angular sector which might jeopardize the subsequent tracking stage. We observe that in order to widen the unambiguous sector, the surveillance antennas should be mounted at a shorter distance but this would yield degraded localization capability if a few antenna elements are exploited. In Section 6.5.2, we devise appropriate techniques aiming at further enhancing the performance of the system both in terms of target detection and localization. Specifically, a proper strategy is proposed to mitigate the effect of ambiguous DoA estimation while guaranteeing remarkable accuracy for targets within the antenna main lobe.

Before concluding this section, Table 6.6 investigates the impact of the disturbance cancellation stage on the obtained performance against drone targets. Specifically, Table 6.6 compares the estimated detection rate over the available 50 CPIs obtained when the cancellation stage is not applied, when a conventional ECA is adopted and when exploiting the suggested ECA-S approach for the same case studies of Figures 6.10(a)–6.12(a). The results clearly show that the disturbance cancellation stage is an essential block for the detection of such targets. We observe that the use of ECA and ECA-S allows a significant enhancement in target detection compared to the case of no disturbance cancellation. Moreover, the ECA-S yields a further improvement over the conventional ECA since it has been shown to be more robust to the time-varying characteristics of the disturbance [43,44]. As expected, the advantage of ECA-S becomes more evident when lower SNR targets are considered. This is the case of Data set 3 (the Phantom flew totally outside of the surveillance antenna main beam) where the ECA-S yields an increase of the number of correct detections of about 25% with respect to the conventional ECA.

Table 6.6 Comparison of the detection results obtained for the case studies of Figures 6.10(a)–6.12(a) with different cancellation strategies

	No disturbance cancellation		**ECA**		**ECA-S**	
	Mavic (%)	Phantom (%)	Mavic (%)	Phantom (%)	Mavic (%)	Phantom (%)
Dataset 1	50	6	96	68	100	70
Dataset 2	0	0	78	66	78	70
Dataset 3	–	0	–	52	–	66

6.4 Neyman–Pearson detector approximation and clutter modelling

In radar applications, the most extended detector criterion is the Neyman Pearson one, which maximizes the P_D maintaining the P_{FA} lower than or equal to a given value [48,49]. If $\tilde{\mathbf{z}}$ is the complex value observation vector, and $f(\tilde{\mathbf{z}}|H_0)$ and $f(\tilde{\mathbf{z}}|H_1)$ are the likelihood functions under null and alternative hypotheses, respectively, a possible implementation of the NP detector consists in comparing the Likelihood Ratio (LR), $\Lambda(\tilde{\mathbf{z}})$, to a detection threshold selected according to P_{FA} requirements, $\eta_{lr}(P_{FA})$, as is expressed in the following [49]:

$$\Lambda(\tilde{\mathbf{z}}) = \frac{f(\tilde{\mathbf{z}}|H_1)}{f(\tilde{\mathbf{z}}|H_0)} \underset{H_0}{\overset{H_1}{\gtrless}} \eta_{lr}(P_{FA}) \tag{6.10}$$

This approach requires complete knowledge of the likelihood functions, and significant detection losses appear when the statistical characteristics of the observation vector under both or any of the hypotheses differ from those assumed in the LR detector design [50]. Because of that clutter characterization is a key issue in radar-detection research activities for the formulation of the NP optimum detector, and the design of suboptimum approaches to approximate it.

In the following sections, clutter models described in radar literature to characterize the null hypothesis are reviewed, the statistical analysis techniques used to propose theoretical statistical models fitting real data are presented, and the LR detector based on clutter characterization is formulated.

6.4.1 Radar clutter characterization

In real radar scenarios, clutter signals present temporal and spatial variability in their statistical model and the associated parameters. Although clutter statistics can be characterized by environment radar measurements, this is a very complex task in bistatic geometries. In open radar literature, not much information about statistical models for bistatic clutter is accessible due to the limited number of operating systems and the multiple parameters to consider. They were mainly performed at typical radar frequencies, and the results cannot be easily extrapolated to communication systems frequencies usually exploited by PRs [51].

In this section, the study of the statistical properties of clutter and its dependences on the environment and system geometry is carried out for DVB-T-based PRs. The proposed statistical models will be key results for the formulation of the LR detector to improve the detection capabilities of PRs for drone monitoring.

6.4.1.1 Clutter modelling

The Gaussian model is applicable when the echo from each resolution cell can be modelled as that from a number of independent, random scatterers, with no one individual scatterer producing an echo of a magnitude commensurate with the resultant echo from all scatterers [52]. Under small grazing angle conditions, as the radar resolution increases, it is capable of discovering the non-homogeneity of the surface, making the Gaussian model unsuitable [52,53].

In [51], a wide set of theoretical models was considered for characterizing a complex random variable, $\tilde{\mathbf{z}}$: exponential, Weibull, *K*-distribution, log-normal, gamma and gamma mixture distribution for the intensity ($R = |\tilde{\mathbf{z}}|^2$); normal, logistic and Gaussian Mixture Distribution for the in-phase ($I = \Re\mathrm{e}(\tilde{\mathbf{z}})$) and in-quadrature ($Q = \Im\mathrm{m}(\tilde{\mathbf{z}})$) components.

Statistical analysis is based on the estimation of the Probability Density Function (PDF) and Cumulative Distribution Function (CDF), the calculation of the parameters of the considered distributions using the method of the moments and the use of goodness-of-fit tests based on the estimation of the distance between the Empirical CDF (ECDF) and the theoretical one [51].

Non-parametric tests such as the two-sample Kolmogorov–Smirnov (KS-test2) and the two-sample Cramér–von-Mises (CM-test2) are considered [51,54]. Given two identically distributed random variables, denoted as $\mathbf{x} = [x_1, \ldots, x_{N1}]$ and $\mathbf{y} = [y_1, \ldots, y_{N2}]$ composed of N_1 and N_2 independent observations, respectively, the KS-test2 and the CM-test2 are implemented as follows [51,54,55]:

- Evaluate the ECDF of both observation vectors: $\hat{F}_X(\mathbf{x})$ and $\hat{F}_Y(\mathbf{y})$.
- Compute the statistic KS-distance, d_{KS}, and CM-distance, d_{CM}, using (6.11), where *sup* is the supremum of the set of distances, and *H*(*x*) is the empirical distribution function of the two samples together [54,55].

$$\begin{aligned} d_{KS} &= \sup_x \left| \widehat{F}_X(\mathbf{x}) - \widehat{F}_Y(\mathbf{y}) \right|; \\ d_{CM}^2 &= \frac{N_1 \cdot N_2}{(N_1 + N_2)} \int_{-\infty}^{+\infty} \left| \widehat{F}_X(\mathbf{x}) - \widehat{F}_Y(\mathbf{y}) \right|^2 \cdot dH(x) \end{aligned} \tag{6.11}$$

- Compare d_{KS} and d_{CM} with a threshold selected according to the significance level, α, and the two sample sizes, N_1 and N_2. If the distances exceed the thresholds, the null hypothesis will be rejected [55]. Alternatively, if the *p-value* is less than the chosen α, which suggests that the observed data are sufficiently inconsistent with H_0, H_0 should be rejected [55].

The third- and fourth-order normalized central moments, denoted as skewness and kurtosis, respectively, are usually used to check if the I and Q components are Gaussian distributed [56]. The skewness parameter, γ_3, measures the degree of asymmetry of the PDF of X about its mean value μ_X, whereas the kurtosis parameter, γ_4, analyses the non-Gaussian nature of the data. Both parameters are obtained using (6.12). For a Gaussian distribution, $\gamma_3 = 0$ and $\gamma_4 = 3$, so these parameters can be considered a measure of the deviation of the data from a Gaussian model.

$$\gamma_3 = \frac{E\left[(X-\mu_X)^3\right]}{E\left[(X-\mu_X)^2\right]^{3/2}}; \quad \gamma_4 = \frac{E\left[(X-\mu_X)^4\right]}{E\left[(X-\mu_X)^2\right]^2} \tag{6.12}$$

6.4.1.2 Statistical analysis of real bistatic radar clutter

The CAF is based on the matched filter implementation and, although it is not optimum for detection when clutter is present, it is frequently used as the observation space for the detection stage for several reasons [57]: it is simpler than the optimum receiver; clutter scattering function may be unknown, it can work almost as well as the optimum receiver in many situations, and it operates as Maximum Likelihood Estimator of the complex amplitude, delay and Doppler shift of the desired targets and the clutter. Taking into account these assumptions, the observation vector $\tilde{\mathbf{z}}$ in (6.10) is composed of a single complex sample of the CAF output, $\tilde{\mathbf{z}} = \tilde{s}[l, m]$, and an exhaustive statistical analysis is carried out throughout all the CAF space.

A local study of different CAF regions is performed. The CAF is split into different Doppler regions following a criterion based on CAF mean level estimation and possible targets Doppler shifts: areas where aircrafts flying at low altitudes could be detected, areas where terrestrial vehicles are expected, and the area close to the zero-Doppler shift, where ground clutter highest contributions are concentrated and drones could be flying. In the region centred in the zero-Doppler line, an exhaustive analysis is performed dividing it into different sub-regions. The Doppler shift values associated with each area depend on the PR geometry and the selected radar scenario.

In [51], ground clutter characterization in PR scenarios at UHF frequencies was carried out to provide deeper knowledge about the impact of ground clutter components on the detection of targets with low Doppler shifts. The data set analysed in [51] was acquired by the IDEPAR demonstrator, a multichannel DVB-T-based PR developed by the University of Alcalá [32]. A rural scenario with similar characteristics considered in this chapter was selected. In the considered case study, a directive antenna was used in the surveillance channel, which was processed without applying any pre-processing stage that could alter clutter statistics. Two integration time values $T_{int} \in \{250, 500\}$ ms were analysed for evaluating the impact of system Doppler resolution in clutter characterization.

Results proved that a Gaussian clutter model could be assumed for higher Doppler shifts, justifying the suitability of conventional detection solutions used in radar literature for aeroplanes and ground targets with high velocity. The statistical parameters were similar for the different regions and depend on the selected T_{int}. For Doppler areas centred in the zero line, a non-homogeneous characterization was required: mixtures of gammas and Gaussians were proposed for modelling the intensity and I/Q components, respectively, statistics of which depend on the T_{int}. A higher number of mixture components are required for the zero Doppler line. This characterization allowed the improvement of detection performance for drones and ground vehicles with low Doppler shift [51,58].

The effects on the ground clutter characterization of one of the most extended PR architecture based on ECA pre-processing stage is analysed in [59] for $T_{int} = 250$ ms. This solution assumes that most of the interference components are rejected by the adaptive filter, so CFAR detectors designed for Gaussian clutter residuals are usually considered. However, in [59], a non-homogeneous characterization based on compound models was required for Doppler shifts close to zero, where ground clutter residuals concentrate. These results proved that although the ECA filter solution provides an improvement in the PR detection capabilities, it does not remove the non-homogeneity of the clutter. In the non-homogeneous area, the CFAR detectors based on Gaussian statistics are not optimum, decreasing the PR performance for the detection of targets characterized by low Doppler shifts on the CAF, such as drones.

6.4.2 Likelihood Ratio detector formulation

For the formulation of the LR detector, the knowledge of the likelihood functions of the complex observation vector is required. Clutter characterization studies in PR scenarios at UHF frequencies were presented in [17,51,58–60] and summarized in Section 6.4.1.2.

In the case study, the complex observation vector was composed of a single complex sample, $\tilde{\mathbf{z}} = \tilde{z}$, so the joint PDF of $\bar{\mathbf{z}} = [\Re e\{\tilde{z}\}, \Im m\{\tilde{z}\}]^T = [z_1, z_2]^T$ under both hypotheses must be estimated, where T denotes the transpose operation. Taking into account the statistical characterization summarized in Section 6.4.1.2, the LR detector could be formulated as follows [51]:

- ***Gaussian areas:*** I and Q components are Gaussians and independent, so the likelihood function of $\bar{\mathbf{z}}$ can be expressed as the product of its marginal. Under this assumption, the LR implementation consists in comparing $|\bar{\mathbf{z}}|^2$ with a detection threshold selected according to P_{FA} requirements [57].
- ***Non-Gaussian areas:*** I and Q components of the non-Gaussian areas could be modelled using a mixture of Gaussian distributions. In this case, both components were orthogonal, but we could not assume that they were independent, so the knowledge of the joint PDFs is required. The Expectation–Maximization algorithm was used for estimating their joint distributions as mixtures of 2D Gaussians [61]. The likelihood function for N_{gauss} mixture of Gaussians is detailed in (6.13), where $\bar{\mu}_i$ and $\mathbf{C}_i$ are the mean vector and the covariance matrix of the ith component, respectively, $|\mathbf{C}_i|$ denotes the determinant

of $\mathbf{C}_i$, and k_i are the mixture coefficients ($k_1 + \cdots + N_{N_{gauss}} = 1$) [51].

$$f(\overline{\mathbf{z}}) = \sum_{i=1}^{N_{gauss}} \frac{k_i}{2\pi\sqrt{|\mathbf{C}_i|}} \cdot \exp\left(-\frac{(\overline{\mathbf{z}} - \overline{\mu}_i)^T \cdot \mathbf{C}_i^{-1} \cdot (\overline{\mathbf{z}} - \overline{\mu}_i)}{2}\right) \tag{6.13}$$

For obtaining $f(\overline{\mathbf{z}}|H_1)$, a Gaussian target model with Gaussians I and Q components of zero mean and variance σ^2_{target} was assumed. To calculate $f(\overline{\mathbf{z}}|H_1)$, the target covariance matrix ($\mathbf{C}_{target} = \sigma^2_{target} \cdot \mathbf{I}$, being $\mathbf{I}$ the 2D identity matrix) is added to the covariance matrices of the individual components of the H_0 mixture. The σ^2_{target} parameter could be estimated as a function of the Signal-to-Interference Ratio (SIR) and the total mean vector and covariance matrix of $f(\overline{\mathbf{z}}|H_0)$, which are expressed in (6.14), being $tr(\mathbf{C}_{total})$ the trace of $\mathbf{C}_{total}$.

$$\begin{aligned} SIR &= 10 \cdot \log_{10}\left(2 \cdot \sigma^2_{target}/tr(\mathbf{C}_{total})\right) \\ \mathbf{C}_{total} &= \sum_{i=1}^{N_{gauss}} k_i \cdot \left(\mathbf{C}_i + (\overline{\mu}_i - \overline{\mu}_{total}) \cdot (\overline{\mu}_i - \overline{\mu}_{total})^T\right); \overline{\mu}_{total} = \sum_{i=1}^{N_{gauss}} k_i \cdot \overline{\mu}_i \end{aligned} \tag{6.14}$$

The LR detector for N_{gauss} mixture of Gaussians is implemented using (6.15), where η_{lr} is the detection threshold fixed according to P_{FA} requirements [51].

$$\frac{\sum_{i=1}^{N_{gauss}} \frac{k_i \cdot \exp\left(-\frac{(\overline{\mathbf{z}}-\overline{\mu}_i)^T \cdot \left(\mathbf{C}_i + \sigma^2_{target} \cdot \mathbf{I}\right)^{-1} \cdot (\overline{\mathbf{z}}-\overline{\mu}_i)}{2}\right)}{2\pi\sqrt{\left|\mathbf{C}_i + \sigma^2_{target} \cdot \mathbf{I}\right|}}}{\sum_{i=1}^{N_{gauss}} \frac{k_i \cdot \exp\left(-\frac{(\overline{\mathbf{z}}-\overline{\mu}_i)^T \cdot \mathbf{C}_i^{-1} \cdot (\overline{\mathbf{z}}-\overline{\mu}_i)}{2}\right)}{2\pi\sqrt{|\mathbf{C}_i|}}} \underset{H_0}{\overset{H_1}{\gtrless}} \eta_{lr}(P_{FA}) \tag{6.15}$$

Non-Gaussian areas of the CAF require the implementation of a bank of LR detectors, but due to the formulation in terms of Gaussian mixtures, their implementation and adaptation to potential clutter changes is really easy.

6.4.2.1 Experimentation and validation

The detection capabilities of the LR detector based on clutter characterization are evaluated in a real radar scenario. The IDEPAR demonstrator [32] was deployed in a rural scenario located in the facilities of the Spanish National Institute of Aerospace Technology in 'La Marañosa', ITM (Madrid, Spain), where controlled drone's flights were performed. The measurement campaign was carried out during the trials of the CONDOR project promoted by the Spanish Ministry of Defence for the evaluation of anti-drone systems. A cooperative DJI Phantom 4 drone with a GPS device was flown following trajectories defined for the military trials.

In the measurement campaign, IDEPAR used a commercial Yagi–Uda antenna in both reference and surveillance channels, characterized by an azimuth beamwidth of 24° and a gain of 17 dB. In Figure 6.13, a general view of the selected radar scenario

Figure 6.13 Radar scenario: IDEPAR demonstrator deployed in the INTA facilities. © 2019 IEEE. Reprinted, with permission, from Reference [17]

is depicted together with the PR and IoO positions, the reference and surveillance antennas, the view from the surveillance antenna (the AoI) and the target trajectory. Torrespaña transmitter, located at 16.6 km far from the PR, was selected as IoO due to its high radiated power and omnidirectional radiation pattern.

The demonstrator was configured to acquire data sets of 20 s with $B = 25$ MHz continuous bandwidth at a central frequency of 698 MHz. Only two DVB-T channels were available, giving rise to an effective bandwidth of 16 MHz. $T_{int} = T_{upd}$ was considered to obtain non-overlapping consecutive CPIs. A value of $T_{int} = 250$ ms was used to avoid targets' range and Doppler cell migration (Table 6.1), providing 80 CPIs. Bistatic Doppler and range resolutions of 4 Hz and 12 m, respectively, were obtained. The acquired signal was pre-processed using the ECA algorithm to reduce DPI and clutter contributions. The cooperative Phantom 4 drone followed a two-way trajectory with a minimum distance of 1.722 km and a maximum one of 1.8 km from the PR location. A desired $P_{FA} = 10^{-5}$ was selected as a detection performance requirement.

For designing the LR solution, the clutter characterization at the output of the CAF was performed. Taking into account the methodology and results detailed in Section 6.4.1, three Doppler regions and the corresponding bistatic velocities, $v_{bistatic}$, were defined based on CAF magnitude mean level estimation and the targets expected in the radar scenario: areas where aircrafts flying at low altitudes could be detected (Region 1: $f_d \in [-242, -98)$ Hz, $v_{bistatic} \in (42.12; 104.01]$ m/s for 1A and $f_d \in (98, 242]$ Hz, $v_{bistatic} \in (-42.12; -104.01]$ m/s for 1B); areas where terrestrial vehicles are expected (Region 2: $f_d \in [-98, -38)$ Hz, $v_{bistatic} \in (16.33; 42.12]$ m/s for 2A and $f_d \in (38,98]$ Hz, $v_{bistatic} \in (-16.33, 42.12]$ m/s for 2B) and Region 3 where ground clutter contributions are concentrated and drones could be flying ($f_d \in [-38; 38]$ Hz, $v_{bistatic} \in [-16.33; 16.33]$ m/s). Region 3 was split into different Doppler sub-regions to carry out an exhaustive analysis.

Goodness-of-fit results confirm that Regions 1 and 2 are characterized by non-zero mean Gaussian clutter models, whereas a non-homogeneous characterization is required for Doppler areas close to zero. Compound models composed of mixtures of Gaussians PDFs are selected for modelling the I and Q components of the sub-regions defined in Region 3. In the zero Doppler line and the Doppler shift values close to it, 3 and 4 Gaussians are proposed to characterize the I and Q components, respectively. For the rest of Doppler sub-regions, a lower number of Gaussians is used, being 2 Gaussians enough to model the I/Q components.

The use of the ECA filter modifies the clutter statistics of the selected Doppler lines along whole T_{acq}, increasing the complexity and the computational cost of the proposed solution. For each CPI, a bank of LR detectors is designed using (6.15), where the likelihood functions are estimated taking into account the statistical models proposed in the clutter characterization study and their associated parameters obtained by the method of moments. Monte Carlo simulations with an estimation error lower than 10% are used for obtaining the fixed detection threshold, η_{lr}, for each LR detector. The proposed detection scheme is compared to a conventional solution based on a square-law envelope detector with a fixed threshold to analyse the PR detection performance for small-drone detection.

Detection results obtained using the square-law and the bank of LR detectors are presented in Figure 6.14. Plots from the 80 CPIs of the whole acquisition time are plotted with grey dots. Detections associated with the Phantom 4 drone are remarked in red. The trajectory provided by the GPS device of the drone is also represented in green line. From visual inspection, a higher number of detections associated with the cooperative target are observed for the bank of LR detectors, so

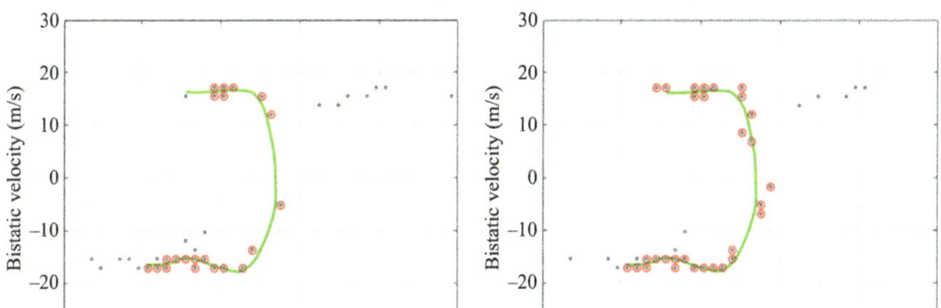

Figure 6.14 Detection results: square-law detector (a) and bank of LR detectors (b). Grey dots: detection plots from the 80 CPIs. Red dots: detection plots related to the cooperative drone. Green line: GPS data

an improvement in the PR detection capabilities is expected. The drone has been detected at a minimum distance of 3.4 km up to a maximum distance of 3.55 km from the PR in the bistatic range domain.

Quantitative results based on the estimation of the P_D and the P_{FA} at the output of the detection stage are obtained following the methodology detailed in [32]. The P_D is estimated at a plot level to avoid the errors associated with the pixel detection grouping techniques. A common ground truth for the cooperative target was generated using the GPS data. Monte Carlo simulation with an estimation error lower than 10% was used to estimate the P_{FA} at a pixel level removing target contributions. For the detection scheme based on a bank of LR detectors, $P_{FA} = 1.95{\cdot}10^{-5}$ and $P_D = 60\%$ for the cooperative target were obtained, whereas $P_{FA} = 2.08{\cdot}10^{-5}$ and $P_D = 42.5\%$ values were estimated for the square-law detector. The proposed solution provides a significant detection performance improvement with respect to the conventional scheme, increasing approximately 18% the P_D with a similar P_{FA}, close to the desired one.

Finally, the tracking system described in [32] was applied at the output of the proposed detection scheme, confirming one trajectory in the radar scenario. In Figure 6.15(a), the estimated trajectory is depicted with the superimposition of the detection plots and the real trajectory provided by the GPS device. The confirmed track is composed of 70 estimated points, giving rise to a $P_D = 87.5\%$ at the output of the tracking stage and increasing the PR performance.

Taking into account the drone motion, the trajectory could be split into three different intervals. In the first 5.25 s, the drone is moving away from the PR at an initial bistatic range of 3.45 km with a bistatic velocity of 17.9 m/s. Then, manoeuvres of deceleration and acceleration (D/A) are performed from 5.5 s to 11.5 s to change the flight direction. Finally, from 11.75 s to 20 s, the drone is approaching the PR with a bistatic velocity of −17.9 m/s.

In Figure 6.15(a), tracking losses are observed at the beginning of the trajectory due to the lack of enough detection plots during the first 2.5 s. However, the tracker is able to estimate the drone trajectory in the D/A interval with a few

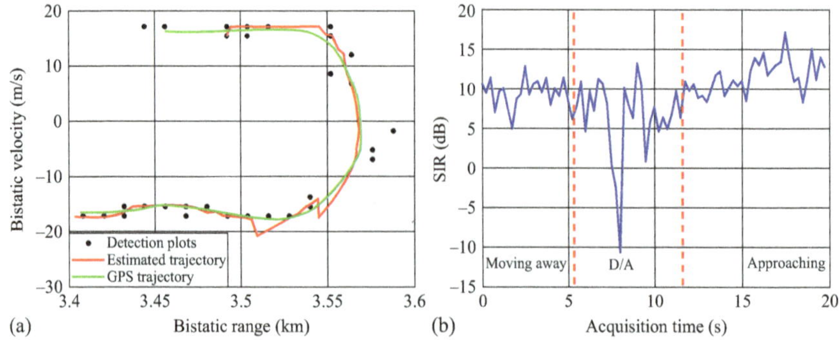

Figure 6.15 Estimated trajectory by the PR processing stage, GPS data and detection plots (a); SIR study at the input of the detection stage (b)

detected plots because the track has been already validated. These missed detections could be related to a reduction in the SIR at the input of the detection stage. In Figure 6.15(b), the SIR estimated for the $[l,m]$ CAF cells associated with the cooperative drone for the whole acquisition time is depicted. The SIR depends on the system geometry as well as on the relative facing of the drone. Nevertheless, an SIR reduction is observed when the bistatic velocity is close to zero (D/A interval) due to the pre-processing filtering technique.

Presented results confirm that the bank of LR detectors, designed according to Doppler-shift-dependent clutter models, is suitable for drone detection and monitoring applications, outperforming the detection capabilities of the conventional solutions based on a square-law detector. These results also show the impact of a proper statistical characterization of the radar scenario, which is useful for the formulation of optimum detectors based on the NP criterion. The bank of LR detectors based in Gaussians mixtures is quite easy to implement and adapt to potential clutter changes. Because of that, the proposed solution is feasible for being implemented in operative systems.

6.5 Multi-channel signal processing strategies

6.5.1 Exploitation of array antennas for target localization

The use of an antenna array in the surveillance channel and the associated Digital Array Signal Processing (DASP) techniques could provide further benefits: generation of notches along the DoA of interferences such as the DPI or other direct signals from interfering IoOs, 3D target parameter estimation (range, Doppler, azimuth) and improvement in the target localization accuracy, angular resolution and angular coverage.

In the literature, solutions based on a two-stage spatial filtering scheme are widely used [62,63], first step of which is focussed on target detection using CFAR techniques, whereas the second one implements DoA algorithms. In this section, modified versions of the previous two-stage spatial filtering approaches proposed in radar literature are presented. Beamforming techniques based on Side Lobe Level (SLL) reduction [64] and optimum weights designed to minimize the variance of the interference using the statistical characterization of real clutter data [17] are analysed. The proposed solutions are evaluated in a real radar scenario for drone detection using DVB-T-based PRs.

6.5.1.1 Array signal processing techniques

Snapshots for array signal processing can be generated in the time domain (outputs of the analogue-to-digital converters associated with each single radiating element) or in the frequency domain (outputs of the CAF processing stage calculated for each single radiating element). If beamforming techniques are applied in the transformed domain, we could take the advantage of the integration gain, system resolution, the sparsity of the targets and interferences present in the input signal, and that each beam space will be only responsible for the azimuth discrimination [62].

In most PR scenarios, CPI is significantly greater than the propagation time across the array, and the product between B and T_{int} is large enough, assuming an almost flat input signal temporal spectra. Under these assumptions, the frequency-domain snapshot model is usually applied [65] and is considered in this work.

As is observed in the general DVB-T-based PR processing scheme presented in Figure 6.6, independent CAFs ($\tilde{s}_i(t)$, $i = 1,\ldots,N$) are generated for each single radiating element of the surveillance array before the DASP stage. In the CAF domain, the snapshots are obtained for each (range bin, Doppler bin) pair, (l,m): $\mathbf{s}_s[l,m] = [\tilde{s}_1[l,m],\ldots,\tilde{s}_N[l,m]]^T$ [64].

DASP techniques allow the generation of multiple beams pointing to a set of azimuth directions, $\mathbf{\Phi} = \{\phi_{sa,1},\ldots,\phi_{sa,N\phi sa}\}$, using appropriate weight vectors, $\mathbf{w}[m, \phi_{sa,i}]$. If the array is located along y-axis, with the XY plane parallel to the tangent plane to the array location, the steering vector ($\mathbf{a}[m, \phi_{sa,i}]$) of a signal impinging from a direction $\phi_{sa,i}$ is defined in (6.16), where f_s is the sampling frequency, f_p is carrier frequency, d_i with $i = 1,\ldots,N$ is the distance of the ith antenna element relative to the origin, and c is the speed of light. The resulting signal after beamforming processing is expressed in (6.17), where superscript H represents the Hermitian conjugate operation.

$$\mathbf{a}\left[m,\phi_{sa,i}\right] = \left[e^{j(2\pi/c)\left(f_s\frac{m}{N_{\text{int}}}+f_p\right)\cdot d_1\cdot\sin\left(\phi_{sa,i}\right)},\ldots,e^{j(2\pi/c)\left(f_s\frac{m}{N_{\text{int}}}+f_p\right)\cdot d_N\cdot\sin\left(\phi_{sa,i}\right)}\right]^T \tag{6.16}$$

$$\mathbf{y}\left[l,m,\phi_{sa,i}\right] = \mathbf{w}\left[m,\phi_{sa,i}\right]^H\cdot\mathbf{s}_s[l,m] \tag{6.17}$$

Conventional data-independent beamforming algorithms were applied in two modified versions of the two-stage spatial filtering scheme presented in [62,63]. In both approaches, a two-stage spatial filtering scheme based on target detection using multiple simultaneous beams and on the estimation of the angle of arrival using DoA techniques was designed for estimating target parameters in the 3D space (range, Doppler, azimuth).

1. **Two-stage spatial filtering processing scheme**

   A two-stage spatial filtering scheme based on [62,63] is considered. The proposed modifications are based on the combination of weighting techniques to control SLL and maximizing directivity, instead of uniform conventional beamforming. Both beamforming approaches were detailed in [64], and the associated performances were analysed. In Figure 6.16, the main functional blocks of the proposed method are depicted [64]:

   (i) The first stage implements beamforming based on orthogonal beams calculated under SLL requirements ($\mathbf{w}_{\text{SLL}}$) to generate a beam space where a 3D CFAR detector is applied. The detector operates using a windowing technique that defines Cells Under Test (CUTs) related to maxima in the beam space and a 3D reference window to improve the interference statistic estimation and to reduce CFAR losses.

   (ii) The second stage corresponds to a high-resolution DoA estimation. A new set of steering angles was defined to increase azimuth estimation accuracy. The Minimum Variance Distortionless Response algorithm

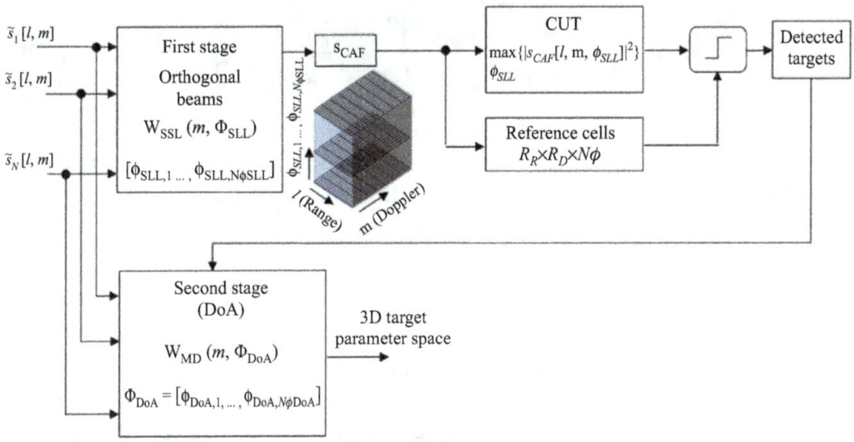

Figure 6.16 Proposed two-stage frequency-domain spatial filtering scheme [64]. © 2017 by the authors

was considered [65], and the weight vector was calculated to maximize the directivity ($\mathbf{w}_{MD}$).

2. **Two-stage spatial filtering processing scheme based on optimum beamforming techniques**

DASP techniques based on a two-stage spatial filtering processing scheme usually apply incoherent CFAR detectors for the first stage. These solutions are optimum in the presence of independent and identically distributed samples with exponential PDF, but their performance decreases significantly under other clutter conditions. To improve conventional non-coherent detection scheme performance in non-homogeneous scenarios, beamforming techniques optimized to minimize the variance of the interference and increase the SIR are proposed to generate the beam space for the first stage of Figure 6.16 [17].

The snapshot generated in the CAF domain, $\mathbf{s}_s[l,m] = [\tilde{s}_1[l,m], \ldots, \tilde{s}_N[l,m]]^T$, is composed of the target return, $r[l,m]$, in the incoming direction $\phi_{sa,I}$, plus the received clutter in each single array element, $\mathbf{c}[l,m] = [c_1[l,m], \ldots, c_N[l,m]]$. That is: $\mathbf{s}_s[l,m] = \mathbf{a}[m,\phi_{sa,i}]\cdot r[l,m]+\mathbf{c}[l,m]$, being $\mathbf{a}[m,\phi_{sa,i}]$ the steering vector expressed in (6.16). Assuming the distortionless criterion, the beamformer output is obtained using (6.18).

$$y[l,m,\phi_{sa,i}] = r[l,m] + \mathbf{w}_{opt}[m,\phi_{sa,i}]^H \cdot \mathbf{c}[l,m] \tag{6.18}$$

The mean square of the output interference is expressed in (6.19), where $\mathbf{M}_c$ is the interference covariance matrix. After minimizing the interference variance using Lagrange multipliers, the optimum weights are calculated using (6.20) [57].

$$E\left[\left|\mathbf{w}_{opt}[m,\phi_{sa,i}]^H \cdot \mathbf{c}[l,m]\right|^2\right] = \mathbf{w}_{opt}[m,\phi_{sa,i}]^H \cdot \mathbf{M}_c[m] \cdot \mathbf{w}_{opt}[m,\phi_{sa,i}] \tag{6.19}$$

$$\mathbf{w}_{opt}\left[m,\phi_{sa,i}\right]^H = \frac{\mathbf{a}\left[m,\phi_{sa,i}\right]^H \cdot \mathbf{M}_c[m]^{-1}}{\mathbf{a}\left[m,\phi_{sa,i}\right]^H \cdot \mathbf{M}_c[m]^{-1} \cdot \mathbf{a}\left[m,\phi_{sa,i}\right]} \tag{6.20}$$

The proposed beamforming is highly dependent on clutter statistics, so a statistical characterization of the observation space, the CAF generated for each single array element, is required. Detection and DoA stages are implemented as in the previous two-stage scheme.

6.5.1.2 Experimentation and validation

The proposed two-stage schemes are evaluated in the rural scenario presented in Section 6.4.2.1 [17]. In this measurement campaign, IDEPAR was configured with the commercial high-directivity antenna described in Section 6.4.2.1 in the reference channel, and a five-element NULA along the y-axis in the surveillance one. The single radiating element had a bandwidth of 60° and a gain of 6.9 dB. The inter-element distances were estimated using a genetic algorithm to obtain a compromise solution between side lobe level and main beamwidth [66], providing an angular resolution of 17°. IDEPAR configuration and processing parameters used in Section 6.4.2.1 were considered. In this case study, the cooperative drone was approaching the PR at an initial distance of 600 m. The five signals acquired by the NULA elements were preprocessed using the ECA algorithm to reduce DPI and clutter contributions. For each CPI and NULA element, a CAF was generated to apply the two-stage scheme:

- In the first stage, orthogonal beams were generated. Steering angles of $\phi_{sa} = [-22.4°,0°,22.4°]$ were selected to cover the azimuth coverage area limited by the single-antenna beamwidth (60°).
- The optimization problem formulated in [64] was solved to obtain $\mathbf{w}_{SLL}$. A clutter characterization of each array element was required to estimate $\mathbf{w}_{opt}$ (6.20).
- A CA-CFAR detector with a reference window of size $R_r \times R_d \times R_a = 64\times16\times3$ was used, being R_r, R_d and R_a the number of range, Doppler and azimuth reference cells. The P_{FA} was set to 10^{-5}.
- In the second stage, a set of 361 steering angles equally spaced between −30° and 30° was generated to perform a high accuracy DoA estimation.

The statistical analysis of each array element was carried out following the strategy detailed in Section 6.4.1 and the Doppler regions defined in Section 6.4.2.1. Results confirmed that a non-homogeneous characterization was required for Doppler shifts close to zero, selecting the mixture of Gaussians for modelling the I and Q components. A Gaussian clutter model was used to characterize the higher Doppler areas. Different interference characterization was obtained for each NULA element due to multipath and DPI effects. The antenna with the higher Doppler shift interval associated with a non-homogeneous clutter was selected to determine the interference model used for calculating optimum beamforming weights. Results proved that clutter statistics depended on the array element and the Doppler shift in the CAF, so optimum beamforming weights were estimated for each Doppler region defined in the statistical analysis and each CPI of the whole acquisition time [17].

To carry out a comparative study between both DASP techniques, the SIR at the output of the first-stage beamformer was estimated for the $[l,m]$ CAF cells associated with the cooperative drone in the whole interval. In Figure 6.17(a), an SIR improvement is observed when the optimum weights are applied, so an increase in the PR detection performance is expected. To prove the benefits of the DASP techniques, the SIR at the input of the first stage was estimated for each array element. In Figure 6.17(b), the SIRs for the best and worst cases are compared to that obtained with the optimum beamformer, showing a significant SIR increase when the DASP technique is considered.

In Figure 6.18(a), the superimposition of the detector outputs in the 80 CPIs is presented, displaying all the false alarms detected through all acquisition time. Two

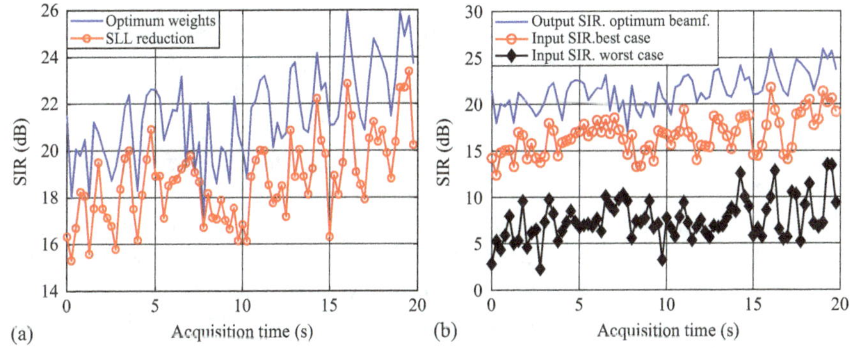

Figure 6.17 Comparative SIR study between proposed DASP techniques (a) and at the input and output of the first stage beamformer (b). (a) © 2019 IEEE. Reprinted, with permission, from Reference [17]

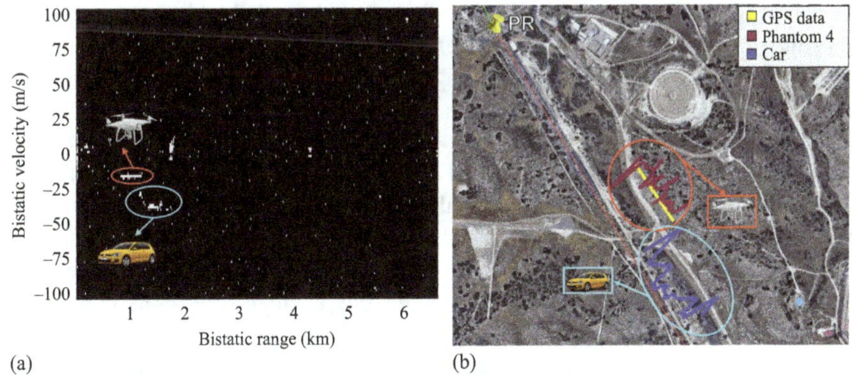

Figure 6.18 Results of the two-stage filtering scheme based on optimum weights: superposition of the detection matrices provided by the detection stage (a) and 2D estimated trajectories in the radar scenario (b). © 2019 IEEE. Reprinted, with permission, from Reference [17]

main groups of detections can be identified: one was associated with the cooperative drone using its GPS data (marked with red), and the other one was related to an unexpected terrestrial target (marked with blue).

Detection performance is evaluated through the P_D and P_{FA} estimation following the methodology detailed in [32,64]. Common ground truths were calculated for the cooperative target using the GPS data to estimate the P_D, which is calculated at the plot level due to errors associated with the pixel detection grouping techniques. The P_{FA} was estimated at pixel level removing the target contributions and using Monte Carlo simulations with an estimation error lower than 10%. For the scheme based on SLL reduction, $P_{FA} = 7.32 \cdot 10^{-5}$ and $P_D = 97.5\%$ for the cooperative target were obtained. $P_{FA} = 6.06 \cdot 10^{-5}$ and $P_D = 98.75\%$ were estimated when optimum weights are applied. Results show that the solution based on optimum weights outperforms the performance of the one designed to control SLL, providing a higher P_D with a lower P_{FA}. In both cases, the estimated P_{FA} is higher than the desired one. A CA-CFAR detector designed assuming Gaussian clutter is applied at the output of the beamforming stage. However, the clutter statistical analysis proved that a non-homogeneous characterization is required for Doppler shifts close to zero. In this area, the CA-CFAR detector is not optimum, decreasing the PR performance.

The 3D target's parameters obtained at the output of the two-stage filtering technique were transformed to 2D map coordinates to be depicted in the radar scenario, Figure 6.18(b). As is observed, the estimated trajectory associated with the cooperative drone fits well with the GPS, and the second one was related to a terrestrial target running in the secondary road. These results validate the proposed two-stage processing schemes and confirm the suitability of DVB-T-based PR for small-drone detection.

6.5.2 Exploitation of the frequency and spatial diversity to improve the detection and localization performance

As it is well known, target detection performance depends highly on the selected transmitter of opportunity and, in particular, it significantly varies with the employed DVB-T channel. The radiating characteristics of the transmitter and the electromagnetic conditions of the propagation channel vary across the wide frequency band allocated to the DVB-T service in the UHF band (from 470 to 862 MHz). In addition, also the target scattering can change with the different considered frequency channels. Eventually, target DoA estimation is characterized by low accuracies when operating at UHF band. Moreover, as mentioned in the previous section, when we operate with a uniform linear array, the requirements of accurate DoA estimation and wide unambiguous angular sector sets competing constraints on the array equivalent length.

To overcome the aforementioned limitations, we can take advantage of the frequency and spatial diversity allowed by the multi-frequency operation and a NULA layout. Specifically, aiming at improving the target detection capability and the reliability of the considered system, we consider the joint exploitation of multiple DVB-T channels emitted by the same broadcast illuminator of opportunity based on the detection schemes proposed in [12]. Among the different MF operation approaches, the decentralized detection scheme (MF DEC) is adopted in the following since it has been shown to be more robust to the presence of ghost targets or unwanted plots formations. Therefore, a detection is declared at a given range/velocity location when L detections out of Q available frequency channels are obtained for the considered range–Doppler bin $[l_0, m_0]$ of the maps separately evaluated at each surveillance channel. Specifically, by defining $b_q[l_0, m_0]$ the binary map at the output of the CFAR detection stage of Figure 6.6, we obtain

$$\sum_{q=1}^{Q} b_q[l_0, m_0] \underset{H_0}{\overset{H_1}{\gtrless}} L \tag{6.21}$$

In this case, a conventional CA-CFAR threshold is separately applied at each single frequency channel in order to achieve the desired P_{FA} after the MF binary integration:

$$P_{FA} = \sum_{i=L}^{Q} \binom{Q}{i} P_{FA}^{SF^i} \left(1 - P_{FA}^{SF}\right)^{Q-i} \tag{6.22}$$

where P_{FA}^{SF} is the probability of false alarm set at each DVB-T channel. Once the target has been detected on the range–velocity plane, the DoA of the corresponding echo has to be estimated with the purpose of localizing it in the X–Y plane. To this purpose, we take the advantage of MF operation in conjunction with a NULA layout to relax the trade-off between DoA estimation accuracy and unambiguous angular sector. In detail, we resort to an ML approach that incoherently integrates the target echoes at multiple carrier frequencies [67]. Specifically, the ML estimate of the target DoA $\widehat{u}_{ML} = \sin(\widehat{\vartheta}_{ML})$ (ϑ being the angle of arrival, measured relative to the array boresight) is obtained as

$$\widehat{u}_{ML} = \underset{u}{\operatorname{argmax}} \sum_{q=1}^{Q} \frac{1}{\sigma_q^2} \left| \boldsymbol{a}_q^H(u) \boldsymbol{x}_q \right|^2 \tag{6.23}$$

where $\boldsymbol{x}_q$ is the $N \times 1$ vector with the complex values extracted at the range–velocity location where the target has been detected, σ_q^2 is the noise power at the qth frequency channel, while $\boldsymbol{a}_q(u)$ is the spatial target steering vector. For a linear array composed by N surveillance antennas at positions $d_n(n = 0, \ldots, N-1)$ measured with respect to the phase reference point, $\boldsymbol{a}_q(u)$ is given by (6.24), where λ_q is the qth wavelength, $q = 1, \ldots, Q$.

$$\boldsymbol{a}_q(u) = \left[e^{j(2\pi/\lambda_q) d_0 u} \ldots e^{j(2\pi/\lambda_q) d_{N-1} u} \right]^H \tag{6.24}$$

This approach benefits from the increased equivalent target SNR provided by the MF integration, which yields enhanced estimation accuracy for unambiguous targets. However, along with this expected advantage, the frequency diversity and the non-uniform spacing of the receiving antenna elements allow to build up an improved ML likelihood function with reduced side lobes level. Therefore, the probability of outliers in target DoA estimation is lowered even operating with few and largely spaced antenna elements [68].

The processing scheme to be applied at each branch of the architecture in Figure 6.6 is sketched in Figure 6.19.

We illustrate the benefits gained for the considered counter-drone operations both in terms of target detection capability and localization accuracy. Specifically, we focus on Data sets 1 and 2 detailed in Table 6.5 due to the availability of both multiple DVB-T frequency channels and surveillance antennas. The same case studies of Figures 6.10 and 6.11 are considered.

First, we consider the use of two surveillance antennas. Table 6.7 reports a quantitative comparison in terms of estimated detection rate and rate of miss-localized plots (due to DoA estimation ambiguities) when separately exploiting the two frequency channels and after the MF operation. In detail, we employ the MF decentralized detection scheme based on a 1-out-of-2 strategy (MF DEC 1/2), while the ML approach in (6.23) is considered for DoA estimation. Specifically, the maximization of (6.23) is performed over a wide angular sector equal to $[-50°:0.01°:50°]$. It is worth noticing that this sector is wider than the unambiguous sector that is guaranteed given the distance among the employed antennas on receive, with both the considered carrier frequencies.

By observing Table 6.7, we notice that the two DVB-T channels yield comparable detection results when the drones flew at a shorter distance (Data set 1). In detail, it is interesting to note that the Mavic drone is detected with a detection rate equal to 100% over the available 50 CPIs. In this case, for both drones, due to the comparable and remarkable detection performance of both single channels, we observe that the MF operation does not yield any significant improvement in terms of target detection capability with respect to the single-frequency operation. In contrast, the possibility to widen the unambiguous angular sector provided by the

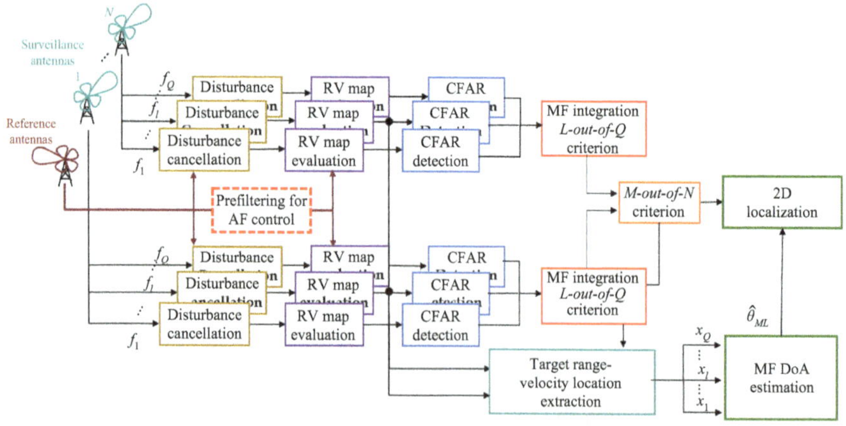

Figure 6.19 Multi-channel multi-frequency DVB-T-based processing scheme for target detection and localization. © 2020 IET. After [20]

Table 6.7 Data sets 1 and 2: comparison of the SF and the MF configurations (over 50 CPIs) using two and three surveillance antennas

		Mavic				Phantom			
		N° of detections		**Ambiguous plots**		**N° of detections**		**Ambiguous plots**	
		2 surv	3 surv	2 surv	3 surv	2 surv	3 surv	2 surv	3 surv
Data set 1	**SF F0**	100%	100%	2%	0%	70%	74%	0%	0%
	SF F1	100%	100%	18%	0%	76%	72%	0%	0%
	MF DEC 1/2	100%	100%	0%	0%	76%	74%	0%	0%
Data set 2	**SF F0**	78%	84%	14%	0%	70%	72%	10%	0%
	SF F1	60%	64%	22%	0%	84%	82%	48%	0%
	MF DEC 1/2	82%	86%	4%	0%	82%	80%	0%	0%

MF approach allows to correctly localize all plots in the Cartesian plane (see Table 6.7 and Figure 6.20(a)) even operating with two surveillance antennas.

When the distance of the drones increases (Data set 2), we observe that the target detection capability shows a larger dependency on the employed DVB-T channel. Moreover, the best performing DVB-T channel varies with the two drones. In this case, the detection performance improvement provided by the MF approach is quite apparent as many additional plots are obtained with respect to the worst SF operation of each target. This clearly shows that the MF approach allows mitigating the temporal variability of the system performance that is typically caused by the time-varying characteristics of both the employed waveforms and the propagation channels, there, including the target scattering and its multipath. In addition, the MF solution avoids the ambiguous estimation of the DoA for targets flying outside the unambiguous angular sector provided by the SF operation. However, there is a non-zero probability that outliers could occur when two antennas are considered. In fact, in the considered test, two ambiguous plots are obtained as highlighted by the green circle in Figure 6.21(a).

Alternatively, to increase at the same time both the DoA estimation accuracy and the unambiguous angular sector, a NULA can be used. The detection rate and rate of miss-localized plots for both DVB-T channels, for the case of three surveillance antennas, are illustrated in Table 6.7. In this case, a 3-out-of-3 strategy is adopted to integrate the detection results at the three available surveillance channels; still the relevant parameters are set in order to obtain a final $P_{FA} = 10^{-7}$. As an example, Figures 6.20(b) and 6.21(b) report the localization results using three surveillance antennas and the single frequency F0. By comparing the results obtained with two and three surveillance antennas, it is evident that in all cases the exploitation of three antennas allows to correctly localize all plots in the Cartesian plane (namely zero ambiguous plots).

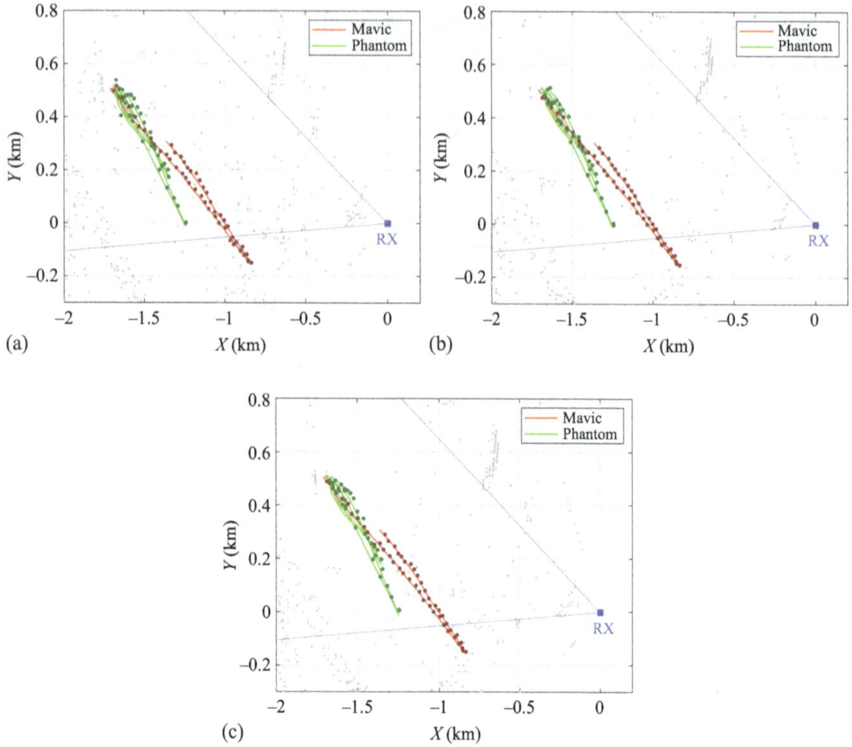

Figure 6.20 Data set 1: localization results for 50 consecutive data files using: (a) MF DEC 1/2 with 2 surv. antennas; (b) SF F0 with 3 surv. antennas; (c) MF DEC 1/2 with 3 surv. antennas. © 2020 IET. After [20]

Moreover, by comparing Figures 6.20(b) and 6.21(b) with Figures 6.10(b) and 6.11(b), we might notice that the use of three antennas allows more accurate localization. This is apparent especially at longer distances from the RX where a reduced DoA estimation accuracy is expected with the SF operation, which, in turn, results in a degraded localization capability due to the projection of the angular errors in the Cartesian domain.

Eventually, a further improvement in terms of localization results might be achieved by jointly exploiting multiple frequency channels and surveillance antennas (Figures 6.20(c) and 6.21(c)). In this case, when three surveillance antennas are used, the MF operation allows only a slight improvement in the localization accuracy with respect to the SF approaches. Notice that the localization improvement does not appear substantial against the considered targets since remarkable localization performance was obtained with the two single DVB-T channels.

Nevertheless, it is worth noticing that the obtained benefits also apply to other targets detected by the PR. As an example, Figure 6.22 shows the localization results of Figure 6.21(c) over the Google Earth map of the local area. As it is evident, in addition to the cooperative drones, the considered system is able to

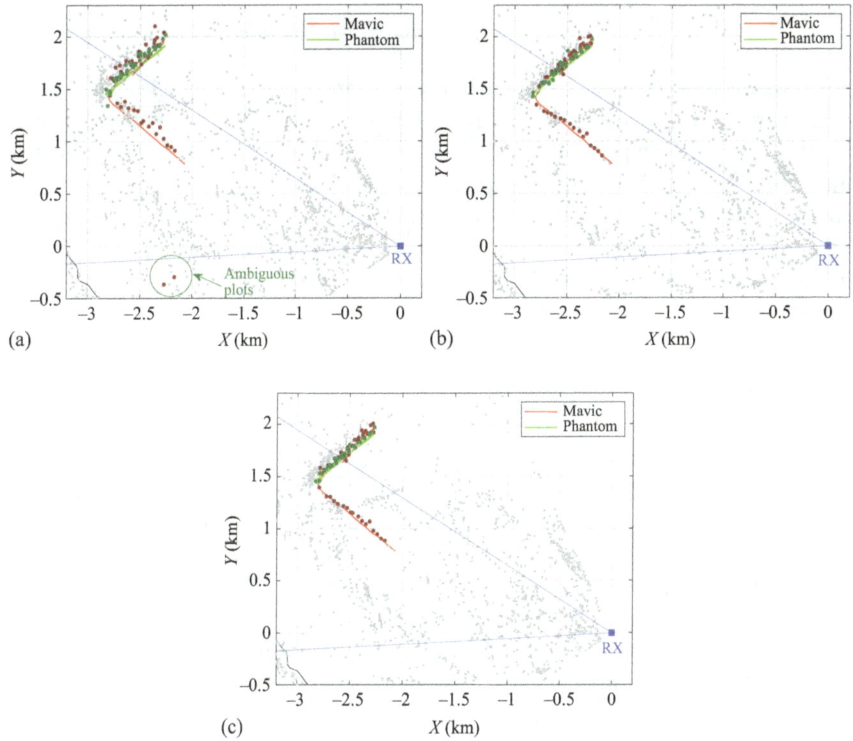

Figure 6.21 Data set 2: localization results for 50 consecutive data files using: (a) MF DEC 1/2 with 2 surv. antennas; (b) SF F0 with 3 surv. antennas; (c) MF DEC 1/2 with 3 surv. antennas. © 2020 IET. After [20]

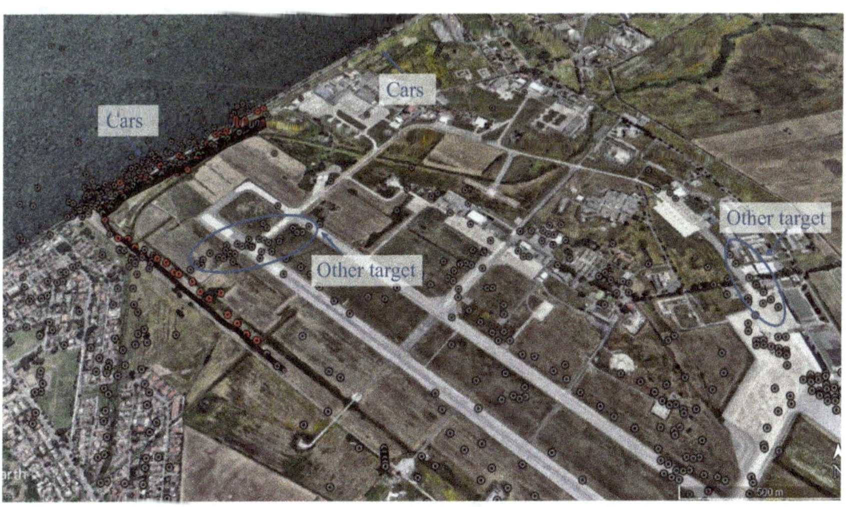

Figure 6.22 Data set 2: localization results of Figure 6.21(c) over the Google Earth map of the local area. © 2020 IET. After [20]

detect many other targets that are likely to correspond to vehicles or aircraft moving on the surface of the airport as well as to birds (comparable with the drones in terms of size). The possibility to widen the unambiguous angular sector greatly simplifies the discrimination among tracks in the Cartesian domain. However, the reported results clearly show that the considered system is able to continuously detect several targets that could be typically encountered in an airport scenario. Therefore, in order to make the DVB-T-based PR sensor able to distinguish among the different type of targets, proper strategies should be considered in the future.

6.6 Conclusions

This chapter tackles the design of DVB-T-based PR sensors and signal processing techniques aiming at enabling silent surveillance of drones in security-sensitive scenarios. Although drones can be valuable friendly collaborative tools for a wide variety of applications, they are also used as vectors of many security threats in modern societies. These are challenging targets from the radar point of view, due to their low reflectivity and high manoeuvrability. The low altitude flights take advantage of trees, ground relief and other man-made elements, to complicate their detection from the scattered energy when they are illuminated by an electromagnetic wave. In this context, PRs are considered an opportunity to enforce surveillance systems in environments where electromagnetic radiations are seriously limited, and a quick deployment and high reconfiguration capabilities are required. The multistatic nature of its operating principle, together with the possibility of using different transmitters with a wide variety of waveforms, frequencies and geometries, makes this research line of great interest.

The study starts with the estimation of expected coverages emphasizing the importance of modelling the propagation losses, including ground effects and a proper model of the BRCS of the target. Studies must be performed for different flight altitudes, specially in environments of complex relief or characterized by numerous vertical structures (urban areas), different frequencies, bandwidths and integration times, parameters that will determine the resolutions of the system and the coherent processing gain required for detection.

Two approaches are presented for the design of the PR detector. The first one focuses on the design of interference cancellation techniques before the coherent processing (CAF); the second one is based on the statistical modelling of the CAF output under hypothesis H_0 and H_1 to formulate the LR detector. In both cases, surveillance systems composed by a single or an array of radiating elements are considered, and different processing schemes and antenna layout are proposed and evaluated.

DASP techniques are used to increase the angular coverage and resolution and to provide the required angular information for 3D (range, Doppler, azimuth) processing. Two-stage approaches are analysed that define different beamforming approaches for detection and DoA estimation.

Also, the inherent possibility of exploiting multiple DVB-T channels emitted by the same transmitter has been explored with the aim to increase the radar range resolution, or to enhance the detection performance by making it robust with respect to the characteristics of the single propagation channel, or to improve the target localization capability by relaxing the constraints on the antenna layout related to the angular ambiguities.

The earlier approaches have been tested against experimental data collected with different PR receivers. The results clearly showed the benefits of the proposed strategies in practical applications and proved the capability of a DVB-T-based PR of detecting and localizing small drones flying at few kilometres from the receiver.

Future activities should address proper approaches aimed at equipping the sensor with the capability to distinguish between different classes of targets.

References

[1] Ritchie M., Fioranelli F., and Borrion H. 'Micro UAV crime prevention: can we help Princess Leia?' in Savona, B. L. (ed.). *Crime Prevention in the 21st Century*. New York, NY: Springer; 2017, pp. 359–376.

[2] https://www.postandcourier.com/news/report-helicopter-crash-on-danie-lisland-may-have-been-caused-by-drone/article_1123767a-12a7-11e8-bf2e-2798e681cd6b.html [Accessed 10 June 2020].

[3] https://www.newscientist.com/article/2190096-drones-are-causing-air-portchaos-why-cant-we-stop-them/ [Accessed 10 June 2020].

[4] https://www.911security.com/blog/airport-security-drone-incident-at-gatwick-airport-may-cost-tens-of-millions-of-dollars [Accessed 10 June 2020].

[5] Shi X., Yang C., Xie W., *et al.* 'Anti-drone system with multiple surveillance technologies: architecture, implementation, and challenges'. *IEEE Communications Magazine*. 2018, 56(4), pp. 68–74.

[6] Howland P.E. 'FM radio based bistatic radar'. *IEE Proceedings – Radar, Sonar and Navigation. (Special Issue on Passive Radar System)*. 2005, 152(3).

[7] Lombardo P. and Colone F. 'Advanced processing methods for passive bistatic radar systems' in Melvin, W. L. and Scheer, J. A. (eds.). *Principles of Modern Radar: Advanced Radar Techniques*. Raleigh, NC SciTech Publishing; 2012, pp. 739–821.

[8] Pignol F., Colone F., and Martelli T. 'Lagrange-polynomial-interpolation-based keystone transform for a passive radar'. *IEEE Transactions on AES*. 2018, 54(3), pp. 1151–1167.

[9] Filippini F., Martelli T., Colone F., *et al.* 'Exploiting long coherent integration times in DVB-T based passive radar systems'. *IEEE Radar Conference*, Boston, MA, USA, 2019.

[10] Saini R. and Cherniakov M. 'DTV signal ambiguity function analysis for radar application'. *IEE Proceedings – Radar, Sonar and Navigation*. 2005, 152(3), pp. 133–142.

[11] Palmer J.E., Harms H.A., Searle S.J., and Davis L. 'DVB-T passive radar signal processing'. *IEEE Transactions on Signal Processing*. 2013, 61(8), pp. 2116–2126.

[12] Martelli T., Colone F., Tilli E., *et al.* 'Multi-frequency target detection techniques for DVB-T based passive radar sensors'. *Sensors*. 2016, 16(10).
[13] Martelli T., Colone F., Tilli E., and Di Lallo A. 'Maritime surveillance via multi-frequency DVB-T based passive radar'. *IEEE Radar Conference (RadarConf)*, Seattle, WA, USA, 2017, pp. 0540–0545.
[14] Liu Y., Wan X., Tang H., *et al.* 'Digital television based passive bistatic radar system for drone detection'. *IEEE Radar Conference*, Seattle, WA, USA, 2017, pp. 1493–1497.
[15] Poullin D. 'Countering illegal UAV flights: passive DVB radar potentiality'. *19th International Radar Symposium (IRS)*, Bonn, 2018, pp. 1–10.
[16] Jarabo-Amores M.P., Mata-Moya D., Gómez-del-Hoyo P.J., *et al.* 'Drone detection feasibility with passive radars'. Proceedings of the 15th European Radar Conf.; Madrid, Spain, Sep 2018, pp. 313–316.
[17] del-Rey-Maestre N., Mata-Moya D., Jarabo-Amores M.P., Gómez-del-Hoyo P.J., and Rosado-Sanz J. 'Optimum beamforming to improve UAV's detection using DVB-T passive radars'. *IEEE International Radar Conference*, Toulon, 2019.
[18] Jarabo-Amores M.P., Mata-Moya D., Gómez-del-Hoyo P.J., del-Rey-Maestre N., and Rosado-Sanz J. 'On the impact of drone airscrews signature on passive radar detection and tracking stages'. *2020 IEEE Radar Conference (RadarConf20)*, Florence, Italy, 2020, pp. 1–6.
[19] Martelli T., Colone F., and Cardinali R. 'Simultaneous short and long range surveillance of drones and aircrafts with DVB-T based Passive Radar'. *IEEE International Radar Conference*, Toulon, 2019.
[20] Martelli T., Colone F., and Cardinali R. 'DVB-T based passive radar for simultaneous counter drone operations and civil air traffic surveillance'. *IET Radar, Sonar & Navigation*. 2020, 14(4), pp. 505–515.
[21] Martelli T., Filippini F., and Colone F. 'Tackling the different target dynamics issues in counter drone operations using passive radar'. *International Radar Conference*, Washington, 2020, pp. 512–517.
[22] Willis N.J. *Bistatic Radar*. Raleigh, NC: SciTech Publishing; 2005.
[23] Velez M.M., Angueira P., De La Vega D., Arrinda A., and Ordiales J.L., 'DVB-T BER measurements in the presence of adjacent channel and co-channel analogue television interference'. *IEEE Transactions on Broadcasting*. 2001, 47(1), pp. 80–84.
[24] Rzewuski S., Borowiec K., Kulpa K., *et al.* 'Supersonic target detection in passive radar'. *16th International Radar Symposium (IRS)*, Dresden, 2015, pp. 89–94.
[25] WinProp Propagation Modelling: A new dimension of wave propagation and radio network planning. https://altairhyperworks.com/product/Feko/WinProp-Propagation-Modeling [Accessed March 2020].
[26] Jarvis A., Reuter H.I., Nelson A., and Guevara E. 'Hole-filled seamless SRTM data V4'. *International Centre for Tropical Agriculture (CIAT)*, 2008. Available from http://srtm.csi.cgiar.org [Accessed March 2020].
[27] Pisa S., Piuzzi E., Pittella E., *et al.* 'Evaluating the radar cross section of the commercial IRIS drone for anti-drone passive radar source selection'.

22nd International Microwave and Radar Conference (MIKON), Poznan, 2018, pp. 699–703.

[28] Guo M., Lin Y., Sun Z., and Fu Y. 'Research on monostatic radar cross section simulation of small unmanned aerial vehicles'. *2018 International Conference on Microwave and Millimeter Wave Technology (ICMMT)*, Chengdu, 2018, pp. 1–3.

[29] Speirs P.J., Schröder A., Renker M., Wellig P., and Murk A. 'Comparisons between simulated and measured X-band signatures of quad-, hexa- and octocopters'. *2018 15th European Radar Conference (EuRAD)*, Madrid, 2018, pp. 325–328.

[30] ANSYS HFSS. 3D Electromagnetic Field Simulator for RF and Wireless Design. https://www.ansys.com/products/electronics/ansys-hfss [Accessed March 2020].

[31] Sevgi L. Complex Electromagnetic Problems and Numerical Simulation Approaches. Piscataway, NJ, USA: Wiley-Interscience; 2003.

[32] Jarabo-Amores M.P., Bárcena-Humanes J.L., Gómez-del-Hoyo P.J., *et al.* 'IDEPAR: a multichannel digital video broadcasting-terrestrial passive radar technological demonstrator in terrestrial radar scenarios'. *IET Radar, Sonar and Navigation*. 2016, 11.

[33] Kulpa K., Malanowski M., Misiurewicz J., and Samczynski P. 'Passive radar for strategic object protection'. *IEEE International Conference on Microwaves, Communications, Antennas and Electronics Systems (COMCAS)*, 2011, pp. 1–4.

[34] Kuschel H., Ummenhofer M., Lombardo P., Colone F., and Bongioanni C. 'Passive radar components of ARGUS 875 3D'. *IEEE Aerospace and Electronic Systems Magazine*. 2014, 29(3), pp. 15–25.

[35] Kuschel H., Heckenbach J., O'Hagan D., *et al.* 'A hybrid multi-frequency Passive Radar concept for medium range air surveillance'. *Microwaves, Radar and Remote Sensing Symposium*, Kiev, 2011, pp. 275–279.

[36] Poullin D. and Flecheux M. 'Passive 3D tracking of low altitude targets using DVB (SFN Broadcasters)'. *IEEE Aerospace and Electronic Systems Magazine*. 2012, 27(11), pp. 36–41.

[37] Poullin D. and M. Flécheux 'A multistatic 3D passive system based on DVB-T'. *International Radar Conference*, Lille, 2014, pp. 1–6.

[38] O'Hagan D.W., Capria A., Petri D., *et al.* 'Passive bistatic radar (PBR) for harbour protection applications'. *Proc. of the IEEE National Radar Conference*, Atlanta, GA, USA, 2012, pp. 446–450.

[39] Langellotti D., Colone F., Lombardo P., Sedehi M., and Tilli E. 'DVB-T based passive bistatic radar for maritime surveillance'. *IEEE Radar Conference*, Cincinnati, OH, USA, 2014.

[40] Langellotti D., Colone F., Lombardo P., Tilli E., Sedehi M., and Farina A. 'Over the horizon maritime surveillance capability of DVB-T based passive radar'. *European Radar Conference (EuRAD 2014)*, Rome, Italy, 2014.

[41] Martelli T., Filippini F., Pignol F., Colone F., and Cardinali R. 'Computationally effective range migration compensation in PCL systems for maritime surveillance'. *IEEE Radar Conference (RadarConf18)*, Oklahoma City, OK, USA, 2018, pp. 1406–1411.

[42] Colone F., O'Hagan D.W., Lombardo P., and Baker C.J. 'A multistage processing algorithm for disturbance removal and target detection in passive bistatic radar'. *IEEE Transactions on Aerospace and Electronic Systems*. 2009, 45(2).

[43] Colone F., Palmarini C., Martelli T., and Tilli E., 'Sliding extensive cancellation algorithm for disturbance removal in passive radar'. *IEEE Transactions on Aerospace and Electronic Systems*. 2016, 52(3), pp. 1309–1326.

[44] Martelli T., Cardinali R., and Colone F. 'Detection performance assessment of the FM-based AULOS® Passive Radar for air surveillance applications'. *International Radar Symposium (IRS)*, Bonn, 2018, pp. 1–10.

[45] Colone F., Langellotti D., and Lombardo P. 'DVB-T signal ambiguity function control for passive radars'. *IEEE Transactions on AES*. 2014, 50(1), pp. 329–347.

[46] Daun M. and Koch W. 'Multistatic target tracking for non-cooperative illumination by DAB/DVB-T'. *OCEANS 2007 – Europe*, Aberdeen, 2007, pp. 1–6.

[47] Daun M. 'Deghosting in passive air surveillance systems'. *11th International Radar Symposium (IRS)*, Vilnius, 2010, pp. 1–8.

[48] Neyman J. and Pearson E. 'On the problem of the most efficient tests of statistical hypotheses'. *Philosophical Transactions of the Royal Society of London. Series A*. 1933, 231(9), pp. 289–337.

[49] Trees H.V. *Detection, Estimation, and Modulation Theory*. Second Edition. Hoboken, NJ, USA: Wiley; 2013.

[50] Aloisio V., di Vito A., and Galati G. 'Optimum detection of moderately fluctuating radar targets'. *IEEE Processing on Radar, Sonar and Navigation*. 1994, 141(3), pp. 164–170.

[51] del Rey-Maestre N., Jarabo-Amores M.P., Mata-Moya D., Bárcena-Humanes J.L., and Gómez-del Hoyo P.J. 'Machine learning techniques for coherent CFAR detection based on statistical modeling of UHF passive ground clutter'. *IEEE Journal of Selected Topics in Signal Processing*. 2018, 12(1), pp. 104–118.

[52] Skolnik M. *Introduction to Radar Systems*. Third Edition. New York, NY: McGraw-Hill Science; 2002.

[53] Conte E., Longo M., and Lops M. 'Modelling and simulation of non-Rayleigh radar clutter'. *IEE Proceedings F Radar and Signal Processing*. 1991, 138(2), pp. 121–130.

[54] D'Agostino R. and Stephens M. *Goodness of Fit Techniques*. New York, NY, USA: Marcel Dekker; 1986.

[55] Anderson T. 'On the distribution of the two-sample Cramér-von-Mises criterion'. *The Annals of Mathematical Statistics (Institute of Mathematical Statistics)*. 1962, 6(3), pp. 255–259.

[56] Farina A., Gini F., Greco M., and Verrazzani L. 'High resolution sea clutter data: statistical analysis of recorded live data'. *IEE Proceedings – Radar, Sonar and Navigation*. 1997, 144, pp. 121–130.

[57] Trees H.V. *Detection, Estimation, and Modulation Theory-Part III: Radar-Sonar Processing and Gaussian Signals in Noise*. Hoboken, NJ, USA: Wiley; 2001.

[58] del-Rey-Maestre N., Jarabo-Amores M.P., Mata-Moya D., *et al.* 'Performance of conventional and LR based passive radar detectors in ground traffic applications'. *International Conference on Control, Artificial Intelligence, Robotics and Optimization (ICCAIRO)*, Prague, Czech Republic, 2018, pp. 23–26.

[59] del-Rey-Maestre N., Jarabo-Amores M.P., Bárcena-Humanes J.L., Mata-Moya D., and Gómez del Hoyo P.J. 'ECA filter effects on ground clutter statistics in DVB-T based passive radar'. *26th European Signal Processing Conference (EUSIPCO)*, Rome, Italy, 2018, pp. 1227–1231.

[60] Gómez-del-Hoyo P.J., del-Rey-Maestre N., Mata-Moya D., Jarabo-Amores M.P., and Benito-Ortiz M.C. 'Coherent detection and 3D tracking stages of a DVB-T based passive radar for terrestrial traffic monitoring'. *IOP Conference Series: Materials Science and Engineering*, 2019.

[61] McLachlan G. and Peel D. *Finite Mixture Models*. Hoboken, NJ, USA: Wiley; 2000.

[62] Moscardini C., Conti M., Berizzi F., Martorella M., Capria A. 'Spatial adaptive processing for passive bistatic radar'. *Proceedings of the IEEE Radar Conference*, Cincinnati, OH, USA, 2014. Volume 1, pp. 1061–1066.

[63] Bournaka G., Heckenbach J., Baruzzi A., Cristallini D., and Kuschel H. 'A two stage beamforming approach for low complexity CFAR detection and localization for passive radar'. *IEEE Radar Conference*, Philadelphia, PA, USA, 2016.

[64] del-Rey-Maestre N., Mata-Moya D., Jarabo-Amores M.P., *et al.* 'Passive radar array processing with non-uniform linear arrays for ground target's detection and localization'. *Remote Sensing. Special Issue Radar Systems for the Societal Challenges*. 2017, 9, pp. 1–29.

[65] Trees H.V. *Optimum Array Processing: Detection, Estimation and Modulation Theory*. Hoboken, NJ, USA: Wiley; 2013.

[66] Rosado-Sanz J., Rey-Maestre N., Mata-Moya D., *et al.* 'Advantages of non-uniform linear arrays based on COTS elements in passive radar applications'. *Int. Microwave & Radar Conf. (MIKON)*, Poznan, 2018, pp. 199–203.

[67] Filippini F., Martelli T., Colone F., *et al.* 'Target DoA estimation in passive radar using non-uniform linear arrays and multiple frequency channels'. *IEEE Radar Conference*, Oklahoma City, OK, USA, 2018, pp. 1290–1295.

[68] Filippini F., Colone F., and De Maio A. 'Threshold region performance of multi-carrier maximum likelihood direction of arrival estimator'. *IEEE Transactions on Aerospace and Electronic System*. 2019, 55(6), pp. 3517–3530.

Chapter 7

Multiband passive radar for drones detection and localization

Pierfrancesco Lombardo[1], Fabiola Colone[1], Carlo Bongioanni[1], Octavio Miguel Cabrera Morrone[1], Francesca Filippini[1], Tatiana Martelli[1] and Ileana Milani[1]

7.1 Introduction

For many surveillance systems, a multisensory architecture is essential when aiming at providing accurate and reliable surveillance of critical infrastructures and protected areas. This is certainly the case for airports, harbours, prisons and chemical facilities as well as for public places such as parking areas, train stations or stadiums, especially due to the new and emerging threat represented by commercial drones and unmanned aerial vehicles (UAVs) [1–3]. In particular, their small size, low radar reflectivity and flexibility of manoeuvre make them especially difficult to detect, localize and classify. The potentials of passive radar (PR) in counter drone operations, by exploiting different illuminators of opportunity (IOs), such as DVB-T, DAB, GSM, WiFi, GNSS (Global Navigation Satellite Systems) and DVB-S that were introduced in previous chapters, have been recently investigated in several works [4–18]. The reported results demonstrate that such sensors could be effectively employed for detecting and localizing small and medium drones up to a few kilometres from the PR receiver when DVB-T transmissions are exploited. However, in urban areas and highly populated environments, where the use of the non-invasive and green PR technology represents an attractive solution for enhancing the security level [19], it is important to consider the possibility to exploit a multisensory system by integrating PR with multiple operational frequency bands.

The integration of PRs operating at different frequency bands can be obtained using two approaches:

1. A hardware integration can be conceived that provides a single multiband sensor, which exploits a common multichannel receiver, common data

[1]Department of Information Engineering, Electronics and Telecommunications (DIET), Sapienza University of Rome, Rome, Italy

memory and common processing techniques for different signals of opportunity and therefore greater frequency bandwidth.

2. At functional level, where each sensor operates with independent receiving structures but the outputs of the multiple sensors are combined to provide improved performance.

In the following sections, we proceed as follows. First, in Section 7.2, we briefly review the most attractive sources of opportunity for an anti-drone PR with their characteristics and signal processing requirements. Three specific sources of opportunity are identified that appear to be well suited for a multiband operation due to the complementarity of their characteristics. Therefore, we illustrate the concept of the multisensory system on the basis of the parasitic exploitation of such signals and discuss possible implementations and benefits. The different components of the multisensory system are then separately addressed in Sections 7.3–7.5 which are devoted respectively to anti-drone PR based on DVB-S satellite broadcast signals, DVB-T terrestrial TV and WiFi, in order to analyse the specific characteristics of their sensors as well as their performance. This analysis also allows to identify hardware and signal processing requirements for an effective implementation of the integrated multiband PR system.

7.2 Exploitation of different waveforms of opportunity

Different techniques have been proposed in the literature to detect and localize drones by exploiting PR systems based on different RF waveforms of opportunity [19–21]. Due to the small size of the target of interest and to the high variability of their motion characteristics, it is essential to have high range and Doppler frequency resolution. Thus, we only consider IOs with frequency bandwidth larger than 7–8 MHz and the possibility of providing long coherent integration times (CITs).

A wide frequency bandwidth is offered by satellite TV broadcast signal operating with the DVB-S standard. The bandwidth of a single DVB-S channel, typically of 32 MHz, allows a quite fine bistatic range resolution of about 5 m, which is much higher than other IOs. Since satellites transmitting DVB-S operate in the Ku-band (with wavelengths in the order of few centimetres), they potentially allow a good accuracy and a high-velocity resolution, which is easy to exploit due to the continuous wave (CW) characteristic of the broadcast signals. The widespread coverage of the TV broadcast makes them appealing as a ubiquitous source of opportunity. On the other hand, the main limitation is the target–satellite distance resulting in very low received signal-to-noise ratio (SNR) values, restricting its operation to short-range applications [16–18,22,23]. In practice, a coverage variable in the range of 40–200 m is reasonably available by exploiting the DVB-S source, when using antenna beams between 5 and 45 degrees. This makes it very valuable providing the accurate localization of drones, after their presence have been detected inside a potential surveillance area.

The GNSS waveforms can also be used with similar characteristics as the DVB-S broadcast illuminators, showing in some cases even a wider frequency bandwidth and a ubiquitous multistatic coverage [15]. However, GNSS are characterized by a similar low power level in proximity of the Earth surface and require the sensor to track the Medium Orbit satellite transmitters that imply higher hardware complexity and a continuously variable geometry, which, in turn, changes continuously range and velocity resolution, as well as detection and estimation performance. For this reason, we consider the exploitation of GNSS IOs only for a possible additional improvement.

Terrestrial broadcast TV can also be of interest, showing the same CW transmission characteristic of DVB-S signals, but with a narrower frequency bandwidth (7–8 MHz) and lower carrier frequency (VHF/UHF), which provides both lower resolution in range and lower sensitivity to the target velocity. In contrast, the received power level is typically higher, which is expected to provide operational capability at longer ranges [4–10,24–27]. Among its characteristics, the low-frequency operation is capable of providing the presence of resonance conditions with the structure of commercial drones that might cause significant increases of their radar cross section (RCS). Coverages of a few kilometres can be obtained by exploiting this source of opportunity, which appears especially useful to search for the presence of potential drones in an assigned surveillance area.

Slightly greater frequency bandwidth is provided by the transmitted waveforms for local area networking applications (WiFi, LTE, etc.) which have been proliferating at a rapid rate for both commercial and private uses. Unlike the case of the TV broadcast transmissions, these sources have low power level (being intended for local networking), a burst characteristic and their use of any specific frequency channel is subject to a multiple access protocols, so that the source location can change from burst to burst. Appropriate operation has been demonstrated in the short-range for the WiFi transmissions, which are widely accessible sources of opportunity [13,14,28–30], but not ubiquitous as the geostationary satellite sources of the DVB-S. The coverages achievable using these sources are somewhat wider than the DVB-S but largely lower than the DVB-T sources. Among their attractive features, it is noted that these sources can operate with small-sized and low-cost antenna systems that can also be deployed in the form of antenna arrays and exploited for improved angular operation both in azimuth and in elevation.

Table 7.1 summarizes the main characteristics of the different IOs considered for the multisensory system addressed in this chapter.

The different coverage, range resolution, velocity resolution and angular accuracy suggest that three types of PR sensors, respectively, based on DVB-S, DVB-T and WiFi signals of opportunity, can nicely operate as parts of a multiband multisensory system, structured as a multi-tier network of PR.

The concept of the multisensory system is sketched in Figure 7.1. The DVB-T PR sensor, which has the widest coverage but the worse resolution in both range and velocity, can perform the task of detecting the potential presence of a drone in a wider area, so to alert the short-range PR sensors that cover the detected drone approximate location or to provide target cueing, in case their antenna system can

Table 7.1 Summary of the main characteristics and complementary aspects of the illuminators of opportunity considered for the multisensory system

	DVB-S	DVB-T	WiFi
Coverage	Limited (few tens of metres)	Very good (several kilometres)	Reasonable (few hundreds of metres)
Resolution	Very good (range: 5 m, velocity: ~0.1 m/s)	Reasonable (range: 20 m, velocity: ~1 m/s)	Good (range: 9–14 m, velocity: ~0.5 m/s)
Accuracy	Very good (Ku-band)	Limited (UHF band)	Good (S or C bands)
Availability/ additional opportunities	Global coverage	Improved RCS for commercial drones	Potential for joint passive radar/passive source location

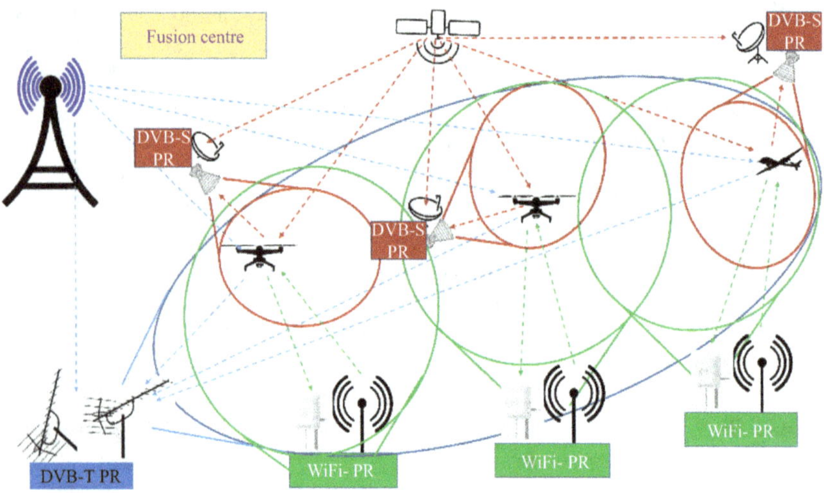

Figure 7.1 Multisensor system based on multiband passive radar components

be steered towards the area of interest. Typically, the WiFi-based PR can provide a good coverage, to follow the drone motion at short range, as well as to accurately measure its 3D position by exploiting its antenna system. However, the WiFi signal coverage is not complete everywhere, whereas the availability of DVB-S signals is quite flat over extremely wide areas; therefore (using a sufficiently dense network of receivers), the DVB-S PR can cover potentially all the region and provide increased range resolution. Once the detection is confirmed, both WiFi-based and DVB-S-based PR can be used to provide a localization with much more accuracy than the DVB-T radar alone. The separation of the search and tracking modes is even more effective, if the DVB-S- and WiFi-based PRs are designed to provide a steerable surveillance antenna system that can be really cued by the DVB-T PR towards the preliminarily detected target.

Moreover, as illustrated in Section 7.3, the DVB-S PR provides a very good velocity accuracy, which can be exploited to implement a high-quality target tracker. In addition, we must mention that the high-velocity resolution can be used to study the drone-induced micro-Doppler that potentially allows the drone-type classification. As indicated in Figure 7.1, the fusion logic described earlier is implemented in the fusion centre, where the data processed by all PR sensors are transmitted (communication links not described to avoid a too busy sketch).

While not explicitly shown in Figure 7.1, we observe that a practical system is typically equipped with at least a couple of antennas for the estimation of the target direction of arrival (DoA). The two antennas for azimuth estimation are shown in Figure 7.2 (whereas a third one or more could be used to also estimate the drone's elevation), where only the sensors focussing on a specific drone target are reported. Figure 7.2 also shows that the multiband PR system can nicely be integrated with a device-based passive coherent location sensor that receives the waveforms autonomously emitted by the target (purple dashed lines in the figure) to localize it. Device-based passive coherent location can provide target localization either by exploiting two angle measurements from two receivers or exploiting angle and a time difference of arrival (TDoA) measurements. This can be a quite effective integration, since many commercial drones exploit the ISM band of 2.4 GHz to transmit to the controller/operator both position information and images of video collected during its operation. Moreover, due to the large similarity in the frequency range of operation and in the processing techniques, this sensor can fully be

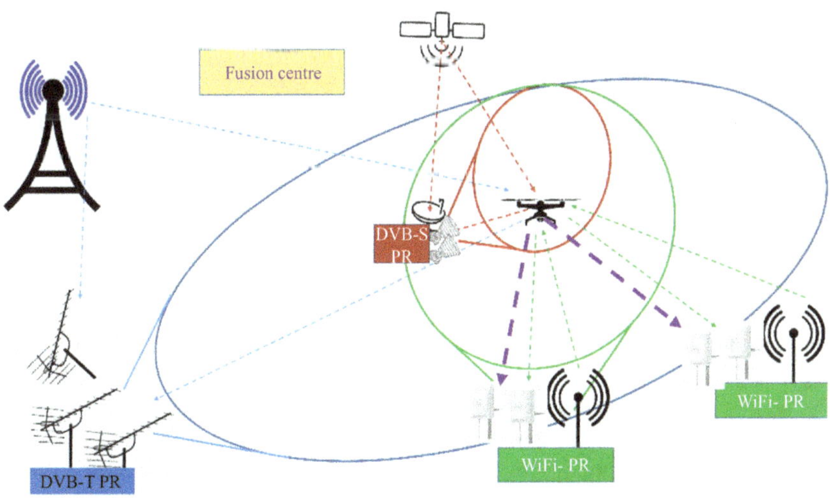

Figure 7.2 Multisensor system based on multiband passive radar components incremented with a device-based passive location system in the ISM frequency range used by the WiFi signals

integrated with the hardware of the WiFi-based PR. Adding this type of sensors can provide a global multiband system with largely improved accuracy and continuity of localization for the drone targets.

While all the three types of PR, respectively, exploiting DVB-S, DVB-T and WiFi signals, can be built around a software-defined radio (SDR) receiver with three or four coherent channels, and they can potentially be integrated in a single multiband PR sensor, with quite different characteristics, the following considerations should carefully be considered at the design stage.

1. Their antenna subsystems are significantly different.
2. While SDRs are available to cover the frequency range, including DVB-T and WiFi frequency bands, to receive the satellite DVB-S signal, a first conversion must be implemented from Ku- to L-band keeping (or reconstructing by processing) phase coherence among channels.
3. DVB-T and DVB-S, despite using both CW signals, require the mostly different hardware and have the mostly different coverage, whereas WiFi and DVB-S signals, which have the mostly similar coverage, operate with different hardware and with all differences between burst and CW transmission modes.
4. Due to the different geometry in terms of transmitter, receiver and target location, other than the differences in the used antenna beams, the requirements for the signal processing chain are quite different among the three systems. This has a significant impact on the processing techniques for the acquisition of the reference signal and for the cancellation of the direct signal and its multipath.

In the following sections, we present the receiver hardware block diagrams and the relevant signal processing schemes that are required to cope with the individual characteristics listed earlier.

7.3 Passive radar based on DVB-S

Most of the current digital TV broadcast satellites exploit the DVB-S signal standard in the Ku frequency band from 10 to 12 GHz. The transmitted waveform at each carrier is a composite signal multiplexing multiple TV programs that are combined into an overall signal with 30-MHz bandwidth, which result in a range resolution of 5 m. This characteristic, together with a very high velocity sensitivity available due to its Ku-band operation, makes the DVB-S waveform extremely attractive [16–18,22,23]. However, the long distance to the source of opportunity implies a very low power density in proximity of the Earth surface, which tends to limit the capability to detect small targets. Moreover, the Ku-band operation and the wide signal bandwidth might have an impact on the sensor's hardware cost and on the computational complexity of the required signal processing.

In this section, the previous limitations are addressed, and appropriate solutions are identified to enhance the potentials of DVB-S PR systems as one of the subsystems of the multisensor system and to make it cost-effective. Specifically, low-cost hardware implementations and low-complexity signal processing schemes are devised to recover the target energy by exploiting long CITs and polarization diversity.

7.3.1 DVB-S-based PR processing schemes

In the DVB-S case, the emitting source of opportunity is typically very far from the receiver location (order of thousands of kilometres) and very far from the Earth surface. Therefore, directive antennas are typically used since the receive signal level tends to be low. Consequently, the direct signal represents a very small component of the signal collected by the surveillance antennas and the corresponding receiving channels and its multipaths are negligible. This implies that

1. a very simple direct signal/clutter cancellation technique can be used and often this can be even avoided without degradation of the overall performance;
2. the reference waveform cannot be extracted from the surveillance channel samples, whereas a narrow beam reference antenna must be used, steered towards the satellite source of opportunity;
3. a long CIT must be used to compensate for the very low signal levels backscattered from the drone, which requires to include advanced techniques to compensate for the target range and Doppler frequency migration.

The developed DVB-S-based PR processing scheme for target detection [17] is illustrated in Figure 7.3. First, the signals collected by means of the reference and surveillance antennas undergo a disturbance cancellation stage for the removal of the undesired contributions in the surveillance channels, i.e. direct signal, clutter and multipath echoes. For this purpose, we resort to the extensive cancellation algorithm (ECA) [20,21,31]. Then, the output signal from ECA filter and the

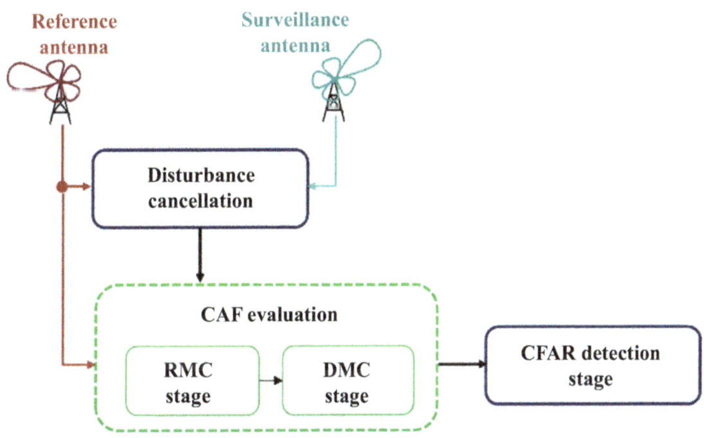

Figure 7.3 DVB-S-based processing scheme for target detection exploiting long coherent integration times. © 2020 IEEE. Reprinted, with permission, from Reference [17]

reference signal are used to evaluate the bistatic range–Doppler cross ambiguity function (CAF).

To increase the CIT, in order to benefit from the coherent integration gain, the target migration effects must be effectively compensated [17]. Consequently, for a moving target with both radial and tangential velocity components, the CAF evaluation should also include a range migration compensation (RMC) and a Doppler migration compensation (DMC) stage. For the RMC stage, we use the keystone transform (KT) based on the implementation presented in [32]. We recall that the KT technique is widely used in active radar to compensate for the range migration. It can be applied to PR systems employing CW transmissions by resorting to a batch approach [20], in order to recreate the classical slow-time/fast-time framework of a pulsed radar operating at a given pulse repetition frequency. We also recall that the use of the suboptimal batch approach yields some SNR loss with respect to the use of an optimum algorithm for the CAF evaluation.

Following the RMC stage, the Doppler migration is corrected at each range cell r, operating on the range-migration-compensated data in the slow-time domain, $x[r, m]$ [17,33]. For a given Doppler rate α, the required correction is obtained by

$$\chi[r,\ v] = FFT\left\{x[r,\ m]\exp\left(-j\frac{\pi}{\lambda PRF^2}\alpha m^2\right),\ m = 0, \ldots,\ M-1\right\} \qquad (7.1)$$

Since the target Doppler rate is unknown at the detection stage, a bank of filters is exploited for an appropriate set of Doppler rate values. A grid of N_α Doppler rate values $\{\alpha_0, \ldots,\ \alpha_{N_\alpha - 1}\}$, spaced by $\alpha = \lambda / T_{CIT}^2$ (where λ is wavelength and T_{CIT} is employed CIT), according to the depth of focus criterion. The 2D outputs from (7.1) obtained for different values of α are collected in a 3D array $\chi[r,\ v,\ n]$, $r = 0, \ldots,\ N_R - 1$, $d = 0, \ldots, N_D - 1$, $n = 0, \ldots N_\alpha - 1$, where N and N_D are the array extensions in the bistatic range and Doppler dimension, respectively. Subsequently, a 2D map is built starting from the 3D array $\chi[r,\ v,\ n]$ by maximizing its intensity across the Doppler rate dimension, separately at each range–Doppler bin:

$$y[r, v] = \max_n\{|\chi[r, v, n]|^2\} \qquad (7.2)$$

Afterwards, a 2D Cell Average–Constant False Alarm Rate (CA-CFAR) scheme is applied to detect potential targets. Notice that due to the previous non-linear operations, the CFAR threshold cannot easily be evaluated in closed form. Therefore, in this work, the threshold to be used in the detection scheme, in order to obtain the desired probability of false Alarm (P_{fa}) value, has been evaluated via Monte Carlo simulation.

7.3.2 Experimental drone detection and localization with DVB-S-based PR

An experimental DVB-S PR has been set up, based on COTS, and fielded in different acquisition campaigns to test its effectiveness for infrastructure protection and specifically when operating against drones. To obtain a clean reference signal, a high gain, offset parabolic reflector antenna is used for the reference channel, with a feedhorn installed in the reflector focus as illustrated in Figure 7.4(a). A low-noise block (LNB) is connected to the feed in order to provide the first low-noise amplification of the received signal and its downconversion from Ku- to L-band. Low-cost LNBs are available for satellite TV reception that can be used to receive either in horizontal or in vertical polarization. As the parabolic reflector antenna has a very narrow beam, only the feedhorn was used as antenna for the surveillance channels in order to continuously monitor a reasonably large area. As shown in Figure 7.4(b), an OMT (orthomode transducer) was connected at the feedhorn output in order to simultaneously collect both the polarizations on receive for the purpose addressed in Section 7.3.4. All LNBs employed are provided with the same reference of 10-MHz signal, so that coherent downconversions are performed at surveillance and reference channels (Figure 7.4(b)). Following the work in [18], a four-channel SDR board is used to downconvert the signals from the reference channel and up to three surveillance channels and to digitize all of them. In Figure 7.4(b), we also show the presence of the diplexers COTS, which allow to properly distribute the 10-MHz reference, the energy source for the LNBs and the L-band outputs.

A DJI Mavic Pro drone of size approximately 20-cm long and 8.3-cm wide (propellers excluded) was used as a cooperative target to test the experimental receiver and evaluate the performance of the proposed signal processing scheme. The drone, which is characterized by a complex motion model, was piloted to fly away from the receiver up to approximately 25 m, and the digitized data files were processed according to the scheme in Figure 7.3. The coherent integration gains were evaluated when different CAF algorithms and CIT values were used. Specifically, for the considered analysis, the CIT has been extended up to 3 s. First, the conventional CAF (namely no compensation case) is evaluated by resorting to the direct FFT algorithm [20] which is the most efficient solution for the CAF formation when DVB-S signals are exploited. Then, we considered the case of a CAF evaluation where the RMC stage alone has been added and finally, the CAF

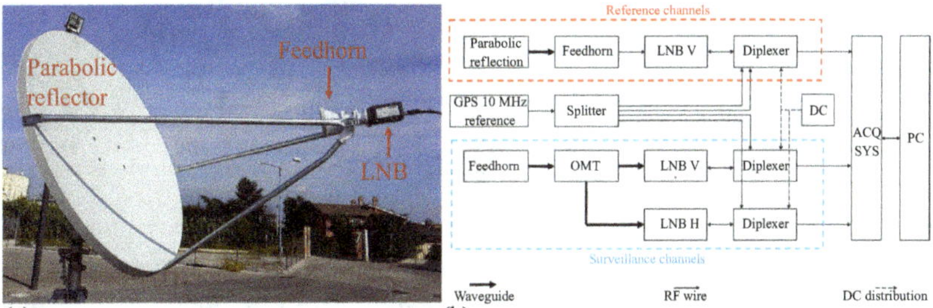

(a) (b)

Figure 7.4 (a) Offset parabolic reflector and (b) DVB-S passive receiver architecture. © 2020 IEEE. Reprinted, with permission, from Reference [17] evaluation where both RMC and DM walk effects have been corrected. As an example, Figure 7.5 shows a zoom of the range–velocity maps obtained using the V polarization for a single data file around the drone location [32 m, 1.6 m/s].

All the maps have been normalized to the noise power (estimated in a region of the CAF where it is unlikely for targets to be present) in order to measure the target's SNRs. It is worth noting that in the DVB-S case, when close range targets are considered (this is the case), the bistatic range is comparable with the receiver-to-target distance since the transmitter-to-target distance is approximately equal to the baseline [17].

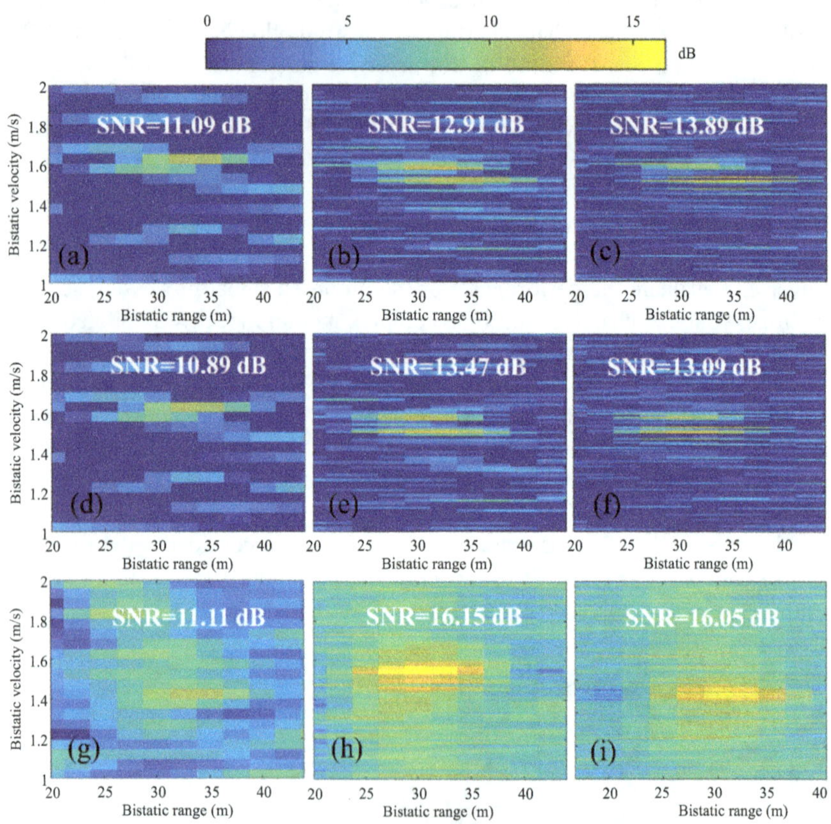

Figure 7.5 Comparison of detection maps to CIT and the cell migration compensation technique. Drone-case: zoom of the range–velocity maps using V polarization at a single scan with different algorithms: (a)–(c) conventional CAF with CIT = 0.5, 2 and 3 s, respectively; (d)–(f) RMC with CIT = 0.5, 2 and 3 s, respectively; (g)–(i) RMC+DMC with CIT = 0.5, 2 and 3 s, respectively. © 2020 IEEE. Reprinted, with permission, from Reference [17]

When the conventional CAF map is evaluated with a CIT = 0.5 s (see Figure 7.5(a), (d) and (g)), the drone is received with low SNR value (approximately 11 dB) and it is likely to not exceed the threshold stage for typical P_{fa} values. When we extend the CIT up to 2 and 3 s and the conventional CAF is adopted, the SNR improvement is limited due to the target migration effects that become apparent, especially in the Doppler dimension (Figure 7.5(b) and (c)). From Figure 7.5(e) and (f), we observe that the compensation of range migration alone yields a negligible advantage on the target peak. In contrast, using the additional DMC stage (Figure 7.5(h) and (i)), the drone peak appears well focused and a significant SNR improvement is obtained with respect to the conventional CAF.

However, the comparison between parts (h) and (i) of Figure 7.5 shows that the extension of the CIT duration from 2 up to 3 s yields no integration gain. This might be caused by additional migration effects due to higher order terms that appear both in the range and Doppler variation laws. Despite this, the reported results clearly demonstrate that the proposed solutions allow extending the CIT at least up to 2 s against drones, even though characterized by an unpredictable motion model.

Moreover, from Figure 7.5 we clearly observe that a very high velocity resolution is available with the long CIT, which allows to obtain detailed characteristics on the target properties and potentially opens the way for a micro-Doppler analysis.

7.3.3 Phase-locked vs non-phase-locked receiver architectures

An effective operation has been obtained by means of the DVB-S PR sensor based on the coherent receiver architecture in Figure 7.4(b) [34]. To obtain a fully coherent scheme, high-quality LNB devices must be employed that can use the same external LO or at least the same 10-MHz reference signal. We refer to this architecture as the phase-locked architecture. In contrast, the use of commercial, non-phase-locked LNB could allow lower cost sensors to be utilized. Due to the Ku-band frequency allocation of the satellite signals, the phase noise of the LO is expected to be more critical than for other PR sources. To investigate this point, we considered a non-coherent reception scheme where the LNBs use independent LOs to obtain the conversion from K-band to L-band, while a four-channel SDR provides the baseband digitized signals. This scheme requires an appropriate processing module to recover phase/frequency synchronization among receiving channels.

For a direct comparison of the two different architectures, we implemented an experimental setup that allowed simultaneous tests against exactly the same input signals (see Figure 7.6). Specifically, a two-way balanced RF waveguide splitter was directly connected to the reference and surveillance antennas as illustrated in Figure 7.6(a) so that the same signals are made available at the input of the two receiver architectures. Figure 7.6(b) shows the reflector parabola, with the feedhorn and waveguide splitter feeding both LNBs used for the reference channels.

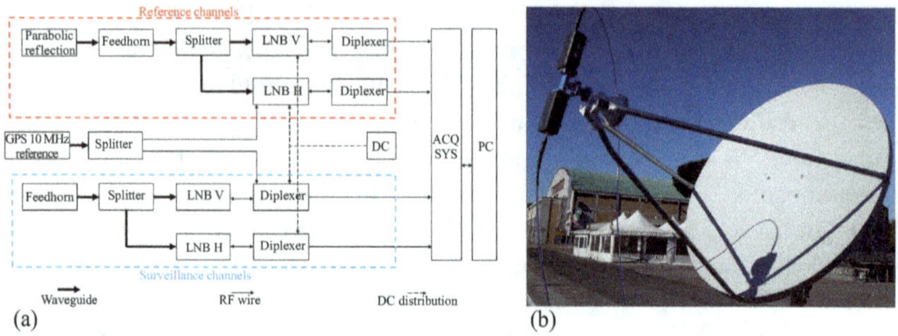

Figure 7.6 (a) Simultaneous coherent and non-coherent architecture and (b) Parabolic reflector with the feedhorn and waveguide splitter feeding both LNBs used for the reference channels. © 2021 IEEE. Reprinted, with permission, from References [17,34]

In Figure 7.6(a), the same components are adopted as in Figure 7.4(b) but connected in a different way. The phase-locked architecture exploits the reference and surveillance signals at the output of the LNBs disciplined by the 10-MHz LO reference. In contrast, the non-phase-locked architecture takes as inputs the signals from the LNBs operating independently by using their own reference.

To compare the performance of these two architectures, an experiment was carried out in which a car moved away from the radar while the four L-band signals were acquired. As previously mentioned, the four signals were arranged in two sets: one set of reference and surveillance in phase-locked setting and another set of reference and surveillance in non-phase-locked setting. The peculiar bistatic geometry obtained by exploiting the geostationary illuminator makes the cancellation stage not necessary. This allowed the use of simple offline processing, including only the preliminary filter stage, the calculation of the CAF and then a CA-CFAR detector applied to consecutive time windows of T_i seconds for each of the time frames. While this can be applied directly to the signals collected with the coherent or phase-locked architecture, the non-coherent or non-phase-locked acquisition scheme requires the preliminary estimation of the frequency offset between the two downconverted signals at the reference and surveillance antennas. To recover the coherency between the two channels in the non-coherent architecture, we could assume that, over short time periods, the use of different reference oscillators only yields a frequency offset ω_o. This would indicate that after applying conventional PR techniques, we would obtain Doppler–range maps shifted in frequency by an amount equal to $\omega_o/2\pi$, which is a negligible loss in terms of coherent integration. The time periods that guarantee the validity of the approximation earlier are largely dependent on the stability of the employed oscillators. In this regard, it is interesting to understand the practical bounds on the CIT set by the approximation earlier when commercial LNBs are employed. Moreover, despite the fact that the frequency offset does not affect the coherent integration gain, it might jeopardize the estimation of the target speed. In principle, this frequency offset can be

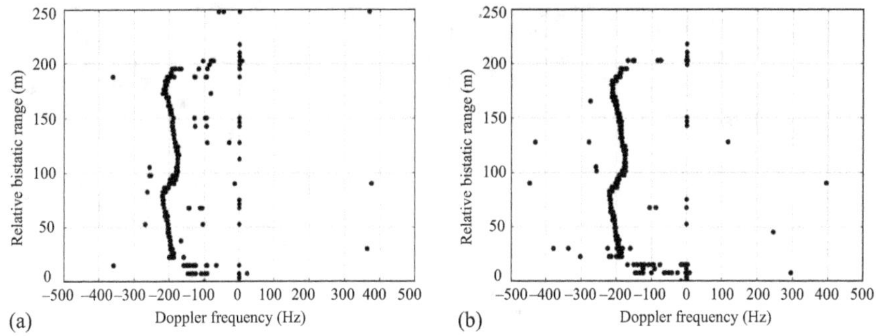

Figure 7.7 Detection maps with a $P_{fa} = 10^{-6}$ for (a) phase-locked and (b) non-phase-locked architectures. © 2021 IEEE. Reprinted, with permission, from Reference [34]

estimated on the basis of the received signals themselves, if we assume that they include the dominant contribution (e.g. the direct signal coming from the transmitter of opportunity). This is possible using a matched filter or CAF function [12,14] assessed at zero range

$$|\psi(m,\omega_0)| = \left|\sum_{k=0}^{N} r_b^*[n-m]s_b[n]\exp(-j\omega_o t)\right| \tag{7.3}$$

and find which offset frequency maximizes it. The lengths of the discrete signals $r_b[n]$ and $s_b[n]$ are represented by N. This procedure for the estimation of ω_o has been exploited to correct the Doppler–range maps.

Figure 7.7(a) and (b) shows the raw detections obtained with the phase-locked and non-phase-locked architectures, respectively, for the considered experimental test. Specifically, in each figure the outputs of the detection stage are shown for consecutive partially overlapped signal frames of $T_i = 0.5$ s each, displaced by 0.3 s. As can be seen, with both architectures it was possible to obtain a clean detection map, at least for a short-range application. This suggests a way to use the low-cost, non-phase-locked architectures for short-range applications. To understand the limits of the non-coherent architecture, Figure 7.8 shows the increase of direct signal to noise, $\Delta DNR = DNR_{Ti}/DNR_0$, with DNR_0 the DNR for an integration initial time, as a function of the integration time for both architectures. The use of DNR to perform this study is highly useful as we can consider the direct signal as a target in range and zero Doppler, which will not present cell migrations thus enabling an increased integration time when necessary until the effects of the lack of gain for the non-phase-locked system become evident. It is seen that after 1 s of integration in the non-phase-locked architecture there is no integration gain, at least in the analysed data set. This analysis directly links the lack of integration gain to the strict hypothesis that the phase error for non-phase-locked architecture

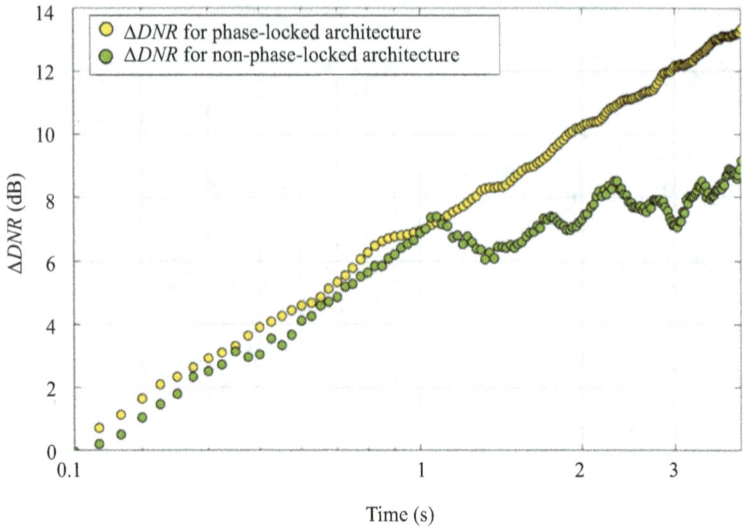

Figure 7.8 ΔDNR for phase-locked and non-phase-locked architectures. © 2021 IEEE. Reprinted, with permission, from Reference [34]

may be considered linear. Therefore, the non-coherent scheme is usable for small integration times [34].

7.3.4 Exploiting polarizations for drone detection

As previously mentioned, to exploit both H and V polarizations, we use an OMT that allows to receive simultaneously the outputs at both polarizations from the same antenna, using a feedhorn with circular flange (as the C-120) and two LNBs (see Figure 7.9(a)). We consider the general case of a coherent DVB-S-based PR receiver, equipped with P differently polarized surveillance channels. First, the available surveillance signals are separately processed according to the main stages of a conventional PR processing scheme, sketched in Figure 7.9(b) [16]. Afterwards, the P outputs can be jointly exploited for target detection purposes.

Specifically, after disturbance cancellation, we use frames of length $L = 2{\cdot}10^6$ samples. The P outputs of this disturbance cancellation stage are then used to evaluate the CAF to obtain one bistatic range–Doppler map for each polarimetric channel, say $\psi(f_D, R_B)_p, p = 0, \ldots, P-1$, where (f_D, R_B) represent the generic Doppler and range pair. A DJI Mavic Pro has been used as cooperative target, flying away from the receiver up to approximately 25 m.

Two different data files are considered in the following, data file 1 and data file 2, which account for two different signal fragments of 0.5 s each. The corresponding results are reported in Figures 7.10 and 7.11, respectively. Each subfigure represents the test statistic over the bistatic range–Doppler plane before the detection stage, mapped into the nominal P_{fa} that would allow the corresponding

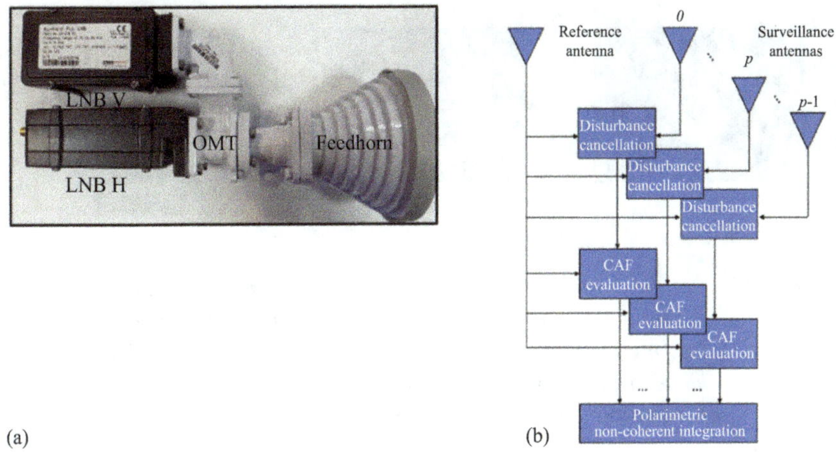

Figure 7.9 Feedhorn with (a) OMT and LNBs for H and V polarization, (b) polarimetric passive radar processing scheme. © 2020 IEEE. Reprinted, with permission, from Reference [16]

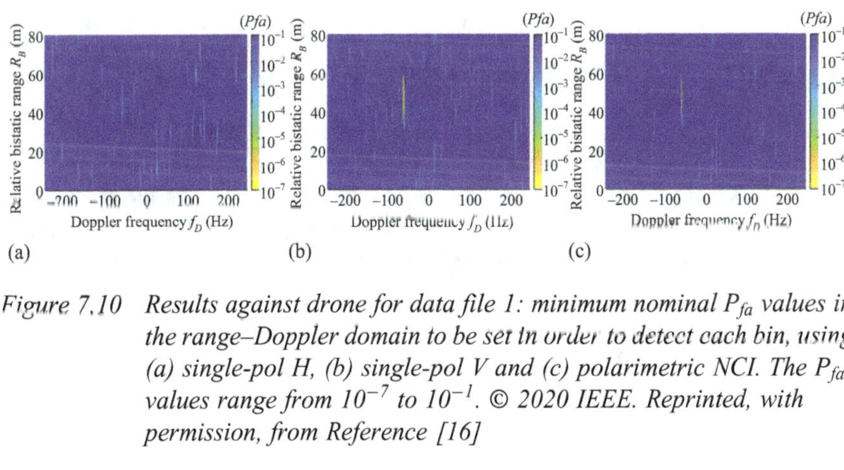

Figure 7.10 Results against drone for data file 1: minimum nominal P_{fa} values in the range–Doppler domain to be set in order to detect each bin, using (a) single-pol H, (b) single-pol V and (c) polarimetric NCI. The P_{fa} values range from 10^{-7} to 10^{-1}. © 2020 IEEE. Reprinted, with permission, from Reference [16]

threshold exceeding. In other words, each pixel has been scaled to represent the minimum value of nominal P_{fa} to be set for that pixel to yield a detection.

Specifically, Figure 7.10(a) and (b) is obtained for data file 1 when separately using the signals collected at vertical (V) and horizontal (H) polarization outputs, respectively. Similarly, Figure 7.11(a) and (b) shows the single-pol maps obtained at data file 2, corresponding to a different time window of the same size. The difference in the plots depends on the complex structure of the target and its motion/rotation and

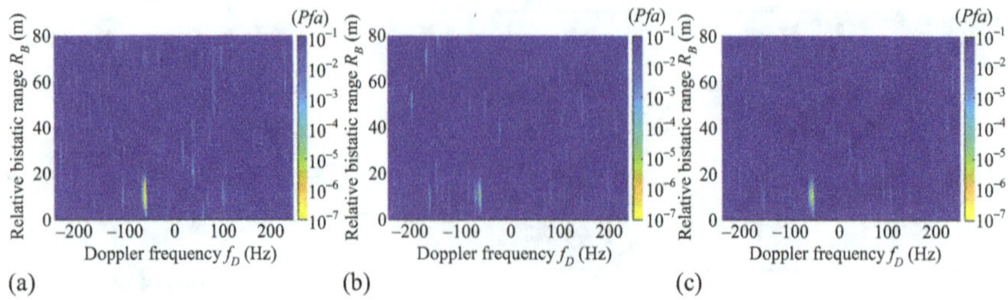

Figure 7.11 Results against drone for data file 2: minimum nominal P_{fa} values in the range–Doppler domain to be set in order to detect each bin, using (a) single-pol H, (b) single-pol V and (c) polarimetric NCI. The P_{fa} values range from 10^{-7} to 10^{-1}. © 2020 IEEE. Reprinted, with permission, from Reference [16]

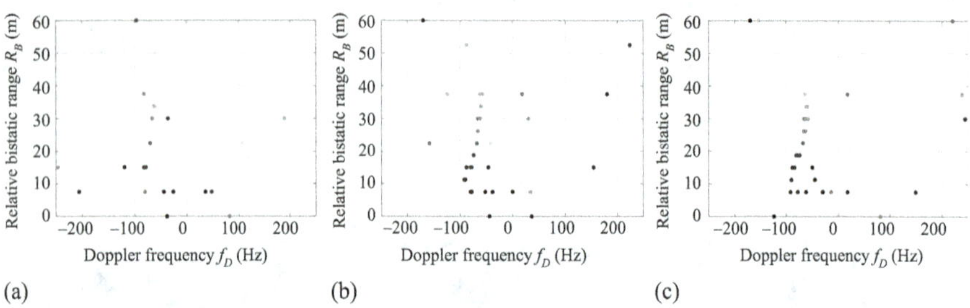

Figure 7.12 Results against drone: raw detections for 40 consecutive data windows with $P_{fa} = 10^{-4}$, using (a) single-pol H, (b) single-pol V and (c) polarimetric NCI. Different grey shades map the temporal information, starting from the darkest one. © 2020 IEEE. Reprinted, with permission, from Reference [16]

the signal polarization reflection that may vary in time (see also [15,16]). In fact, data files 1 and 2 show quite different behaviours, which makes the a priori selection quite hard. Notice that the available data set does not even allow to identify the polarimetric channel showing the best performance on average. Therefore, to improve robustness in detection, a simple non-coherent integration (NCI) of the signals processed at the two channels, i.e. $P = 2$ (H and V), can be used:

$$\psi_{NCI}(f_D, R_B) = \left|\psi(f_D, R_B)_V\right|^2 + \left|\psi(f_D, R_B)_H\right|^2 \tag{7.4}$$

Figures 7.10(c) and 7.11(c) show $\psi_{NCI}(f_D, R_B)$, respectively, for data files 1 and 2. It is not difficult to observe that the NCI approach provides stable results, always close to the best single-polarized channel.

Finally, Figure 7.12 shows the raw detections obtained for 40 consecutive data files as grey plots in a common bistatic range–Doppler plane, for $P_{fa} = 10^{-4}$. The different grey shades map the time window at which the detections are extracted across the acquisition time, starting from the darkest one. In particular, Figure 7.12 (a) and (b) shows the results obtained using the H and V polarization, respectively, while Figure 7.12(c) shows the results obtained using the polarimetric NCI approach.

Figure 7.12 shows that using the polarimetric NCI, the drone detection obtains a higher continuity for its entire trajectory. Table 7.2 shows a comparison in terms of number of detections associated to the drone when the three CAF algorithms and different CIT durations are considered.

The following observations are made [17]:

1. When CIT = 0.5 s, the conventional CAF, the only RMC stage and both RMC and DMC provide comparable detection results, since no range and Doppler migration effects show up during short coherent CITs.
2. For longer CITs, the target echo smears both on multiple Range cells and on multiple Doppler cells and only few additional detections are obtained, unless both RMC and DMC stages are applied. Due to the low drone velocity, the range walk is small and using only the RMC does not show a sensible improvement.
3. Using both RMC and DMC stages, the capability of detecting increases with the CIT duration until CIT = 2.5 s that provides 26 and 38 drone detections (out of 40) when using H and V polarization, respectively. These values are highlighted in red in Table 7.2.
4. A further increase of the integration time up to 3 s yields a slight degradation of the detection performance in V polarization, possibly caused by higher order terms that appear both in the range and Doppler variations.

As can be seen the joint exploitation of long CITs and target migration compensation techniques determines a significant improvement in drone detection capability for the considered system, especially when combined with the use of the polarimetric information (either by selecting the best polarimetric channel or by providing a their noncoherent combination, as previously shown).

Table 7.2 Synthesis of the drone detection results when different algorithms and CIT values are adopted for CAF evaluation. © 2020 IEEE. Reprinted, with permission, from Reference [17]

	Conv. CAF		**RMC**		**RMC+DMC**	
	H	**V**	**H**	**V**	**H**	**V**
CIT = 0.5 s	11	23	11	21	11	23
CIT = 1 s	14	26	16	25	21	30
CIT = 1.5 s	17	28	20	28	25	34
CIT = 2 s	16	28	17	26	25	34
CIT = 2.5 s	14	28	16	29	**26**	**38**
CIT = 3 s	15	29	19	30	26	36

Based on the analysis and discussion reported in this section, we can conclude that cost-effective DVB-S passive receivers could be implemented aiming at the detection of drones and small UAV at short ranges based on the PR principle. Along with appropriate signal processing schemes, long CITs and polarization diversity can be exploited to increase the reliability of the system. Incidentally, we notice that such approaches might also aid the classification task by providing additional information on the detected target echoes.

7.4 Passive radar based on DVB-T

The main features that make DVB-T-based PRs suitable for drone surveillance have extensively been described in Chapter 6, where different aspects were widely covered. In this chapter, we discuss additional peculiarities that make the DVB-T PR sensor an essential component of the multisensory multiband system.

As briefly mentioned in Chapter 6, a specific advantage of the PR operation based on DVB-T signals is the ability to provide simultaneous short- and long-range surveillance. For instance, in [9] the authors demonstrated the simultaneous detection of drones flying near the airport area along with the conventional civil air traffic at longer ranges. On one hand, this possibility enables an effective cooperation with high-resolution PR sensors that are more suitable for local area monitoring, such as DVB-S and WiFi-based sensors. On the other hand, this implies that the system operates at the same time with quite different RCS targets, so that effective solutions should be devised to mitigate the potential masking effects caused by the consequent high dynamic range.

7.4.1 DVB-T-based PR systems for simultaneous short- and long-range monitoring

As shown in Figure 7.13, in order to develop joint short- and long-range surveillance capability, the collected data need to be simultaneously forwarded to two parallel processing chains that are tailored to maximize the performance against the drones at short range and the civil air traffic at farther ranges, respectively. This approach is enabled by the previously discussed characteristics of the DVB-T signals and by the exploitation of SDR receiver architectures. This implies that the signal processing to be entirely digital can be flexibly adapted to meet the requirements of different search tasks.

This aspect has been specifically investigated in [9]. As it is apparent from Figure 7.13, the two processing chains are made up of the same blocks. Nevertheless, the algorithm implemented in each block – together with its relevant parameters – must be properly adapted to the considered application. This might also include the timings to be employed (i.e. coherent integration intervals length, update rates and time windows used for cancellation), which can be asynchronously set for the two processing chains to suit the needs of the different tasks.

The disturbance cancellation stage exploited in the DVB-T PR was addressed in Chapter 6 where the ECA-S (ECA-sliding) approach was shown to be a

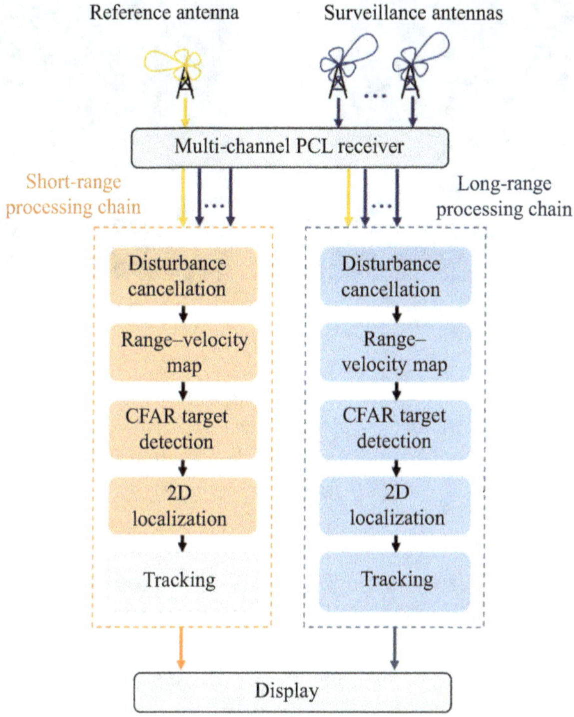

Figure 7.13 DVB-T-based processing scheme for simultaneous detection and localization of drones and aircraft [9]. © 2020 IET

suitable solution for both processing chains. However, based on the considerations in [35], batches of small dimensions should be used against civil air traffic aerial targets to synthesize a wide Doppler cancellation notch to effectively remove the non-stationary disturbance, while longer batches are preferred for the detection of slowly moving objects with low RCS to yield a narrower cancellation notch, thus preserving the target echo contribution. As illustrated in Figure 7.13, once the cancellation stage has been applied, and the reference signals have been properly filtered to remove the high side lobes and spurious peaks appearing in the DVB-T signal ambiguity function (AF) [24], the collected signals undergo the range–velocity map evaluation. For this evaluation, the two different surveillance applications require different CITs T_{int}, as well as different algorithms, to be employed. It should also be considered that the limit to the used CIT is also related to the computational complexity required by the processing technique, as well as the maximum admissible latency for the detections that could potentially need to trigger other sensors. Specifically, in the long-range civil air traffic surveillance application case, aiming at increasing the detection capability of low RCS aircraft as well as at widening the coverage area, long T_{int} should be considered. In contrast,

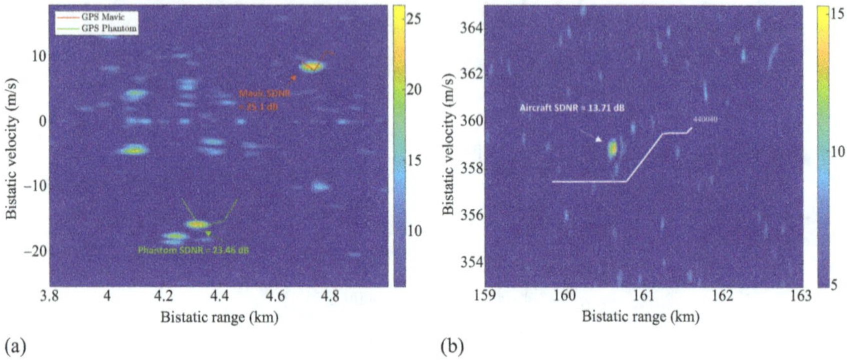

Figure 7.14 Enlarged view of the bistatic range–velocity maps obtained with the processing scheme of Figure 7.13 applied at the same scan against (a) drones at short range and (b) aircraft at long range (ADS-B track in white) [9]. © 2020 IET

when short-range surveillance is sought, due to the limited distances and to the low velocity of the targets of interest, shorter CITs could be employed. It is also worth mentioning that in both cases extended T_{int} would require an effective compensation of the target range and/or Doppler frequency migration effects. The need for the motion compensation will also depend on the considered application. For example, aircraft flying at long ranges are expected to suffer more from a range migration, while small UAVs flying close to the receiver are expected to suffer more from a Doppler frequency walk effect. Effective strategies to deal with these issues are reported in [32,33].

Once the range–velocity maps have been evaluated at all the available surveillance channels, a CA-CFAR detection stage is separately applied to each map to detect targets with a given probability of false alarm P_{fa} and, subsequently, the target is localized in the X–Y plane by estimating its DoA based on the availability of different surveillance antennas. Finally, a tracking stage is applied to reduce the false tracks. For this, it is likely that different techniques should be employed. A standard, low-cost tracking stage can be used for the civil air traffic. However, an advanced algorithm should be considered to account for the complex and unpredictable motion of drones in the short-range surveillance case.

As the first example of the results obtained with the illustrated signal processing stages, Figure 7.14(a) and (b) shows the range–velocity maps obtained for a single data file (scan) from data set 2 described in Section 6.3.2 of Chapter 6 for both the short-range and long-range cases, respectively. Specifically, based on the availability of air truth information (namely GPS or ADSB registrations), we report an enlarged view of the maps around the two targets of interest. All the reported maps have been scaled for the average background level (both clutter residuals and thermal noise) so that each value represents the estimated signal-to-disturbance-plus-noise ratio (SDNR).

In the enlarged view of Figure 7.14(a) we observe that, at the time of the selected scan, two drones are flying in the monitored area with a bistatic range greater than 4 km and opposite bistatic velocities (see the red and green GPS trajectories). Two high peaks with SNR values of 25.1 and 23.46 dB are clearly visible around the Mavic and Phantom GPS tracks, respectively. Simultaneously, a well-focused peak with SNR = 13.71 dB is present at further distance and at very high velocity (enlarged view of Figure 7.14(b)), around the bistatic location [160 km, 359 m/s], which is likely to correspond to the aircraft return (see ADSB track in white). Incidentally, we notice that many other peaks appear in both figures. These are caused by undesired contributions, such as side lobes of the AF, disturbance residuals, as well as other noise sources, or they can correspond to other targets moving in the considered area in the short-range case. The latter issue can become a major problem in applications such as airport surveillance; therefore, the following subsection is devoted to address this problematic issue.

Before concluding this section, a more extensive example is provided in Figure 7.15(a) and (b) that reports the simultaneous short-range and long-range detection results over the bistatic range–velocity plane obtained for the data set 2 used in Chapter 6, Section 6.3.2. Specifically, Figure 7.15(a) reports the same result of Figure 6.11(a) in Chapter 6 where the capability of the DVB-T PR to detect drones up to few kilometres from the receiver is clearly demonstrated. Figure 7.15(b) shows the detection results obtained against the same data set, when the parallel processing scheme used was optimized for wide area surveillance. In each figure, the grey dots represent all the raw detections of the passive sensor across consecutive CIT within the considered acquisition time, the coloured tracks in Figure 7.15(a) are those provided by the GPS receiver on board the cooperative drones, while in Figure 7.15(b) in black we report the available ADSB data for civil air traffic. Eventually, the red plots in Figure 7.15(b) represent the output of the tracking stage. Examining Figure 7.15(b), we notice that the system is able to detect most of the aircraft flying in the surveyed area. Specifically, targets at bistatic ranges up to 150 km are detected with excellent continuity. Notice that the target track

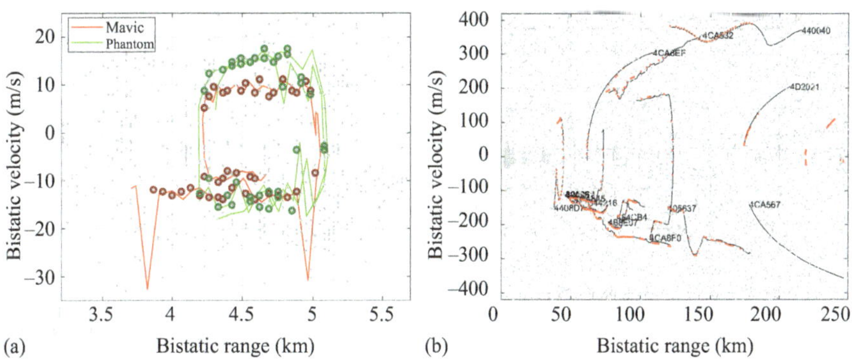

Figure 7.15 Detection results and ground truth tracks for data set 2 at (a) short ranges and (b) long ranges [9]. © 2020 IET

4CA567 is not detected as it falls within the surveillance antenna backlobe. Moreover, many other targets are detected with good continuity at further distances.

7.4.2 Tackling the different target dynamic issues

As previously mentioned, the capability to detect mini UAVs and drones, which are characterized by very small RCS values, might be strongly limited in high-density target scenarios where different target classes coexist. This is the case of an airport terminal area where targets of different sizes may be present.

In principle, the detection of weak target echoes is not prevented by the presence of thermal noise since long CITs could be exploited to increase the target SNR available in the range–Doppler map. In contrast, the weak drone returns could be masked by stronger targets located in the same range cell or in neighbouring range–velocity cells. Note that the power received from drones can be of several dBs lower than larger targets, such as civil air traffic aircrafts. Therefore, in this subsection, we aim at overcoming this issue by taking the advantage of a CLEAN-like multistage algorithm [36] to remove the strongest target contributions, in addition to the use of long CITs.

Figure 7.16 shows the block diagram of the proposed signal processing method. First, as described in Section 7.4.1, the signals collected with the reference and surveillance antennas undergo disturbance cancellation stage and range–velocity map evaluation. We refer to the latter as $\chi[r,v]$, with r and v representing the generic bistatic range–velocity bin. Incidentally, we recall that for targets with a dominant radial velocity, the CAF evaluation might also include an RMC stage [32,37].

Afterwards, a CLEAN-like algorithm [36] is used with the purpose of estimating the dominant target echoes and progressively removing them from the image so that the weak target returns begin to appear.

At the ith iteration of the proposed algorithm, the following steps are performed:

1. At the beginning ($i = 0$), the range–velocity map $\chi^{(0)}[r,v]$ undergoes a first thresholding stage that aims at identifying the target echoes that are stronger than a fixed power threshold μ.
2. Once a strong target echo has been identified at (r_0,v_0), its range cell, i.e. $\chi^{(i)}[r_0,v]$, is fed in input to an inverse fast Fourier transform to move from Doppler frequency to the slow-time domain.
3. The output of this stage undergoes a focusing stage, using the Doppler rate value α^* that results in the maximum target peak, as described in [33].
4. Then, the target contribution is properly filtered and a reverse focusing is applied to recover the current estimate of the target signal in the range-compressed slow-time domain, $x^{(i)}[r,m]$.
5. Finally, the filtered and defocused target contribution is subtracted from the original, and the new cleaned version undergoes an FFT to move from slow-time to Doppler frequency domain.
6. The obtained range–velocity map $\chi^{(i)}[r,v]$ undergoes the thresholding stage described at point 2.

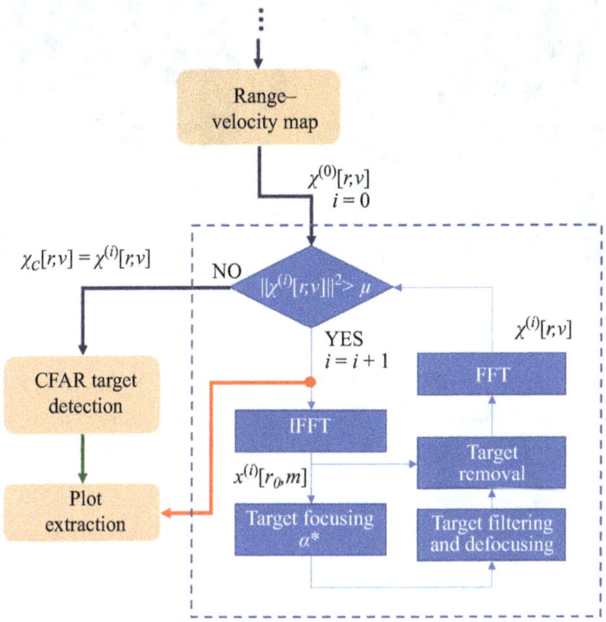

Figure 7.16 Block diagram of the proposed signal processing scheme [10].
© 2020 IEEE. Reprinted, with permission, from Reference [10]

The algorithm should be stopped when the bistatic range–velocity map is likely to contain only weak target echoes, namely when the threshold μ is not exceeded by any range–velocity bin. Note that the threshold value must be properly selected according to the considered surveillance application, namely it will largely depend on the considered target types and on the ranges of interest. We euristically set the threshold to 20 dB above the noise floor in order to guarantee the strongest targets to be easily identified. Once the considered algorithm has been performed, an adaptive target detection stage is applied to the cleaned bistatic range–velocity map $\chi_C[r,v]$, according to the approach described earlier. Finally, the remaining target detections will be input to the plot extraction stage, along with the strongest target ones, resulting from stage three of the algorithm.

Figure 7.17(a) shows the bistatic range–velocity map obtained at a single data file among the data set 1 described in Chapter 6 with $T_{int} = 1$ s. Notice that due to the selected CIT, an RMC stage has been applied, based on the approach described in [32,37]. The map has been scaled for the background level (estimated in a region of the CAF where it is unlikely for targets to be present) in order to measure the target SDNRs. The white line represents the available GPS data for the cooperative drone at the considered time interval. An enlarged view around the drone position is shown in Figure 7.17(b). By observing Figure 7.17(a) and (b), a peak with SDNR = 17.69 dB is visible around the bistatic location [4.81 km, 10 m/s] that is likely to be the return from the drone. Simultaneously, other peaks are visible in

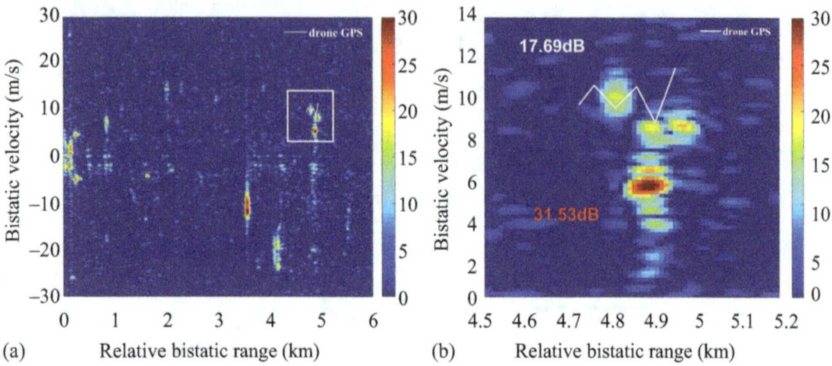

Figure 7.17 Results obtained against real data with $T = 1$ s after RMC: (a) bistatic range–velocity map $\chi(0)[r,v]$; (b) enlarged view of (a). *© 2020 IEEE. Reprinted, with permission, from Reference [10]*

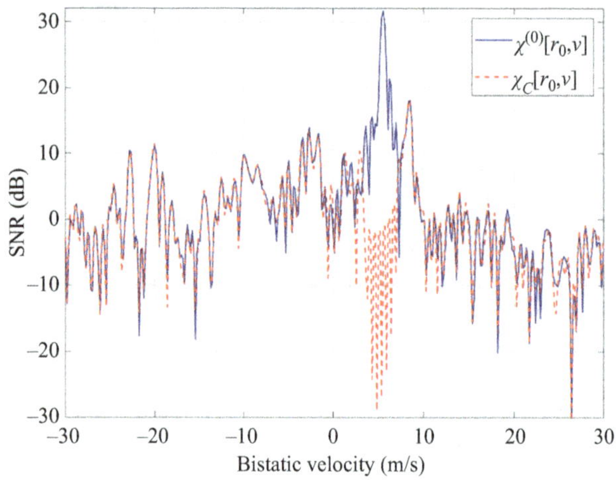

Figure 7.18 Comparison between $\chi^{(0)}[r_0,v]$ and $\chi_C[r_0,v]$ after the first algorithm iteration. *© 2020 IEEE. Reprinted, with permission, from Reference [10]*

Figure 7.17(b), among which the highest one at position [4.88 km, 5.6 m/s] is received with an SDNR = 31.53 dB. In addition, several other peaks are visible in the map which are likely to correspond to other targets moving on the monitored area, e.g. cars, small aircraft, as well as birds. This confirms that airport terminal areas are very likely to show a high-density target scenario and therefore they also show the need for the described advanced strategy to handle this problem.

Figure 7.18 shows the result obtained by applying the algorithm shown in Figure 7.16, using a fixed threshold of 20 dB above the noise power in order to remove

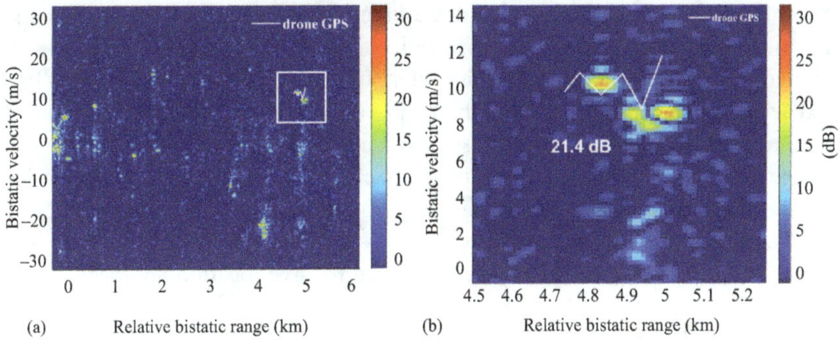

Figure 7.19 Results obtained after the proposed strategy: (a) $\chi_C[r,v]$ after DMC; (b) enlarged view of (a). © 2020 IEEE. Reprinted, with permission, from Reference [10]

the strongest target contribution. Specifically, we compare the Doppler frequency cut obtained in the range bin where the strongest peak around the drone was present, namely at 4.88 km, before and after one iteration of the proposed strategy. We notice that a wide notch is obtained after the proposed algorithm (see dashed red curve) and the strongest target is removed. This is also confirmed by observing Figure 7.19(a) and (b), where we see the cleaned version of the range–velocity map and its enlarged view around the drone, respectively, after undergoing a DMC stage.

From Figure 7.19, we observe that (i) the drone peak is now very visible and can easily be distinguished from the background, largely due to proper compensation of the target motion, (ii) the strongest target is removed from the map using the proposed algorithm, thus enabling the detection of the weaker drone peak (now at 21.4 dB above the noise floor) that was covered by its side lobes, (iii) finally, by comparing Figure 7.19(a) and Figure 7.17(a), many other strong peaks are removed from the map, which is likely to make weaker peaks appear.

Based on the analysis and discussion reported in this section, we can conclude that DVB-T signals, although mainly used for long-range surveillance, can effectively be exploited to counter the threat of drones especially in surveillance applications of sensitive areas such as airport terminal areas. In fact, the possibility of detecting small targets at several kilometres would allow the potential threat to be identified and started to be tracked before other subsystems, although with less accuracy than that of other waveforms. Subsequently, this useful information could be used as input to subsystems that exploit different signals, more appropriate for the surveillance of these targets at very short range as well as their classification.

7.5 Passive radar based on WiFi

As mentioned in Section 7.2, when considering short range surveillance applications, the WiFi transmissions represent an attractive solution since these are widely accessible sources of opportunity that are able to provide reasonable wide bandwidth (i.e. range resolution) and reasonable high-velocity accuracy (due to high exploited frequencies). While these characteristics are in common with the DVB-S-based sensors, the exploitation of WiFi signals also provides additional advantages for the resulting multiband, multisensory system. Among them, the availability of commercial antenna elements of compact size allows the implementation of low cost receiving arrays to estimate target angle both in azimuth and in elevation. In addition, since many commercial drones emit signals in the WiFi band, they open the way to the joint use of PR and device-based passive location approaches.

7.5.1 WiFi-based PR receiver architecture and processing schemes

In the WiFi case, the source of the signal of opportunity is typically close to the receiver location (order of metres or tenths of metres) so that the direct signal and its multipaths are strong within the signal collected by the surveillance antennas and the corresponding receiving channels. This implies that

1. an effective cancellation technique must be applied before building the coherent integrations providing the range–velocity maps to remove both the direct signal and its multipaths/clutter;
2. there are two different ways to obtain a clean reference waveform [28,29,38]:
   (i) If the WiFi router (transmitter) is accessible and it is possible to introduce a directional coupler between the access point and its antenna, a solution is to connect it to a receiving channel (downconverter and A/D). This is obviously a possibility enabled by the short distance that allows a small but clean fraction of the transmitted waveform to be acquired. It is worth mentioning that while providing quite a good copy of the signal of opportunity, this approach still requires a dedicated receiving channel to be used to collect the reference signal. We explicitly note that in this case, by using a single four-channel receiver (commonly available among COTS SDR), the sensor can feature up to three surveillance channels, including one of the receiving channels dedicated to the reference signal.
   (ii) If the WiFi router in not accessible, the reference waveform must be extracted from the signal collected by a receiving antenna. However, in the WiFi geometry the strong level of the direct signal in the samples collected by the surveillance antenna allows the direct extraction of a clean waveform. Specifically, the transmitted signal can be reconstructed by demodulating and re-modulating the received signal according to the IEEE 802.11 standard. This approach may suffer from reconstruction errors but it avoids the need for a dedicated reference channel. In this case using a single four-channel receiver, it is possible to build four parallel surveillance channels. These could be arranged into a double-node system with nodes appropriately displaced to cover a wide desired area but connected to the same SDR and being each node equipped with an interferometric pair of surveillance antennas to enable the estimation of the DoA.

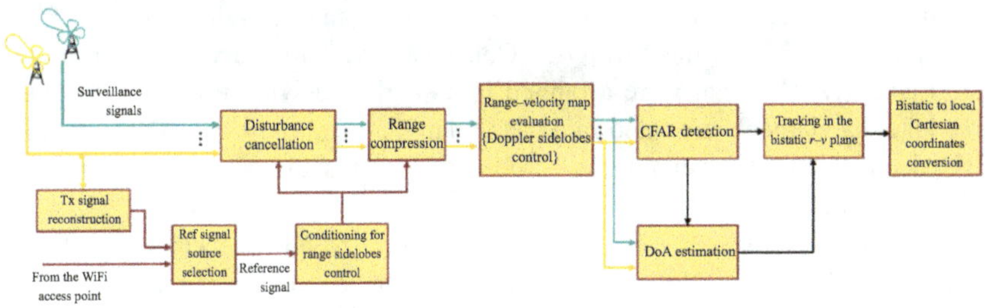

Figure 7.20 WiFi-based PR processing scheme

The general processing scheme for WiFi-based PR is shown in Figure 7.20, when multiple surveillance antennas are employed. For illustration purposes, it encompasses both strategies described earlier for the reference signal recovery. When the transmitted waveform is not directly acquired from the transmitter, the reconstruction of the transmitted waveform from the surveillance channels is the first operation. Hence subsequent stages are independent of the reference signal source selection.

Since the WiFi transmissions are of a pulsed type, the bistatic range–velocity map is evaluated by first cross-correlating the surveillance signals with the available reference signal on a packet-by-packet basis. This stage basically implements a range compression at each signal pulse corresponding to a WiFi packet. Subsequently the obtained results are coherently integrated over a set of consecutive packets. In fact, the range compressed pulses included in the selected CIT undergo an FFT stage in order to obtain the bistatic range–Doppler map. Proper techniques for the range [28] and Doppler [29], side lobes control, are applied since the AF of the WiFi signals is characterized by high side lobes in both range and Doppler dimensions [30].

The processing scheme must typically include the clutter/multipath cancellation stage for the disturbance removal. In particular, the ECA-S has been shown especially effective in the past [35].

After the clutter cancelled, range/velocity maps are evaluated at each surveillance channel, a CFAR threshold is separately applied to them. Finally, a decentralized strategy is exploited to combine the detections of the surveillance antennas belonging to the same receiving node, e.g. target detection is declared only for targets that exceed the threshold at all the surveillance channels. The detection stage also provides a first target localization over the bistatic range–velocity plane. Thereafter, a conventional Kalman tracking algorithm can be applied to reduce the false alarms while yielding more accurate range–velocity measurements. The target localization in local Cartesian coordinates is finally obtained by exploiting the range and DoA measurements provided by multiple surveillance antennas.

7.5.2 Experimental drones detection and 3D localization with WiFi-based PR

Dedicated test campaigns have been conducted to verify the potentials of the considered system to detect and localize a very low RCS drone in 3D. In this experiment, a quad-channel receiver was employed and the reference signal was directly measured from the WiFi AP. Consequently, three surveillance channels were made available that were arranged in a single receiving node, connected to three antenna elements mounted in a triangular shape to allow the 3D localization of the detected target, by exploiting both the filtered bistatic range and the azimuth and elevation angles estimates.

A sketch of the test area together with the WiFi-based PR receiver location is illustrated in Figure 7.21(a). Specifically, the considered area comprised mostly grass surface and was approximately 40-m long and 9-m wide. A very small drone is used as a cooperative target (see Figure 7.21(b)). It is made of carbon fibre and expanded foam and has dimensions 60 cm×60 cm×9 cm.

In this example, the drone moves from point B (Figure 7.21(a)), at about 30 m from the PR sensor, to point A (towards the receiver location) going up and down, namely changing its height across the acquisition time. The corresponding detection results over the range–velocity plane collected during the whole acquisition time are reported with 'x' markers in Figure 7.22(a).

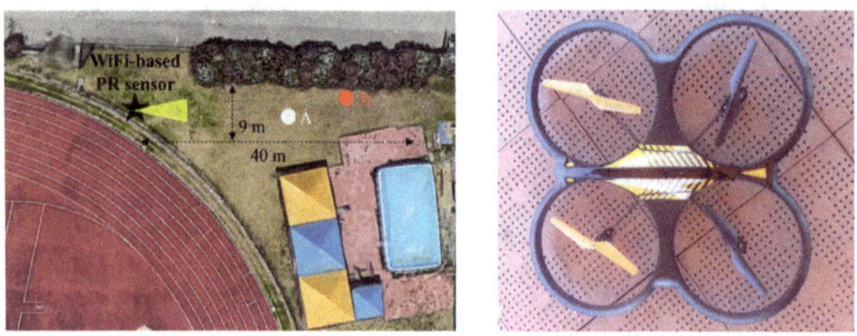

(a) (b)

Figure 7.21 Scenario of the test campaign against drones: (a) sketch of the test area; (b) picture of the used drone [13]. © 2017 IET

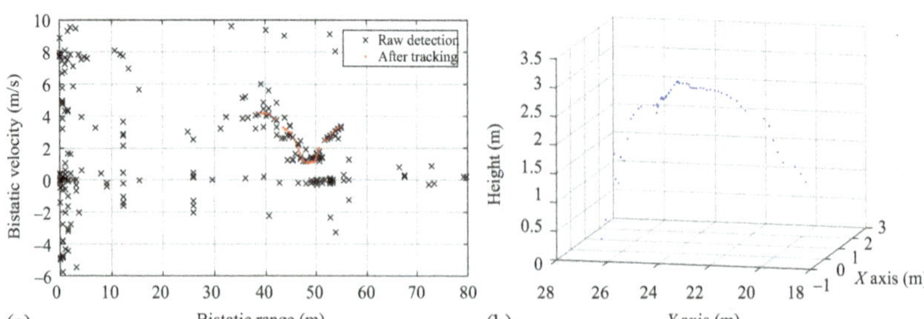

(a) (b)

Figure 7.22 Test with a small RCS drone: (a) detection results over the range–velocity plane; (b) 3D localization results [13]. © 2017 IET

As can be seen, a cloud is visible at bistatic ranges between 40 and 55 m. These detections correspond to the target returns since they are clearly compliant with the test geometry. The isolated markers in Figure 7.22(a) are false alarms due to thermal noise and disturbance residuals that can be discarded easily by a tracking algorithm; to this purpose, we used a Kalman tracking filter operating in the range–velocity plane [39]. Specifically, the sequence of red plots represents the range–velocity estimations associated with the moving target. Then, the phase difference among the surveillance antennas at the target detection point is exploited to evaluate the azimuth and elevation angles. For example, in [20] such sources have been considered in PR systems for small airport surveillance. Additionally, the possibility to retrieve the height of the detected targets by exploiting multiple receiving antennas has been investigated [13]. By employing a set of three antennas with both horizontal and vertical displacement, an interferometric approach can be exploited to estimate both the azimuth and the elevation angles of the target echoes. This allows a full localization in three dimensions: height estimation represents strategic information for an effective classification of the different types of targets/threats. Again, we observe that using a standard four-channel receiver provides a 3D node (one extracted reference channel + three surveillance channels) with the case (a) scheme.

The 3D localization results obtained with such a four-channel node are shown in Figure 7.22(b). We show the raw plot positions; therefore, more continuous tracks could be obtained by applying an additional tracking stage in the Cartesian plane to follow the evolution of the x, y, z coordinates.

As can be seen, the obtained results show that the passive sensor is able to correctly detect and localize in 3D the small flying object along its trajectory. The reported experimental results support the practical applicability of the WiFi-based PR concept for such advanced security applications in private premises and/or public areas as a viable alternative to active sensors, at least in the short range.

7.5.3 WiFi joint operation of passive radar and passive source location

While the WiFi-based PR allows effective detection and localization of the drone, a further improved system could be obtained by combining the multisource PR. In addition, we observe that the WiFi ISM frequency band offers also another interesting opportunity. In fact, many drones have an on-board transmitter, operating in this specific frequency range that is used for communications between drone and controller, to download drone's flight data information, as well as real-time video.

In this case, the WiFi signals allow a target to be localized even through its direct emissions using passive source location (PSL) techniques. The inherent features of the two strategies (PR and PSL) and the results obtained from their comparison on experimental data have shown an interesting complementarity between them. This result opens the way for a joint exploitation of the two localization strategies and to the development of appropriate sensor fusion techniques. For the sake of clarity, we briefly summarize the complementary aspects of PR and PSL in Table 7.3.

Table 7.3 PR and PSL features

Passive radar (PR)	**Passive source location (PSL)**
Device-free localization	Device-based localization
Higher computational cost	Lower computational cost
Effective for moving targets	Potentially inaccurate for moving targets
No detection of stationary targets	Stationary targets can be detected and localized
Closely spaced targets cannot be discriminated	Closely spaced targets can be discriminated due to the device MAC address

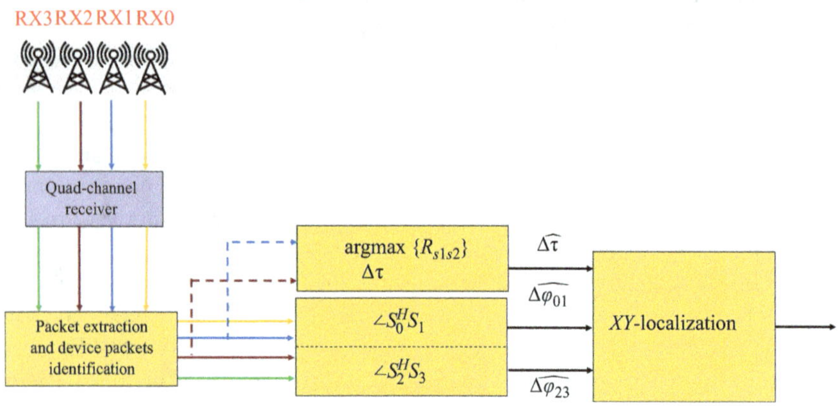

Figure 7.23 PSL processing scheme [14]. © 2020 IEEE. Reprinted, with permission, from Reference [14]

The general processing scheme for WiFi-based PSL is shown in Figure 7.23, where the configuration with four receiving channels is represented (see [14] for details). After the acquisition of the WiFi signals through the available surveillance antennas and the subsequent extraction and identification of the WiFi packets based on their MAC address, the DoA, or equivalently the angle of arrival (AoA), and the TDoA are estimated directly on the received WiFi packets. In particular, the phase difference between the signals collected by two closely spaced antennas is used to estimate the AoA of the target, while the TDoA is obtained by searching the peak of the cross-correlation between the signals received by two distant receiving nodes. As apparent, when using four receiving channels, two AoA measurements and one TDoA could be obtained by properly arranging two receiving nodes each one equipped with two antennas. When three receiving channels are available, the PSL is still able to estimate the target position in the *XY* plane through the combination of one AoA measurement and one TDoA measurement, provided that two nodes are deployed being just one equipped with a couple of antennas on receive. The combination of the available measurements is geometrically obtained through the intersection of a line (AoA) and a hyperbola (TDoA).

In contrast to the PR, the PSL is a 'device-based' localization strategy. It could be potentially inaccurate for moving targets. On the other hand, it is an interesting solution for stationary objects localization, and it allows the unambiguous association of the transmission to a specific target, based on the device MAC address, so that even very closely spaced targets can be discriminated. As illustrated earlier, the PSL estimates the target position directly on the signals transmitted by the drone and received by the surveillance antennas. These signals are therefore characterized by higher SNR conditions, which provide the possibility to reduce the complexity of the processing scheme and, in turn, to decrease the computational cost of the final localization, since it avoids all the steps required by the PR for the extraction of the target echoes, as evident comparing Figure 7.23 with Figure 7.20.

The joint exploitation of PR and PSL processing schemes was tested on experimental data collected in dedicated acquisition campaigns. Details about the considered scenario and the experimental setup are provided in [14]. A commercial drone was used both as target of opportunity for the PR and the PSL systems. Specifically, the DJI Mavic Pro was used. It is a lightweight (about 730 g) drone with small size (about 30 cm×25 cm×8 cm), WiFi 802.11a/b/g/n/ac connectivity, integrated GPS/GLONASS positioning system for flight control and automatic return to take-off position. The streaming video from drone to controller was exploited as device transmissions to perform target detection and localization with the PSL system. Drone localization was performed by means of a proper combination of one range and one AoA measurement for the PR sensor, while the PSL exploits one TDoA and one AoA measurement to reach the same goal. In the first example reported here, the drone moves away from the receiving antennas starting from a distance of about 15 m, then stops and hovers for some seconds at a distance of about 50 m, then it flies back towards the receiving antennas up to a distance of about 30 m.

Figure 7.24 shows the previously mentioned measurements for the experimental test. For each subplot, the ground truth obtained from drone GPS recording is reported in grey. Figure 7.24(a) shows the bistatic range (blue crosses) obtained by the PR sensor after the application of a tracking stage in the range–velocity plane. As can be seen, the PR sensor does not provide a full coverage of the drone flight, specifically in the period of time when the target is in hovering. In Figure 24 (b), the TDoA estimated by the PSL sensor is compared with the ground truth. In this case, the obtained measurements do not depend on the motion status of the drone during its flight. Finally, in Figure 24(c) the AoA estimations from the two sensors are compared: as is apparent, the two sensors provide comparable measurements. Moreover, although there is a very small time interval with no detections (after about 15 s), most of the time one sensor compensates for the lack of measurements from the other one.

As can be seen, the complementarity between the two strategies (PR and PSL) is quite evident from this example on experimental data, especially when they operate on a typical move–stop–move target. In particular, the PR provides a good

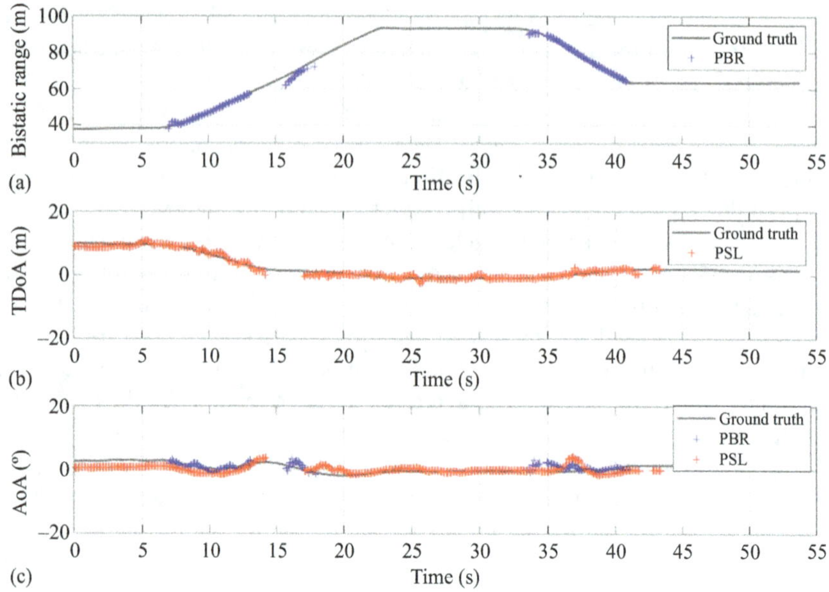

Figure 7.24 Sensor measurements compared with ground truth: (a) bistatic range, (b) TDoA, (c) AoA. © 2020 IEEE. Reprinted, with permission, from Reference [14]

accuracy for moving targets due to the high transmission rate of the AP. In contrast, measurements are absent for static targets. On the other hand, the PSL provides good accuracy localization for both moving and static targets, when the device is active and transmits packets with a high rate. In contrast, PSL measurements can be not enough to track a target if its device does not transmit with regularity. As a consequence, the fusion of PR and PSL measurements is expected to be able to provide significant improvement with respect to the single sensor. The main advantages of the proposed sensor fusion approach are the increase of the positioning accuracy and the decrease of the time where the target is not detected and localized.

Figure 7.25 shows the improvement in localization capability of the sensor fusion technique with respect to the single sensor results, when applying a Kalman filter based on nearly constant velocity model. First of all, the KF is applied to the PR and the PSL sensors, separately. The results are reported in blue for the PR, in red for the PSL and in green for the data fusion techniques in order to investigate the possible improvement coming from the joint exploitation of the two sensors. All the reported curves approximate with reasonable accuracy of the drone's trajectory in the *XY* plane. The green curve is on average the closest one to the real path flew by the drone. This can be better observed in Figure 7.26 that shows the instantaneous positioning errors of the three examined approaches (with respect to

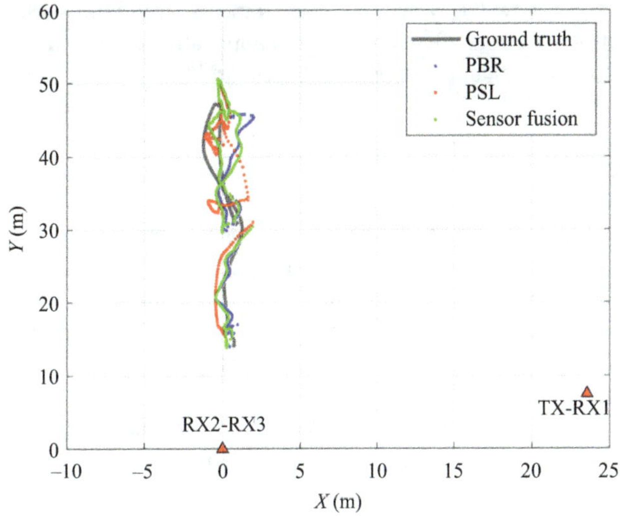

Figure 7.25 Drone XY-localization after Kalman filter application. © 2020 IEEE. Reprinted, with permission, from Reference [14]

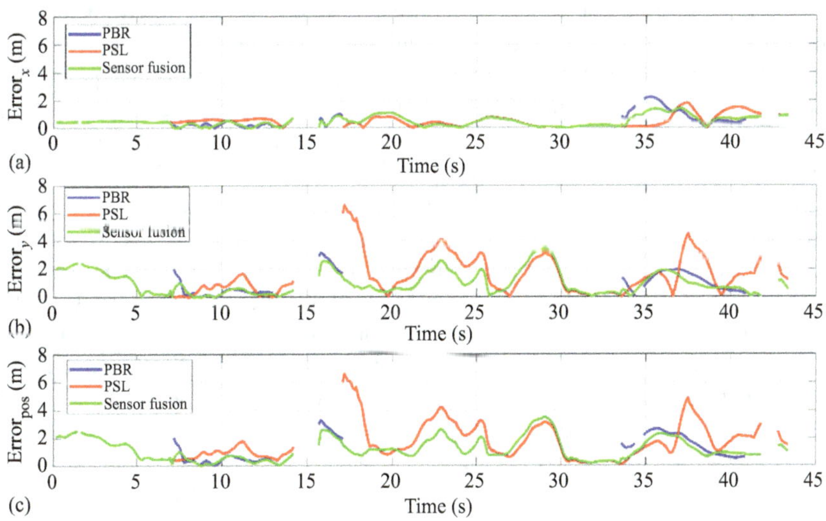

Figure 7.26 Positioning errors after Kalman filter application: (a) x component, (b) y component, (c) overall positioning error. © 2020 IEEE. Reprinted, with permission, from Reference [14]

the ground truth) for the presented experimental test. In particular, the single component errors are reported in Figure 7.26(a) and (b), for the x component and the y component, respectively, while the combination of the two coordinates errors is considered in Figure 7.26(c). In addition to the coverage improvement described before, the results show that when the measurements from both sensors are available, the positioning errors after sensor fusion application are significantly reduced with respect to those obtained with a single sensor.

7.6 Conclusions

In this chapter, the concept of a multiband PR system has been investigated. The multisensory PR architecture has been introduced and appropriate waveforms of opportunity have been selected to provide the system with multitasking capabilities. Specifically, the proposed system exploits DVB-T signals to guarantee the required coverage against drones and UAV, whereas DVB-S and WiFi-based systems offer improved accuracy in terms of target localization and characterization at shorter range. Finally, the WiFi band operation allows to include in the fusion logic also device-based localization techniques based on the radiation spontaneously emitted by commercial drones.

The capabilities of the different subsystems have been demonstrated by means of experimental tests using cooperative targets, which allowed us to verify the benefits conveyed by the selected waveforms of opportunity in the conceived multisensory system. In particular, while the presented analysis and the experimental demonstration confirmed the multiband PR sensor as a way to include the nice features provided by the different available waveforms of opportunity, a perspective improvement is foreseen in the development of the signal processing techniques to fuse the multiple sensors outputs. The basic approach, presented in the earlier discussion, consists in using the DVB-T PR for the drone search function, thus providing the cueing to both WiFi PR and DVB-S PR that operate in parallel to obtain accurate range–velocity bistatic measurements and angle measurements, each one in its own projection geometry. Eventually, the two sensors can be exploited with their high resolution to characterize/classify the target.

Perspective improvements in the fusion technique include algorithms to merge the measurements obtained at different resolutions and accuracies by the three sensors, which also operate intrinsically on different projected bistatic domains on asynchronous signal waveforms, with different update rates. The joint tracking of the drone targets obtained by an optimized fusion of the three-sensor measurements is expected to provide improved accuracy, as well as a significant robustness to (i) availability of the signals of opportunity, (ii) propagation conditions with respect to the individual sensor, (iii) target RCS value for the specific bistatic angle, carrier frequency and polarization, (iv) possible target cancellation by the cancellation filter intended to remove direct signal and interference, when the bistatic geometry shows the absence of relative velocity.

Advanced fusion techniques are also foreseen to be developed for the target characterization and classification, based on the available range and velocity resolution as well as on the polarization. In this line goes also the potential exploitation of the micro-Doppler signatures that might be available especially with the high-frequency DVB-S PR. Also in this case we expect to see significant capability improvements to become available in the future, with the development of appropriate signal processing techniques.

In addition to the foreseen development of the signal processing techniques, joint experimental tests, to be carried out in the near future when the restrictions for the containment of the COVID-19 pandemic will allow, are expected to highlight both the nice features of the multisensory system and the open challenges or needs.

Finally, the near-term evolution of the communication waveforms clearly includes both an improvement in the availability of satellite communication signals, which will include shortly also broadcast directed to handheld/mobile devices and the presence of emerging ubiquitous mobile communication systems with the 5G/6G mobile devices. The former is expected to provide additional satellite signals of opportunity with wider coverage and sensibly higher power densities, allowing to enlarge the surveillance areas of the DVB-S sensor, while keeping its nice features related to the high carrier frequency and to the long CIT available with the stationary emitter. The latter will certainly provide a dense network of emitters of opportunity, operating at different carrier frequencies – up to very high values around 60 GHz in the very short range – with generally variable, but possibly sensibly large frequency bandwidth. Both types of signals of opportunity are high-quality candidates for the inclusion in the multiband PR sensor, or as replacements for the present sensors.

References

[1] De Cubber G., 'Explosive drones: how to deal with this new threat?'. *Proceedings of the 9th International Workshop on Measurement, Prevention, Protection and Management of CBRN Risks*; Belgium, Apr 2019, pp. 1–8.

[2] Ritchie M., Fioranelli F., and Borrion H. 'Micro UAV crime prevention: can we help Princess Leia?' in Savona, B. L. (ed.). *Crime Prevention in the 21st Century*. New York, NY: Springer; 2017, pp. 359–376.

[3] Shi X., Yang C., Xie W., Liang C., Shi Z., and Chen J. 'Anti-drone system with multiple surveillance technologies: architecture, implementation, and challenges'. *IEEE Communications Magazine*. 2018; vol. 56(4), pp. 68–74.

[4] Liu Y., Wan X., Tang H., Li J., Cheng Y., and Zhang H. 'Digital television based passive bistatic radar system for drone detection'. *Proceedings of the 2017 IEEE Radar Conference*; Seattle, WA, USA, May 2017, pp. 1493–1497.

[5] Poullin D. 'Countering illegal UAV flights: passive DVB radar potentiality'. *Proceedings of the 19th International Radar Symposium (IRS)*; Bonn, Germany, Jun 2018, pp. 1–10.

[6] Jarabo-Amores M. P., Mata-Moya D., Gómez-del-Hoyo P. J., *et al.* 'Drone detection feasibility with passive radars'. *Proceedings of the 15th European Radar Conf.*; Madrid, Spain, Sep 2018, pp. 313–316.

[7] del-Rey-Maestre N., Mata-Moya D., Jarabo-Amores M. P., Gómez-del-Hoyo P. J., and Rosado-Sanz J. 'Optimum beamforming to improve UAV's detection using DVB-T passive radars'. *Proceedings of the 2019 IEEE International Radar Conference*; Toulon, France, Sep 2019, pp. 1–6.

[8] Martelli T., Colone F., and Cardinali R. 'Simultaneous short and long range surveillance of drones and aircrafts with DVB-T based Passive Radar'. *Proceedings of the 2019 IEEE International Radar Conference*; Toulon, France, Sep 2019, pp. 1–6.

[9] Martelli T., Colone F., and Cardinali R. 'DVB-T based Passive Radar for simultaneous counter drone operations and civil air traffic surveillance'. *IET Radar, Sonar & Navigation*. 2020; vol. 14(4), pp. 505–515.

[10] Martelli T., Filippini F., and Colone F. 'Tackling the different target dynamics issues in counter drone operations using passive radar'. *Proceedings of the 2020 International Radar Conference*; Washington, DC, USA, Apr–May 2020, pp. 512–517.

[11] Schupbach C., Patry C., Maasdorp F., Boniger U., and Wellig P. 'Micro-UAV detection using DAB-based passive radar'. *Proceedings of the 2017 IEEE Radar Conference*; Seattle, WA, USA, May 2017, pp. 1037–1040.

[12] Knoedler B., Zemmari R., and Koch W. 'On the detection of small UAV using a GSM passive coherent location system'. *Proceedings of the 2016 17th International Radar Symposium (IRS)*; Krakow, Poland, May 2016, pp. 1–4.

[13] Martelli T., Murgia F., Colone F., Bongioanni C., and Lombardo P. 'Detection and 3D localization of ultralight aircrafts and drones with a WiFi-based passive radar'. *Proceedings of the 2017 International Conference on Radar Systems (Radar 2017)*; Belfast, UK, Oct 2017, pp. 1–6.

[14] Milani I., Bongioanni C., Colone F., and Lombardo P. 'Fusing active and passive measurements for drone localization'. *Proceedings of the 2020 International Radar Symposium (IRS)*; Warsaw, Poland, Oct 2020, pp. 245–249.

[15] Ilioudis C. V., Cao J., Theodoru I., *et al.* 'GNSS based passive radar for UAV monitoring'. *Proceedings of the 2019 IEEE Radar Conference*; Boston, MA, USA, Apr 2019, pp. 1–6.

[16] Cabrera O., Bongioanni C., Filippini F., Sarabakha O., Colone F., and Lombardo P. 'Detecting drones and human beings with DVB-S based COTS passive radar for short-range surveillance'. *Proceedings of the 2020 International Radar Conference*; Washington, DC, USA, Apr–May 2020, pp. 37–42.

[17] Martelli T., Cabrera O., Colone F., and Lombardo P. 'Exploitation of long coherent integration times to improve drone detection in DVB-S based passive radar'. *Proceedings of the 2020 IEEE Radar Conference (RadarConf20)*; Florence, Italy, Sep 2020, pp. 1–6.

[18] Filippini F., Cabrera O., Bongioanni C., Colone F., and Lombardo P. 'DVB-S based passive radar for short range security application'. *Presented at 2021 IEEE Radar Conference*; Atlanta, GA, USA, 2021.

[19] Howland P. E. 'Editorial: passive radar systems'. *IEE Proceedings on Radar, Sonar and Navigation*. 2005; vol. 152(3), pp. 105–106.

[20] Lombardo P. and Colone F. 'Advanced processing methods for passive bistatic radar' in Melvin, W. L. and Scheer, J. A. (eds.). *Principles of Modern Radar: Advanced Radar Techniques*. Raleigh, NC: SciTech Publishing; 2012, pp. 739–821.

[21] Griffiths H. D. and Baker C. J. *Introduction to Passive Radar*. Norwood, MA: Artech House; 2017.

[22] Pisciottano I., Cristallini D., and Pastina D. 'Maritime target imaging via simultaneous DVB-T and DVB-S passive ISAR'. *IET Radar, Sonar and Navigation*. 2019; vol. 13(9), pp. 1479–1487.

[23] Cabrera O., Bongioanni C., Colone F., and Lombardo P. 'Non-coherent DVB-S passive radar demonstrator'. *Proceedings of the 2020 International Radar Symposium (IRS)*; Warsaw, Poland, Oct 2020, pp. 228–231.

[24] Colone F., Langellotti D., and Lombardo P. 'DVB-T signal ambiguity function control for passive radars'. *IEEE Transactions on Aerospace and Electronic Systems*. 2014; vol. 50(1), pp. 329–347.

[25] Kuschel H., Heckenbach J., O'Hagan D., and Ummenhofer M. 'A hybrid multi-frequency passive radar concept for medium range air surveillance'. *Proceedings of the 2011 Microwaves, Radar and Remote Sensing Symposium*; Kiev, Ukraine, Oct 2011, pp. 275–279.

[26] Poullin D. and Flecheux M. 'Passive 3D tracking of low altitude targets using DVB (SFN broadcasters)'. *IEEE Aerospace and Electronics Systems Magazine*. 2012; vol. 27(11), pp. 36–41.

[27] Martelli T., Colone F., Tilli E., and Di Lallo A. 'Multi-frequency target detection techniques for DVB-T based passive radar sensors'. *Sensors*. 2016; vol. 16(10), pp. 1–25.

[28] Colone F., Falcone P., Bongioanni C., and Lombardo P., 'WiFi-based passive bistatic radar: data processing schemes and experimental results'. *IEEE Transactions on Aerospace and Electronic Systems*. 2012; vol. 48(2), pp. 1061–1079.

[29] Falcone P., Colone F., and Lombardo P., 'Potentialities and challenges of WiFi-based passive radar'. *IEEE Aerospace and Electronic Systems Magazine*. 2012; vol. 27(11), pp. 15–26.

[30] Colone F., Woodbridge K., Guo H., Mason D., and Baker C. J. 'Ambiguity function analysis of wireless LAN transmissions for passive radar'. *IEEE Transactions on Aerospace and Electronic Systems*. 2011; vol. 47(1), pp. 240–264.

[31] Colone F., O'Hagan D. W., Lombardo P., and Baker C. J. 'A multistage processing algorithm for disturbance removal and target detection in passive bistatic radar'. *IEEE Transactions on Aerospace and Electronic Systems*. 2009; vol. 45(2), pp. 698–722.

[32] Pignol F., Colone F., and Martelli T. 'Lagrange-polynomial-interpolation-based keystone transform for a passive radar'. *IEEE Transactions on Aerospace and Electronic Systems*. 2018; vol. 54(3), pp. 1151–1167.

[33] Filippini F., Martelli T., Colone F., and Cardinali R. 'Exploiting long coherent integration times in DVB-T based passive radar systems'. *Proceedings of the 2019 IEEE Radar Conference*; Boston, MA, USA, Apr 2019, pp. 1–6.

[34] Cabrera O., Bongioanni C., Colone F., and Lombardo P. 'Comparing phase-locked and non-phase-locked architectures for dual-channel DVB-S passive

radar'. *Proceedings of the 2020 17th European Radar Conference (EuRAD)*; Utrecht, the Netherlands, Jan 2021, pp. 350–353.
[35] Colone F., Palmarini C., Martelli T., and Tilli E. 'Sliding extensive cancellation algorithm for disturbance removal in passive radar'. *IEEE Transactions on Aerospace and Electronic Systems*. 2016; vol. 52(3), pp. 1309–1326.
[36] Jenho T. and Steinberg B. D. 'Reduction of sidelobe and speckle artifacts in microwave imaging: the CLEAN technique'. *IEEE Transactions on Antennas and Propagation*. 1988; vol. 36(4), pp. 543–556.
[37] Martelli T., Filippini F., Pignol F., Colone F., and Cardinali R. 'Computationally effective range migration compensation in PCL systems for maritime surveillance'. *Proceedings of the 2018 IEEE Radar Conference (RadarConf18)*; Oklahoma City, OK, USA, Apr 2018, pp. 1406–1411.
[38] Colone F., Martelli T., Bongioanni C., Pastina D., and Lombardo P. 'WiFi-based PCL for monitoring private airfields'. *IEEE Aerospace and Electronic Systems Magazine*. 2017; vol. 32(2), pp. 22–29.
[39] Falcone P., Colone F., Macera A., and Lombardo P. 'Two-dimensional location of moving targets within local areas using WiFi-based multistatic passive radar'. *IET Radar, Sonar & Navigation*. 2014; vol. 8(2), pp. 123–131.

Chapter 8
GNSS-based UAV detection

Christos V. Ilioudis[1], *Carmine Clemente*[1] *and John J. Soraghan*[1]

Among the potential illuminators of opportunity (IO) that can be used in a passive radar (PR) aimed at unmanned aerial vehicle (UAV) monitoring, Global Navigation Satellite Systems (GNSS) represent an interesting alternative providing a global coverage opportunity, known signal structure and scalability. This chapter investigates the capabilities of a GNSS-based PR for UAV detection by looking at relevant aspects such as coverage, power budgets and processing schemes. Finally, the chapter experimentally assesses the capability of this type of system to detect UAVs.

8.1 Introduction

A PR exploits IO in order to perform radar tasks. Recently, the use of PRs for UAV monitoring has attracted significant interest in both the research and industry sectors [1–5]. A common drawback of PR solutions is that the surveillance coverage is reliant on the presence of a suitable IO; thus, it requires ad hoc system design and calibrations before deployment. To enable portability in such systems, GNSS IOs present an interesting alternative providing a global coverage opportunity, known signal structure and scalability. GNSS-based PR systems have been proposed employing either spaceborne or earth-based receivers. In [6,7], the effective use of GNSS as an IO has been demonstrated for ocean reflectometry applications. An airborne GNSS passive bistatic SAR was proposed in [8] for imaging using a synchronisation algorithm to generate the reference signal at the receiver. In a similar approach, the authors in [9,10] proposed a GNSS PR system design for maritime target detection.

A particular case of bistatic radars is the forward scattering (FS) configuration occurring when the target crosses (or is very close to) the line of sight between the illuminator and the receiver [11]. A benefit of the FS configuration is that it results in an enhancement of the radar cross section (RCS) of the target, meaning that FS

[1]Department of EEE, Centre for Signal and Image Processing (CESIP), University of Strathclyde, Glasgow, United Kingdom

radars can be particularly attractive for targets with low reflectivity, as their FS-RCS is mainly dependent on the silhouette of the target and not by its material [12,13]. Due to their particular geometry and lack of range resolution, FS radar was initially investigated for the design of electromagnetic fences [14] and their detection capabilities [15–17]. Recently, advanced signal processing techniques for target localisation and velocity estimation [18–20] and classification [21,22] have being proposed. Furthermore, their bistatic configuration makes them suitable for PR operations [23] with various illuminators such as GNSS [24–26], WiFi [27–29], GSM [30] and broadcast signals [31].

In this chapter, the concept of a GNSS-based PR for UAV detection, localisation and motion estimation is presented for both backscattering (BS) and FS configurations. Specifically, a link-budget analysis to determine the maximum detectable range using different GNSS constellations is conducted for two different case studies. An appropriate signal model and system design are provided, while target parameter estimation based on the ambiguity function (AF) is discussed for a multiple-input–single-output (MISO) configuration. Additionally, the feasibility of the system is validated through experimental acquisitions.

8.2 GNSS-based PR coverage

A PR can only operate under the presence of one or more IOs as they do not comprise dedicated transmitters. The main drawback of using such an approach is that the overall performance of the system is highly dependent on the IOs operational characteristics. In the case of GNSS, this means that the non-stationarity of the IOs needs to be taken into account. Particularly, for GNSS IOs the position and availability of illuminators shall be considered to assess the performance of the system, while the relative motion needs to be taken into account in the signal processing chain. In this section, two GNSS constellations, Global Positioning System (GPS) and Galileo, are examined in order to assess their availability and maximum detectable range that they can offer for UAV targets detection.

8.2.1 Back and forward scattering RCS of UAVs

In order to understand the expected performance of the PR, a coarse estimate of the UAV's RCS should be available. However, the RCS will vary depending on the topology of the PR. The examined system topology is illustrated in Figure 8.1. While the depicted configuration is composed of multiple satellites, it is useful to initially examine a bistatic or single-input–single-output system comprising a single satellite IO Tx, a ground passive receiver with two channels Hx and Rx and an airborne target Tg, with their distances being denoted as D_S satellite to receiver, D_T satellite to target and D_R target to receiver. Initially the signal is emitted from the satellite to earth. A portion of it is directly received by Hx, while another part first reaches Tg and is then scattered at different directions. Depending on the geometry, the passive receiver Rx will capture the echoes from the target from different scattering directions. Specifically, based on the bistatic angle θ, three distinct RCS

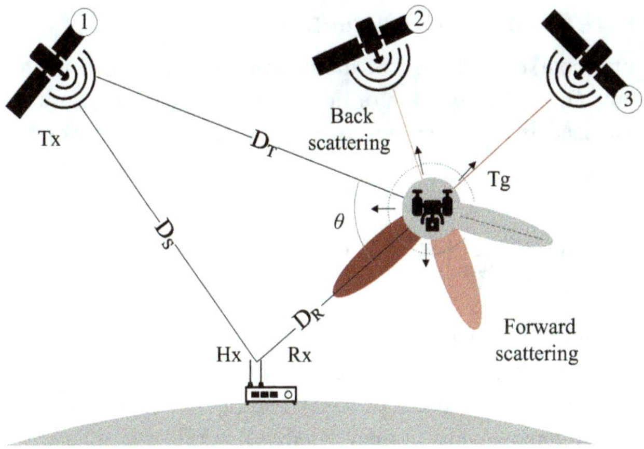

Figure 8.1 Passive GNSS radar topology comprising three satellite IOs Tx, a UAV target Tg and a two-channel receiver Hx and Rx

regions are distinguished: pseudo-monostatic BS at $\theta < 20°$, BS at $20° < \theta < 140°$ and FS at $\theta > 140°$. Compared to the other RCS regions, FS does not offer target ranging; however, it can significantly increase the target's RCS in the forward direction [11]. Considering more than one IOs results in a MISO system. In such a configuration, the receiver can utilise returns from multiple satellites with the target observed in BS or FS region depending on the considered IO.

The BS–RCS for miniature UAV targets has been widely discussed in [32,33]. As the BS–RCS of such complex targets is difficult to estimate for every bistatic angle and target orientation, in this work the BS–RCS is modelled by the RCS of a sphere:

$$\sigma_{\mathrm{BS}} = \pi r^2 \tag{8.1}$$

where r is the radius of the sphere. As shown in [33], spheres with radius close to 8 cm can well approximate the BS–RCS of a Parrot drone target at the 2.4 GHz WiFi band. An overview of the FS-RCS phenomenology is provided in [34], while its applicability for miniature UAV detection was investigated in [35]. The maximum FS-RCS in the optical region can be calculated from [36]

$$\sigma_{\mathrm{FS,\ max}} = \frac{4\pi A^2}{\lambda^2} \tag{8.2}$$

where A is the physical area of the target. This maximum value can only be observed if the receiver antenna falls into the FS main lobe which is centred across the line of sight between the transmitter and target facing the opposite direction from the transmitter, see Figure 8.1, while its width is proportional to λ and inversely proportional to the target's typical dimension [34]. The values of (8.1) and (8.2) are used to estimate the RCS of a miniature UAV which helps to predict the expected performance of the PR.

8.2.2 Passive bistatic radar equation

In order to get a coarse estimation of the maximum operational range, an investigation on the bistatic radar equation of the system is necessary. Similar to [35], the figure of merit used is the signal-to-noise ratio (SNR), which for a single pulse is [37]

$$\rho = \left(\frac{P_T G_T}{4\pi D_T^2}\right)\left(\frac{\sigma}{4\pi D_R^2}\right)\left(\frac{\lambda^2 G_R L_s}{4\pi}\right)\frac{G_P}{P_n} \tag{8.3}$$

where the different parameters are described in Table 8.1 along with their typical values derived from [35,38]. It should be noted that the processing gain is calculated as the time bandwidth product of the signal, i.e. $G_P = T \times B_s$. GNSS satellites operate in a continuous waveform manner, so when referring to a single pulse, the duration of a single code sequence transmitted by the satellite is assumed. When a coherent integration of duration T_I is considered, the improved SNR is given as

$$\hat{\rho} = \frac{T_I}{T}\rho \tag{8.4}$$

where T is the duration of the code used by the satellite. For the specific case of GNSS illumination, multiple satellites can be considered at the same time as IOs. However, the different satellites are expected to be widely distributed around the target; thus, coherency between the different signals is not a realistic assumption and therefore non-coherent integration shall be considered. Non-coherent integration is less effective than the coherent one and the computation of its gain is not straightforward. Generally, the square root of the coherent gain is used as a

Table 8.1 Satellite link budget parameters

	Description		**GPS**	**Galileo**
λ	Wavelength	cm	19.03	19.03
D_T	Satellite-to-target distance (mean)	km	20,200	23,222
P_T	Transmitted power (mean)	dBW	21.7	23.9
G_T	Transmitter gain (mean)	dBi	14.5	16.5
T	Code duration	ms	1	4
B_s	Code bandwidth	MHz	1.023	1.023
G_P	Signal processing gain	dB	30.1	36.1
G_R	Receiver gain	dBi	46	46
L_s	Losses	dB	−3	−3
P_n	Noise power	dB	−128	−128
σ_{BS}	Target FS-RCS	dBsm	−17	−17
σ_{FS}	Target BS–RCS	dBsm	−4.65	−4.65
ρ_{min}	Minimum SNR for detection	dB	8	8

conservative assumption. Assuming N non-coherent received signals from N satellites, the SNR after integration is approximated as

$$\tilde{\rho} = \sqrt{\sum_{i=1}^{N} \rho_i^2} \tag{8.5}$$

where ρ_i, $i = 1, 2, \ldots, N$ is the SNR of the target return from each satellite. The maximum detectable range for a certain set of configurations can be then obtained from (8.3):

$$D_R = \sqrt{\left(\frac{P_T G_T}{4\pi D_T^2}\right)\left(\frac{\sigma}{4\pi\rho_{\min}}\right)\left(\frac{\lambda^2 G_R L_s}{4\pi}\right)\frac{G_P}{P_n}} \tag{8.6}$$

where $\rho_{\min}$ is the minimum SNR required for detection. Additionally, using (8.4) and (8.6) it can be shown how a wider coverage can be achieved when using coherent integration in time:

$$\hat{D}_R = \sqrt{\frac{T_I}{T}} D_R \tag{8.7}$$

Finally, in the case of multiple satellites' integration, the maximum achievable range is approximated as

$$\tilde{D}_R \approx \sqrt[4]{\sum_{i=1}^{N} D_{R,i}^4} \tag{8.8}$$

where $D_{R,i}$ is the maximum achievable range using the ith satellite. It is worth noting that certain variables in (8.6) and (8.7) can be parametrised in the system design process, while others depend on the satellite properties and shall be considered constant. In particular, the transmitted power, gain, wavelength and processing gain have fixed values for each satellite. On the other hand, the receiver gain, integration time, number of satellites and SNR for detection can be set in the system design. The rest of the parameters, such as the target RCS and noise power are unknown or cannot be controlled.

As different constellations have different processing gain and code duration, see Table 8.1, these parameters need to be considered in the integration time selection as from (8.5) it can be shown that Galileo offers lower gain than GPS for the same T_I because of its longer pulse duration. This lower integration gain however is compensated by the higher processing gain Galileo's longer codes offer.

8.2.3 Case studies

In order to provide an idea of the coverage that a GNSS-based PR could provide, in this section two case studies are presented and the link budget model described in

Section 8.2.2 is applied to approximate the maximum distance that a miniature UAV target can be detected.

The GNSS coverage depends on the receiver's geographic position and will also vary during different parts of the day; hence, the analysis is carried out throughout a 24-h period. Table 8.1 summarises the parameters used in the presented analysis, while the power of individual satellites is obtained from [38]. The satellite position model is based on the orbit computation method described in [39]. The method follows a modified version of the SGP4 model applied for standard satellite orbit computation using two-line elements (TLE). The position of each satellite is calculated for the full day (midnight to midnight) at 16/06/2019 based on the TLE files with the same epoch date.

8.2.3.1 Case Study I: Heathrow Airport

Heathrow Airport, Longford TW6, London, the United Kingdom is selected as the first case study. The PR receiver is located at 51.4700°N, 0.4543°W for latitude and longitude, respectively, while the height is set at 25 m above sea level. The first analysis performed considers single illuminators only. The maximum detectable range achieved by each satellite, as calculated from (8.7) using BS–RCS and coherent integration of $T_I = 0.5$ s, is shown in Figure 8.2. Figure 8.2(a) and (b) shows the maximum detectable range throughout 1 day for each satellite in the GPS and Galileo constellations, respectively. Moreover, the last row indexed as 'Max' shows the range when the satellite offering the maximum range in that instant of

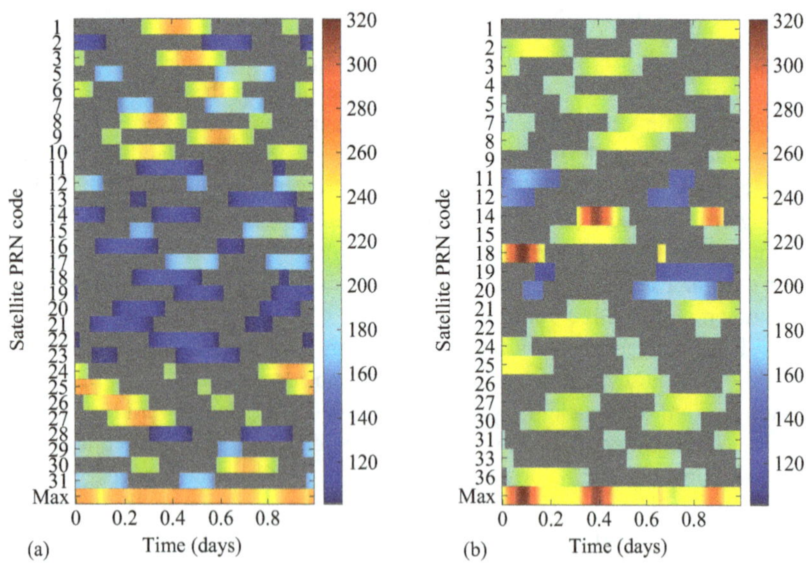

Figure 8.2 Case Study I: maximum detectable range (m) achieved in BS configuration using one satellite for detection: (a) GPS and (b) Galileo

the day is selected. Also, zero range shown in grey colour indicates no satellite visibility. From the results, it can be seen that Galileo can provide longer detectable ranges; however, these are less consistent throughout the day when compared with the GPS illumination.

The next considered case is when multiple illuminators are integrated at the same time, thus using (8.8) for all the visible illuminators at a specific time instant, the results in Figure 8.3 are obtained. From the figure, it can be seen that the maximum achievable range after integrating multiple signal is significantly longer than what individual satellites offer. Moreover, the Galileo constellation offer higher detectable range for most of the day fluctuating between 330 and 410 m for $T_I = 0.5$ s, see left axis in Figure 8.3. Additionally, integrating all visible satellites from both constellations, significantly higher maximum detectable range varying from 390 up to 470 m is achieved. These results can be further improved by increasing the integration time. As it can be seen from the right axis in Figure 8.3, the maximum detectable ranges of 580 m for Galileo and 660 m for combined constellations are achieved.

In order to assess the FS performance, only a subset of the illuminators should be considered. Indeed, as discussed in Section 8.2.1, the FS-RCS only occurs when the target is close to the baseline between the satellite illuminator and the ground receiver. For this reason, we assume that the receiver is pointing West and only satellites positioned West of the airport, particularly between 160° and 200° azimuth angle and between 5° and 50° elevation angle in topocentric coordinates system (TCS), are considered. In Figure 8.4, the maximum detectable range achieved from each satellite in different constellations is shown using an integration time of $T_I = 0.1$ s. Notice that T_I is significantly smaller when FS is considered due to the more restrictive topology. As seen in Figure 8.4, the coverage in

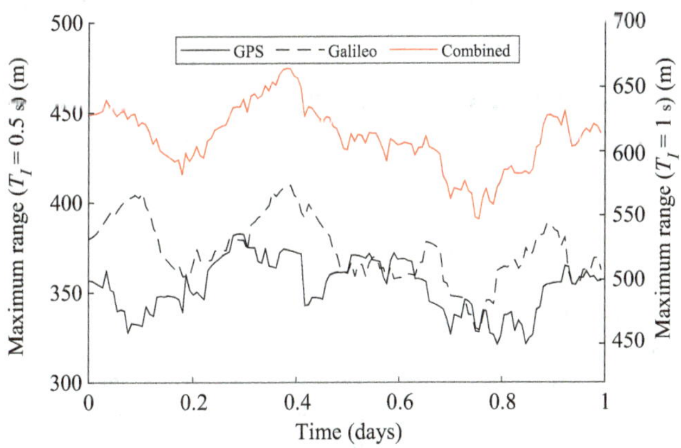

Figure 8.3 Case Study I: maximum detectable range achieved in BS configuration using all the visible satellites in different constellations and a combination of both; left axis for integration time $T_I = 0.5$ s and right axis for $T_I = 1$ s

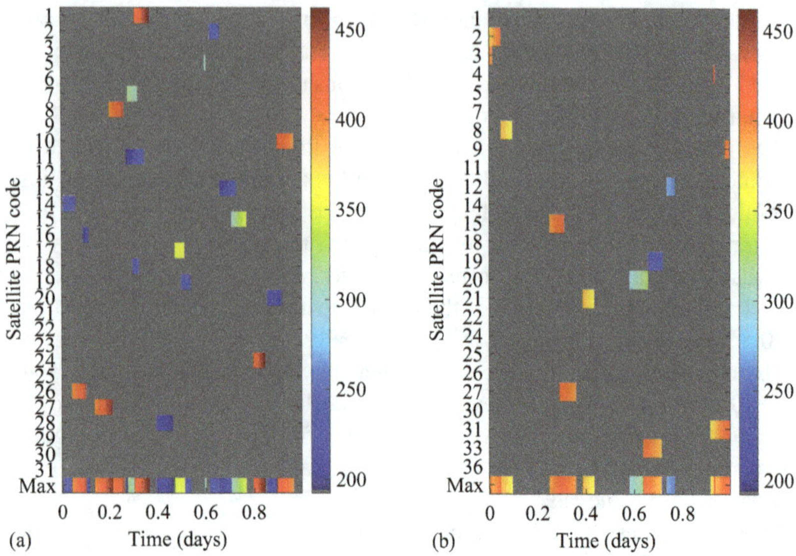

Figure 8.4 Case Study I: maximum detectable range (m) achieved in FS configuration using one satellite for detection: (a) GPS and (b) Galileo

terms of time is significantly less dense compared to the BS in Figure 8.2. On the contrary, the detectable range is longer, even if the integration time is shorter, which is due to the FS-RCS being significantly higher than BS–RCS. As it can be seen, GPS offers longer ranges compared to the Galileo while also being less sparse. In particular, the GPS will have non-zero range for 79.2% of the day compared to Galileo 51.4%.

8.2.3.2 Case Study II: Newark Liberty International Airport

The second case study is the Newark Liberty International Airport, NJ 07114, the United States, located at 40.6895°N, 74.1745°W for latitude and longitude, respectively, with the height of the radar receiver being set at 25 m above sea level. Similar to Case Study I, the maximum detectable ranges achieved by individual satellites and by integrating different signals are shown in Figures 8.5 and 8.6, respectively. Comparing the respective results in Case Studies I and II, it can be seen that the satellite availability profiles appear different. This is expected as the receiver in the Case Study II is placed at different geographical location than Case Study I. Moreover, comparing Figure 8.2 with Figure 8.5 it can be seen that the maximum range of the individual satellites is similar for both the case studies. Furthermore, comparing Figure 8.3 with Figure 8.6, it can be seen that for the Case Study II the maximum detection ranges have generally lower values.

To examine the FS-RCS, only satellites positioned South of the airport, particularly between 240° and 280° azimuth angle and between 5° and 50° elevation

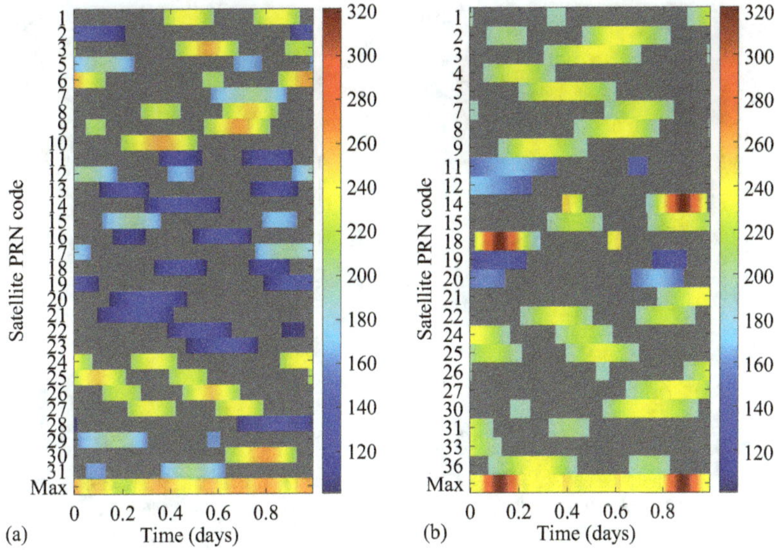

Figure 8.5 Case Study II: maximum detectable range (m) achieved in BS configuration using one satellite for detection: (a) GPS and (b) Galileo

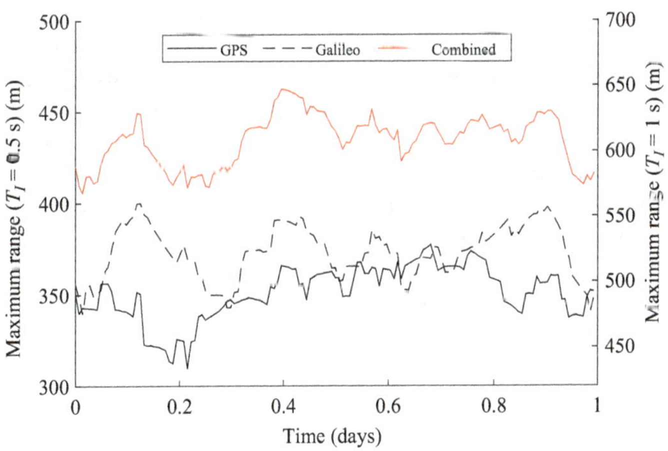

Figure 8.6 Case Study II: maximum detectable range achieved in BS configuration using all the visible satellites in different constellations and a combination of both; left axis for integration time $T_I = 0.5$ s and right axis for $T_I = 1$ s

angle in TCS, are considered. In Figure 8.7, the maximum detectable range is shown to be achieved from each satellite using an integration time of $T_I = 0.1$ s. Comparing Figure 8.7 with Figure 8.4, it can be seen that available satellites are

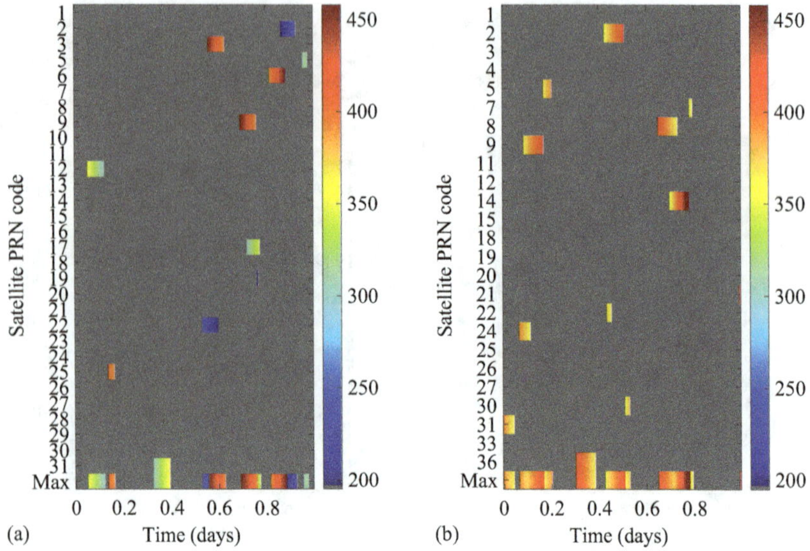

Figure 8.7 Case Study II: maximum detectable range (m) achieved in FS configuration (South) using one satellite for detection: (a) GPS and (b) Galileo

much more sparse in Case Study II than Case Study I. In particular, GPS and Galileo will have non-zero range for 48.6% and 52.8% of the day, respectively, suggesting that a South-facing FS radar would not be particularly effective.

For a better comparison between the two different geographic positions, Case Study II is also examined only for satellites appearing in the West direction, with the same azimuth and elevation restrictions as in Case Study I. From Figure 8.8, it can be seen that the coverage has been significantly improved for the GPS and slightly reduced for Galileo giving non-zero ranges for 68.7% and 47.9%, respectively. Additionally, the maximum detectable range for Galileo is generally lower compared to when only satellites from the South direction are used.

Considering both case studies, it can be seen that while the FS configuration can offer significantly higher detectable ranges when individual satellites are considered, the low availability of satellites in the desired configuration makes it of difficult realisation. On the contrary, while BS configurations offer lower maximum range for individual satellites, the availability of the satellites is much higher allowing for multiple reflections to be integrated. Indeed, non-coherent integration of multiple satellites in BS can offer similar maximum ranges to the FS configuration, while if longer coherent integration in time, not available in FS, is considered the resulting ranges can be even higher. A combination of both configurations using two processing streams, one for BS which can stay always online and one for FS operating when available satellites are present, would be the

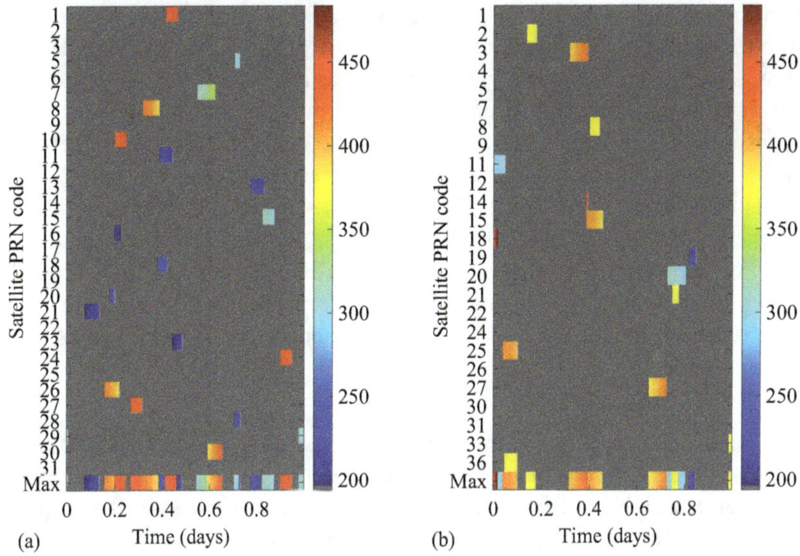

Figure 8.8 Case Study II: maximum detectable range (m) achieved in FS configuration (West) using one satellite for detection: (a) GPS and (b) Galileo

most advantageous approach in terms of detection performance. Furthermore, integration of other available GNSS systems could further improve the maximum detectable range and availability.

8.3 Source signal reconstruction

A fundamental requirement of a PR is the knowledge of the signals transmitted by the IOs, in order to apply time difference of arrival (TDOA) and allow for the estimation of the bistatic range as well as improved SNR. In this section, the appropriate signal model and reconstruction algorithm for the GNSS-based PR are introduced.

8.3.1 Signal model

The generic form of a GNSS signal transmitted by the ith satellite can be written as [8]

$$s_i(t) = m_i(t)g_i(t)e^{j2\pi f_0 t} \tag{8.9}$$

where f_0 denotes the carrier frequency, $g_i(t)$ is Pseudorandom Noise (PRN) code sequence and $m_i(t)$ is navigation message. The component $g_i(t)$ in (8.9) is generated by repeating a binary phase shift keying (BPSK)-modulated sequence composed of N_c chips, each of which has duration equal to T_c. Furthermore, the information carried by $m_i(t)$ is also coded using a BPSK modulation, with a symbol duration T_m which is significantly longer than T_c, i.e. $T_m \gg T_c$. Since the PRN code used by each satellite is known by the receiver, as it is used for the detection of the signal, it is useful to define its duration as the pulse repetition interval (PRI) of the transmitter which equals $\mathrm{PRI} = N_c T_c$. Based on this PRI, *fast-time* is defined as the time between the beginning and the end of each PRI, i.e. $t \in [0, \mathrm{PRI}]$, while intervals of a PRI, i.e. $u = 0, \mathrm{PRI}, 2\mathrm{PRI}, \ldots, (N_p - 1)\mathrm{PRI}$ are referred to as *slow-time* where N_p is the number of received PRIs. Without loss of generality, the transmitter–receiver channel can be assumed to be stationary in fast-time, while changing in slow-time. The direct path signal from the ith GNSS satellite at the passive receiver, after removing the carrier frequency, can be expressed as [35]

$$r_i(t,u) = m_i(t,u)g_i(t,u)e^{j[2\pi f_i(u)(t+u)+\phi_i(u)+\psi_i(u)]} \tag{8.10}$$

where f_i and ϕ_i are the signal's frequency and phase shifts, respectively, caused by the distance and relative velocity between the satellite and the receiver, and ψ_i is a phase error caused by non-free-space propagation phenomena as well as hardware imperfections. Notice that since the channel varies in slow-time, the delay, frequency shift and phase error are modelled in slow-time. Furthermore, the different components are given as

$$m_i(t,u) = m_i(t - \tau_i(u)) \tag{8.11}$$

$$g_i(t,u) = g_i(t - \tau_i(u)) \tag{8.12}$$

$$\phi_i(u) = 2\pi(f_0 + f_i(u))\tau_i(u) \tag{8.13}$$

where $\tau_i(u)$ is signal's delay.

According to (8.10), the signal reflected by the target can be written as

$$\hat{r}_i(t,u) = \hat{m}_i(t,u)\hat{g}_i(t,u)e^{j[2\pi\hat{f}_i(u)(t+u)+\hat{\phi}_i(u)+\hat{\psi}_i(u)]} \tag{8.14}$$

where $\hat{f}_i(u)$ and $\hat{\psi}_i(u)$ are the frequency shift and phase error that the signal experiences in the satellite–target–receiver path. The parameters $\hat{m}_i(t,u)$, $\hat{g}_i(t,u)$ and $\hat{\phi}_i(u)$ can be calculated using (8.11), (8.12) and (8.13), respectively, by replacing $f_i(u)$ with $\hat{f}_i(u)$ and $\tau_i(u)$ with $\hat{\tau}_i(u)$ which is the delay the signal experiences in the transmitter–target–receiver path.

8.3.2 GNSS signal characteristics

In this section, the AF of the GNSS waveform is investigated. Here the AF definition introduced in [40] is used where $\mathscr{A}(\tau, f)$ is the output of the matched filter when the input signal, PRN code in this case, experiences a delay τ and frequency shifted f relative to the nominal values for which the matched filter is designed. In Figure 8.9(a), the AF of the space vehicle number (SVN) 1 GPS satellite signal is shown, overlaid by the zero-Doppler and zero-delay cuts. It should be noted that all the codes in one constellation exhibit similar AF and therefore examining only one example is sufficient for the current analysis. In Figure 8.9(b), the AF of the SVN 1

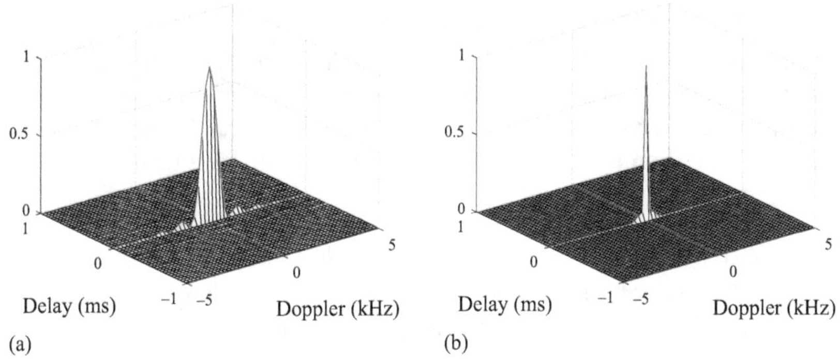

Figure 8.9 Example AF of (a) GPS and (b) Galileo code signal

Table 8.2 Radar parameters

	Description		GPS	Galileo
PRF	Pulse repetition frequency	Hz	1,000	250
$f_{\mathrm{d,un}}$	Maximum unambiguous Doppler	Hz	500	125
ΔD	Range resolution	m	146.5	146.5

Galileo satellite signal is shown. Comparing Figure 8.9(a) and (b), it can be observed that both signals offer a 'thumbtack'-shaped AF with high response around the centre and low values elsewhere. This is a characteristic generally well accepted by radar waveform designers as it provides proper localisation of the target and low ambiguities. Furthermore, the Galileo AF shows a finer Doppler resolution due to the longer duration of the waveform. The AF parameters of the examined signals are summarised in Table 8.2. This table provides a comparison between the waveforms of the two constellations. In particular, the pulse repetition frequency of the GPS signal is 1 kHz, while for Galileo this value drops by a factor of 4 due to its longer duration. As a consequence, GPS has a larger unambiguous Doppler range than Galileo. While the measured Doppler will depend on the geometry of the transmitter–receiver–target system [23], to place these numbers into perspective it will assume the case where the satellite is behind the receiver and the target is moving straight towards them. The experienced Doppler in this case is maximum and is given by

$$f_{\mathrm{d,bi}} = \frac{2V_{\mathrm{g}}}{\lambda} \tag{8.15}$$

with V_{g} representing the target radial velocity, thus for $V_{\mathrm{g}} = 30\ \mathrm{m/s}$ the measured Doppler is 315 Hz. This means that Galileo might exhibit Doppler ambiguities in such a configuration. Lastly, the GPS and Galileo satellites offer the same range resolution, due to the same bandwidth of their pulses.

8.3.3 Signal reconstruction algorithm

The approach proposed in [8] for synchronisation is selected in order to perform the signal reconstruction algorithm. The basic operations of the synchronisation approach are depicted in Figure 8.10. The system utilises two channels: a reference channel used to capture the direct signals form the available satellites and a surveillance channel that captures the target's reflections. The input from the reference channel is passed through a standard GNSS receiver chain where the signals from different satellites are first detected by correlating the reference channel signal with different PRN sequences. This process is commonly referred to as *signal acquisition* and provides initial estimations regarding the code delay and Doppler shift of the signal. This information is later passed to a PRN code tracker which provides a finer and continuous estimation of the signal delay, Doppler and phase parameters. At this stage, the estimated parameters can be used to reconstruct noise-free reference signals from the different IOs that can be then correlated with the surveillance signal in order to perform TDOA and target detection. The whole procedure can be formalised as follows by using the signal model derived in Section 8.3.1. The signal at the reference and surveillance channels can be expressed as

$$r_R(t,u) = \sum_{i=1}^{N_i} \sqrt{a_{i,R}(u)}r_i(t,u) + \sum_{i=1}^{N_i} \sqrt{\hat{a}_{i,R}(u)}\hat{r}_i(t,u) + n_R(t,u) \tag{8.16}$$

$$r_S(t,u) = \sum_{i=1}^{N_i} \sqrt{a_{i,S}(u)}r_i(t,u) + \sum_{i=1}^{N_i} \sqrt{\hat{a}_{i,S}(u)}\hat{r}_i(t,u) + n_S(t,u) \tag{8.17}$$

where $a_{i,R}(u)$, $\hat{a}_{i,R}(u)$ represent the power losses in the reference channel owing to the satellite–receiver and satellite–target–receiver paths, and $a_{i,S}(u)$ and $\hat{a}_{i,S}(u)$ are the respective losses in the surveillance channel. If $a_{i,R}(u) \gg \hat{a}_{i,R}(u)$ is satisfied, the tracker can provide an estimate of the delay $\tilde{\tau}_i(u) \approx \tau_i(u)$, frequency shift $\tilde{f}_i(u) \approx f_i(u)$ and phase shift $\tilde{\phi}_i(u) \approx \phi_i(u)$ that the signal experiences in the satellite–receiver path. Additionally, the tracker outputs an estimate of the

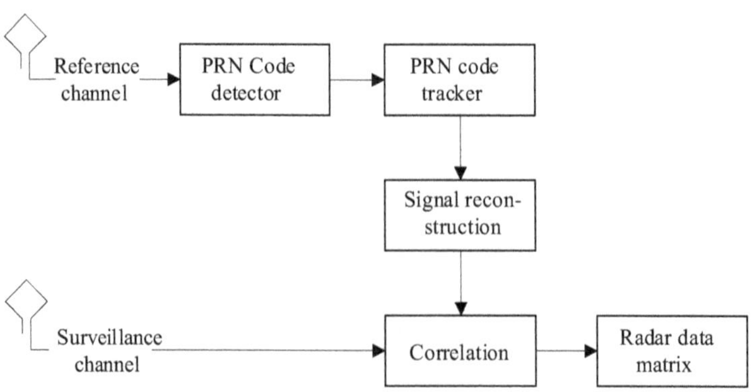

Figure 8.10 Signal reconstruction algorithm

navigation signal $\tilde{m}_i(u)$ containing binary phase information. Using these estimates, the signal from the ith satellite can be reconstructed as

$$\tilde{r}_{i,D}(t,u) = \tilde{m}_i(u)\tilde{g}_i(t,u)e^{j[2\pi\tilde{f}_i(u)(t+u)+\tilde{\phi}_i(u)]} \tag{8.18}$$

where $\tilde{g}_i(t,u) = g_i(t - \tilde{\tau}_i(u))$ is time-shifted PRN code. Following the reconstruction of all the satellite signals in (8.18), these can be used to filter the signal from the surveillance channel. By design, the signals transmitted from different satellites are quasi-orthogonal, thus the cross-correlation between the reconstructed signal (8.18) and the surveillance channel (8.17) can be expressed as

$$Y_i(k,u) = \int_0^{\mathrm{PRI}} \tilde{r}^{\dagger}_{i,D}(t-k,u)r_S(t,u)\mathrm{d}t = y_i(k,u) + \hat{y}_i(k,u) + n_i(k,u) \tag{8.19}$$

where $(\cdot)^{\dagger}$ denotes the complex conjugate operation, and $y_i(k,u)$, $\hat{y}_i(k,u)$, $n_i(k,u)$ are the correlation output components associated with the direct signal, target returns and noise, respectively. For an accurate reconstruction of the signal, and if the phase error in both channels is approximately the same, i.e. $\tilde{\psi}_i(u) \approx \psi_i(u)$, $y_i(k,u)$ and $\hat{y}_i(k,u)$ can be expressed as [35]

$$y_i(k,u) = \sqrt{a_{i,S}(u)}\mathscr{A}(k,0) \tag{8.20}$$

$$\hat{y}_i(k,u) = \sqrt{\hat{a}_{i,S}(u)}\mathscr{A}(\Delta\tau_i(u) + k, \Delta f_i(u)) \tag{8.21}$$

where $\mathscr{A}(\tau,f)$ is AF as described in Section 8.3.2, $\Delta\tau_i(u) = \tau_i(u) - \tilde{\tau}_i(u)$ and $\Delta f_i(u) = f_i(u) - \tilde{f}_i(u)$ represent the bistatic delay and Doppler shift, respectively. From (8.20) and (8.21), it can be seen that the output from the filtered signal $Y_i(k,u)$ comprises two main components: one at zero-delay and zero-Doppler and one at $\Delta\tau_i(u)$ and $\Delta f_i(u)$. If $\Delta\tau_i(u) \neq 0$ or $\Delta f_i(u) \neq 0$, it is therefore possible to estimate the target's range and velocity after filtering the direct signal component. In the literature, various direct signal suppression techniques have been proposed and the reader is referred to [23,41,42] for further discussion.

8.4 Target parameters estimation

Following the TDOA processing, target detection is possible, and equally importantly, position and target motion parameters can be estimated. In this section, the target localisation problem is dealt with by employing a MISO AF approach. This allows the use of the return from multiple satellites and exploits their spatial diversity. It should be noted that the proposed approach addresses only BS returns as TDOA cannot be observed in FS. Target parameter estimation methods suitable for FS have been proposed in literature [18,19]; however, those cannot be directly implemented in the proposed AF approach but could be combined in an information fusion framework. An alternative approach combining matched filter outputs for BS and the first Fresnel zone for FS responses was proposed in [43] assuming a binary ambiguity for the target.

8.4.1 Target localisation

Assuming the acquisition of one PRI in the presence of a target, we define $\mathbf{r}_i$ as the $N \times 1$ vector containing the discrete samples of one PRN code transmitted by the ith satellite, reflected by the target and captured at the receiver. The total received signal can be defined as

$$\mathbf{r} = \sum_{i=1}^{N_i} \mathbf{r}_i + \mathbf{n} \tag{8.22}$$

where N_i is the number of available satellites and $\mathbf{n}$ is additive white Gaussian noise. Assuming that the target introduces negligible Doppler shift in fast-time, i.e. the target motion is modelled as phase shifts in slow-time, the non-coherent MISO AF of the received signal is defined as

$$\mathbf{A}(x, y, z) = \sum_{i=1}^{N_i} |\mathbf{r}_i^{\dagger}(x, y, z)\mathbf{r}| \tag{8.23}$$

where x, y and z are the spatial parameters in an X–Y–Z three-dimensional space, $\mathbf{r}_i(x, y, z)$ is the expected signal from a target in the (x, y, z) coordinates associated with the ith satellite, and $(\cdot)^{\dagger}$ is complex conjugate operation. It is worth mentioning that here a perfect reconstruction of the satellite signal is assumed, meaning that delay and Doppler shifts due to the satellite position and motion are accounted in $\mathbf{r}_i^{\dagger}(x, y, z)$. Examining (8.23), it can be easily seen that the MISO AF is defined as the non-coherent sum of the AFs for the different satellite signals $\mathbf{A}_i(x, y, z)$, i.e.

$$\mathbf{A}(x, y, z) = \sum_{i=1}^{N_i} |\mathbf{A}_i(x, y, z)| \tag{8.24}$$

While coherent summation of the different AFs could theoretically provide superior results, as mentioned in Section 8.2.2, since coherency between the different satellite signals might not be achievable in most cases, non-coherent integration is adopted in this study.

For a better understanding of the system's behaviour, consider the following simplified example. Assume a source placed at the Heathrow airport and the configuration of the available GPS satellites shown in Figure 8.11, where the satellites' coordinates have been scaled down by a factor of 10^5 for better illustration. A target is placed within the surveillance area with the received signal being modelled based on Section 8.1.3 with the reference signal assumed to be perfectly reconstructed. To calculate the reflected target's power received from each satellite, (8.3) is used by replacing P_n with 1. Furthermore, no noise is assumed, while the in-between sample delays are modelled using fractional delay by linear interpolation [44]. In Figure 8.9, the X–Y cross sections of the MISO AF is show for $z = 20$ m and the

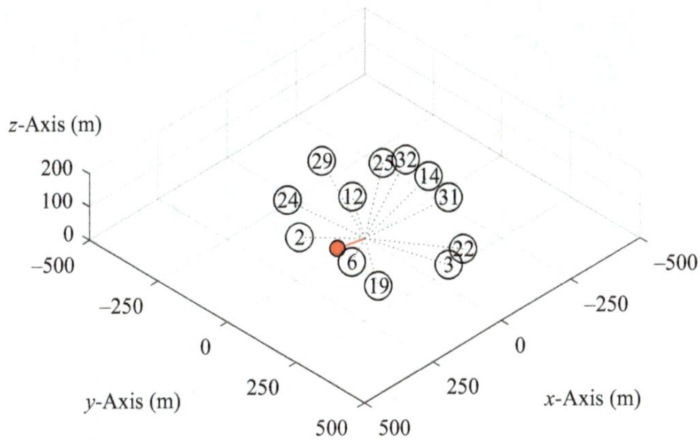

Figure 8.11 Examined geometry with the receiver placed at (0, 0, 0), the coordinates of each satellite being scaled down and the target denoted by the red dot

target placed at different positions. Each MISO AF is normalised by its maximum value and the resolution bin step in all dimensions is 10 m starting form −500 up to 500 m in X and Y axes and 0–250 m in Z. As it can be seen the response of the MISO AF is highly dependent on the target's position. Specifically, when the target is placed away from the receiver, see Figure 8.12(a) and (c), the response appears similar to a circle arc, while when placed on top of the receiver the response assumes a more symmetric shape. Furthermore, when there is a mismatch in height between the actual target position and the AF cut, see Figure 8.12(d), the response has lower intensity.

The MISO AF of the target when placed at (250, 0, 20) at different hours of the day is shown in Figure 8.13. It can be seen, that throughout the day the response changes due to the different satellites' position. It is also worth noting that the spatial resolution in the radial direction is not constant and higher than the one described in Table 8.2. This is caused by the bistatic resolution in the different AFs $\mathbf{A}_i$ varying based on the respective bistatic angle θ. In summary, the MISO AF processing performance, including which satellite to use at a given time of the day, would need to be analysed on a case by case basis depending on the receiver positioning.

8.4.2 Velocity estimation

As the target's Doppler is translated in different phase shifts observed in slow-time, the formula of MISO AFs derived in (8.24) is not sufficient in order to estimate the target's velocity. In monostatic and bistatic configurations, the Doppler processing usually involves a Fourier transform across the slow-time samples of each fast-time bin. As in MISO, the Doppler experienced from different transmitter–receiver pairs cannot be converted to velocity along the same axis, a different approach needs to

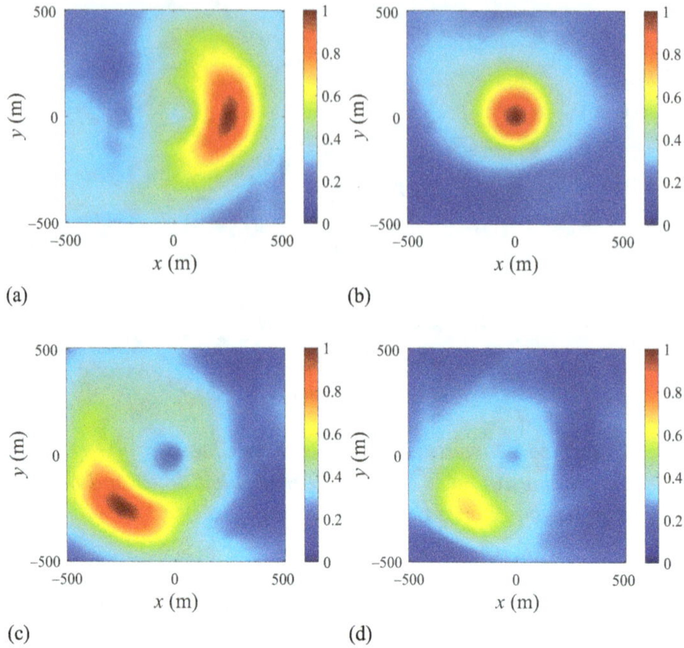

Figure 8.12 MISO AF cross section at z = 20 m for different target position and the receiver and satellites topology shown in Figure 8.11: (a) (250;0;20), (b) (0;0;20), (c) (−250;−250;20), (d) (−250;−250;250)

be pursued. Similarly to the location estimation, the velocity is estimated along all three *X*–*Y*–*Z* axes and by extending (8.24) under the assumption of slow-time processing results in

$$\mathbf{A}(x, y, z, v_x, v_y, v_z, m_0) = \sum_{i=1}^{N_i} \left| \sum_{m=1}^{N_m} \mathbf{r}_i^{\dagger}(x, y, z)\mathbf{r}(m + m_0)e^{-2\pi f_{D,i}(x,y,z,v_x,v_y,v_z)m} \right| \tag{8.25}$$

where v_x, v_y and v_z denote the velocity along the three axes, m_0 is the starting slow-time sample of the examined interval, N_m is the number of slow-time samples composing a coherent time interval (CPI), m is the index of the slow-time samples, and $f_{D,i}(x, y, z, v_x, v_y, v_z)$ is the expected bistatic Doppler in the ith satellite signal divided by the PRF. To better understand the process, the scenario described in Figure 8.11 is assumed with the target located at (250, 0, 20) and with velocity only in the y direction $v_y = 5$ m/s. In Figure 8.14, the MISO AF cut described in (8.25) at the location of the target and for $v_z = 0$ is shown for a CPI of 0.1 s. Under closer inspection, it can be noted that the MIMO AF comprises different iso-Doppler lines deriving from the AFs of the different IOs, with a peak at the estimate target's velocity.

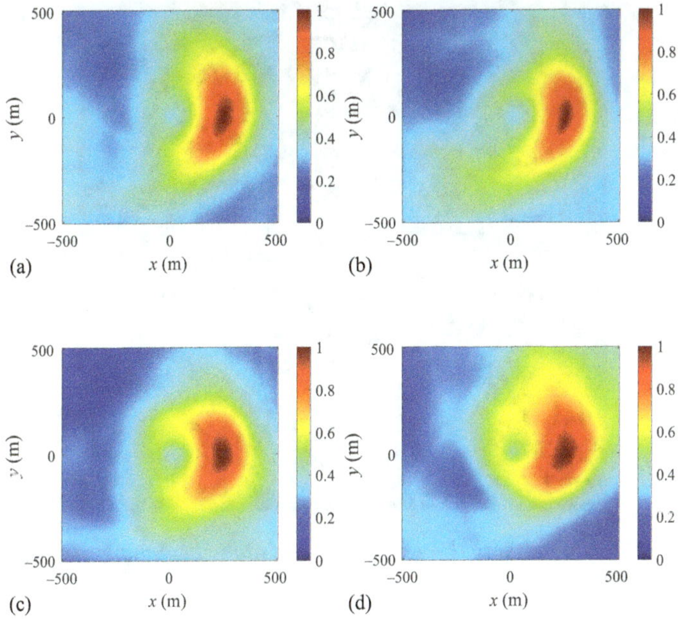

Figure 8.13 MISO AF cross section at z = 20 m for the target placed at (250, 0, 20) at different hours of the day: (a) 00.00, (b) 06.00, (c) 12.00, (d) 18.00

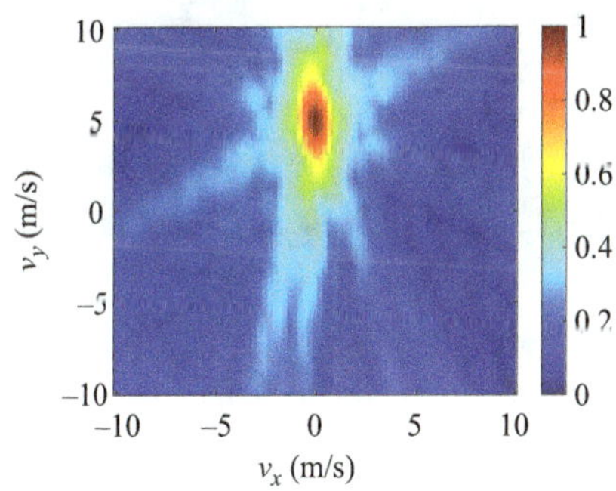

Figure 8.14 MISO AF cross section at the position of the target (250,0,20) and for $v_z = 0$ for a target with velocity only on the y-axis $v_y = 5\ m/s$

8.5 Experimental analysis

To validate the system in real conditions, an experimental campaign was held in the Caplaw Model Flying Club, Johnstone, Scotland. The target used was the DJI Phantom 4 drone, while the area and the satellites' positions are shown in Figure 8.15 with the tracked satellites shown in green colour and blue indicating the

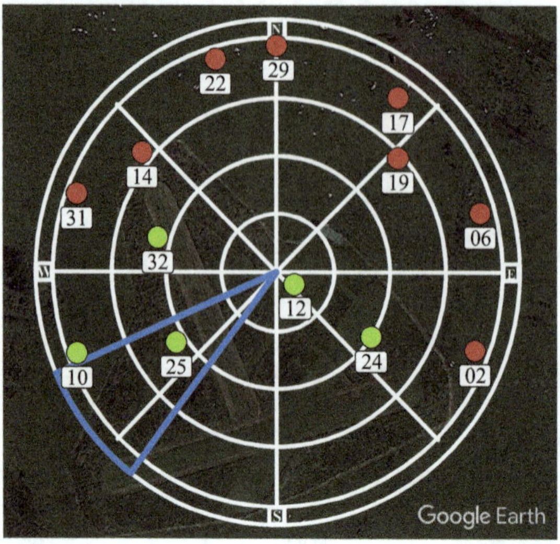

Figure 8.15 Testing area and satellite position at the beginning of the acquisition [45]

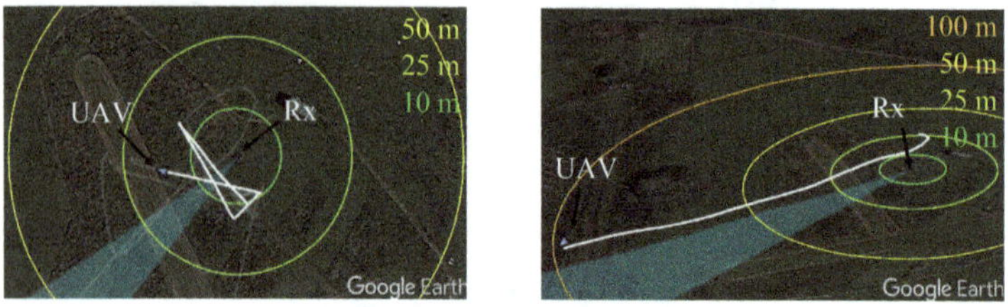

Figure 8.16 Target flight path and different ground range indicators from the receiver position [45]: (a) Scenario 1 and (b) Scenario 2

antenna beam. The mission comprises two main target motions: crossings in front of the receiver as illustrated in Figure 8.16(a) and descending away from the receiver illustrated in Figure 8.16(b). In each acquisition, the time of the recording and the location of the receiver and the satellites were extracted from the captured GPS signal.

8.5.1 Scenario 1: crossing in front of the receiver

The target route in Scenario 1 is shown in Figure 8.16(a). The distance of the target form the receiver during the crossing was approximately 10 m. In Figure 8.17, the

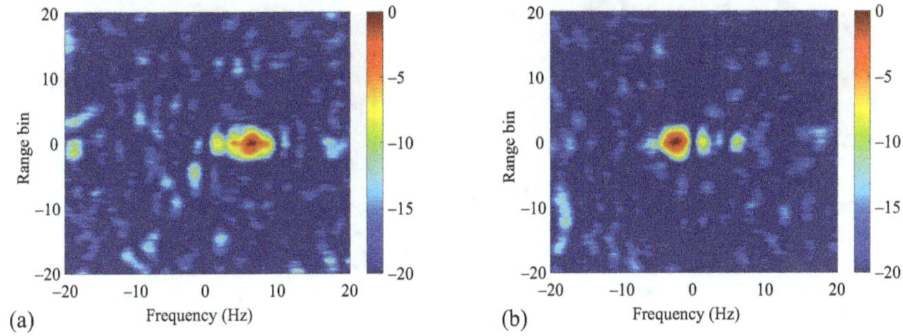

Figure 8.17 Range–Doppler maps (a) right before and (b) after the target crosses the receiver's beam centre; CPI = 0.5 s, PRN 25

range–Doppler maps generated from the returns of satellite PRN 25 are shown for one of the crossings. The CPI used was 0.5 s providing a frequency resolution of 2 Hz. Moreover, in order to remove stationary clutter, the mean of the slow-time signal is subtracted from all range bins. As it can be seen, the target initially appears with a positive frequency, passes from 0 and then goes to negative. This is expected as the decrease in frequency reflects the approaching and receding behaviour of the drone. Additionally, since the satellite is in the FS geometry the Doppler will be 0 when the target crosses the baseline. To better illustrate the target's behaviour, the spectrograms generated using the first range bin of four available satellites are shown in Figure 8.18. As it can be seen for PRN 25, the same behaviour show in Figure 8.17 is apparent for five crossings. Investigating the returns of the different satellites, it can be seen that the target appears in different Doppler shifts due to the different bistatic angles. Additionally, it is noticed that the noise floor on the different spectrograms significantly varies with PRN 25 and 10 (see Figure 8.18(a) and (b)) offering better SNR than PRN 32 and 24 (see Figure 8.18(c) and (d)).

To combine the signals from five available satellites, the non coherent MIMO AF approach is applied. The examined area is from -500 to 500 m in the X and Y axes centred at the location of the receiver and height $z = 5$ m. Moreover, the examined grid in the XY plane has a step of 10 m. To remove the clutter and improve the SNR, a bandpass filter is applied with cut-off frequencies of 2 and 100 Hz in all spatial resolution cells for each satellite. Additionally, to apply non-coherent integration in time a moving mean of 0.2 s window is applied on the time domain of each spatial cell for each satellite separately. In Figure 8.19, the MIMO AF is shown in the absence and presence of the target. All the values have been normalised by the maximum value of the MIMO AF from the beginning of the recording up to the first crossing. As it can be seen, the presence of the target is indicated by a distinguishable lobe. The peak of the lobe appears at $x = -10$ m and $y = -10$ m, giving a detection distance at 15 m from the receiver.

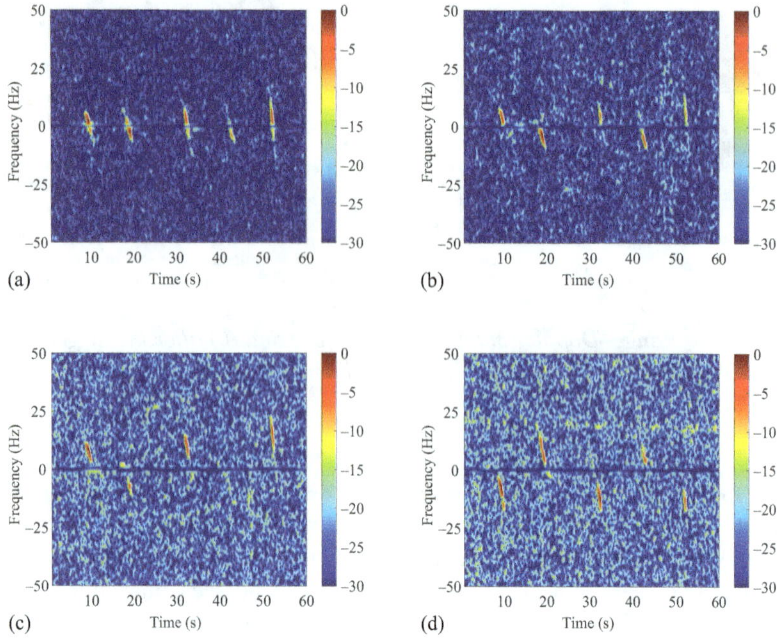

Figure 8.18 Spectrograms generated by filtering the received signal using the signal from GPS satellites: (a) PRN 25, (b) PRN 10, (c) PRN 32, (d) PRN 24, while the target crosses the antenna beam five times

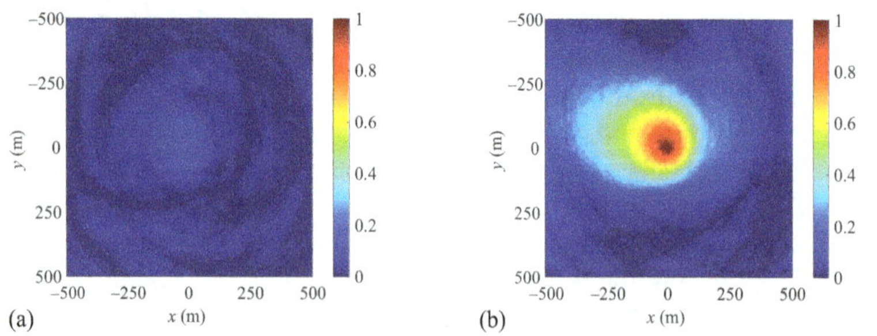

Figure 8.19 Non-coherent MIMO AF (a) target absent (b) target present. AF generated combining five available satellites: PRN 25, 10, 32, 24, 10

8.5.2 Scenario 2: descending away from the receiver

To examine the second motion of the target (see Figure 8.16(b)), the range–Doppler map generated for satellite PRN 25 is shown in Figure 8.20 using a CPI of 0.5 s. As it can be seen the target appears in lower negative frequency as it moves away from the receiver. For better illustration, the spectrograms generated from four available satellites are shown in Figure 8.21. Similar to the crossing motion in Figure 8.18,

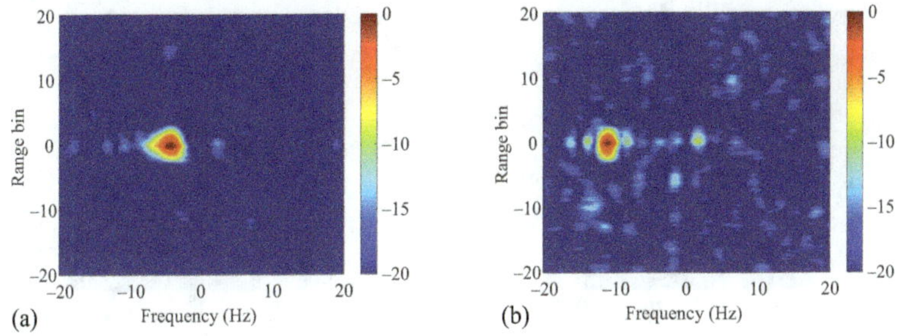

Figure 8.20 Range–Doppler maps (a) right before and (b) after the target crosses the receivers beam centre; CPI = 0.5 s, PRN 25

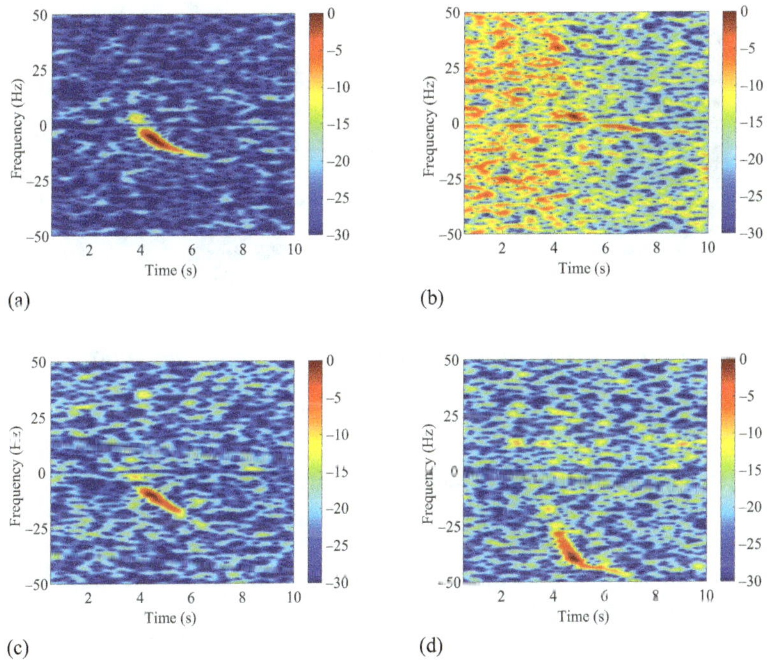

Figure 8.21 Spectrogram generated by filtering the received signal using the signal from GPS satellites: (a) PRN 25, (b) PRN 10, (c) PRN 32, (d) PRN 24, while the target descending away from the antenna beam

the Doppler profile of the target appears in different frequency shifts due to the unique bistatic configuration of each receiver–satellite pair. It should be noted that in this case, the target is not visible in the PRN 10 returns. This can be due to high interference/losses in the satellite–target–receiver channel or poor signal reconstruction. Moreover, the noise floor for the different satellite returns follow the same trend as in Figure 8.18.

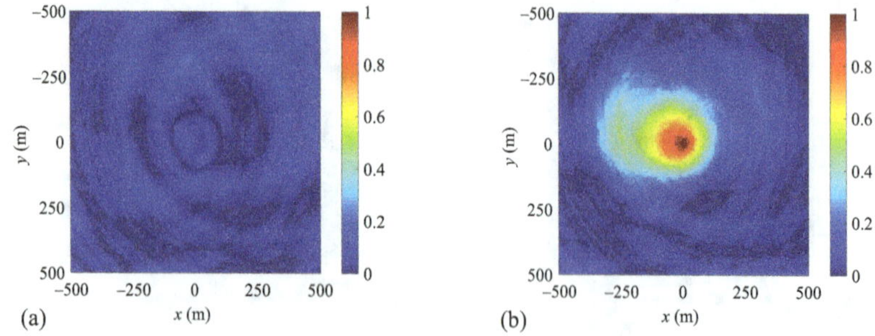

Figure 8.22 Non-coherent MIMO AF (a) target absent (b) target present. AF generated combining five available satellites: PRN 25, 10, 32, 24, 10

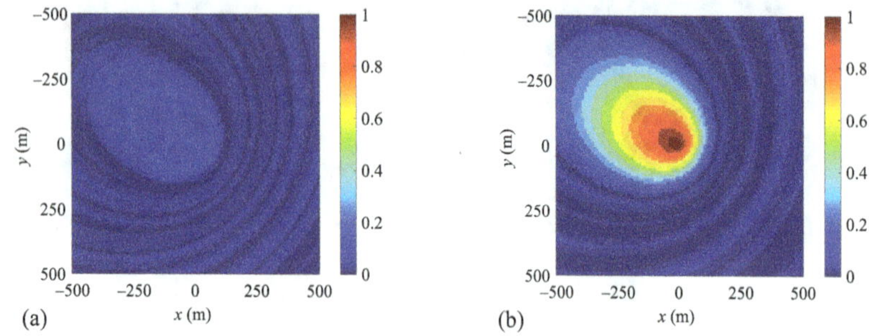

Figure 8.23 Non-coherent MIMO AF (a) target absent (b) target present. AF generated using PRN 25 satellite

To integrate the returns from multiple satellites, the non-coherent MIMO AF is used similarly to Scenario 1. In Figure 8.22, the MIMO AF generated using five available satellites is shown before and after the target enters the scene. Similar to Scenario 1, the target can be indicated by a rise in intensity at the location of the target. The maximum is observed at $x = -10$ m and $y = 0$ m; however, the target is still visible when located at $x = -20$ m and $y = -20$ m giving a maximum value of 0.35 and range 28 m. For comparison, the results when using only one satellite are illustrated in Figure 8.23. As it can be seen, when one satellite is used the AF is characterised with iso-range ellipses. Comparing Figures 8.22(b) and 8.23(b), it is noted that using multiple satellites offers better target localisation capabilities.

8.6 Conclusion

The chapter presented a feasibility study of UAV monitoring using GNSS-based PR systems. In particular, an estimate of the maximum detectable range for small UAV targets was investigated along with the availability of illuminators under BS and FS assumptions. Using two different case studies, it was shown that Galileo satellites can offer higher maximum ranges compared to GPS. Additionally, using all the satellites in one constellation guarantees persistent 24-h availability in BS configuration, while using all the available satellites offers longer detectable ranges. Examining the FS performance for certain orientations, it was demonstrated that while FS-RCS can offer longer maximum detectable range than BS–RCS, the availability of the satellites in FS configuration is quite sparse. The design of the proposed system was derived, describing the appropriate signal model and signal reconstruction algorithm necessary to perform correlation operations. Furthermore, in order to estimate the target's location and velocity a MISO approach using non-coherent integration of different satellite returns was proposed. The behaviour of the MISO AF was investigated under different target positions and part of the day scenarios. Lastly, experimental results demonstrate that such a system and processing approach facilitates UAV detection.

References

[1] Chadwick AD. Micro-drone detection using software-defined 3G passive radar. In: International Conference on Radar Systems (Radar 2017); 2017. p. 1–6.

[2] Jarabo-Amores MP, Mata-Moya D, Gomez-del-Hoyo PJ, *et al.* Drone detection feasibility with passive radars. In: 2018 15th European Radar Conference (EuRAD); 2018. p. 313–316.

[3] Liu Y, Wan X, Tang H, *et al.* Digital television based passive bistatic radar system for drone detection. In: 2017 IEEE Radar Conference (RadarConf); 2017. p. 1493–1497.

[4] Schüpbach C, Patry C, Maasdorp F, *et al.* Micro-UAV detection using DAB-based passive radar. In: 2017 IEEE Radar Conference (RadarConf); 2017. p. 1037–1040.

[5] Knoedler B, Zemmari R, and Koch W. On the detection of small UAV using a GSM passive coherent location system. In: 2016 17th International Radar Symposium (IRS); 2016. p. 1–4

[6] Addabbo P, Giangregorio G, Galdi C, *et al.* Simulation of TechDemoSat-1 delay-Doppler maps for GPS ocean reflectometry. IEEE Journal of Selected Topics in Applied Earth Observations and Remote Sensing. 2017;10 (9):4256–4268.

[7] Ullo SL, Giangregorio G, di Bisceglie M, *et al.* Analysis of GPS signals backscattered from a target on the sea surface. In: 2017 IEEE International Geoscience and Remote Sensing Symposium (IGARSS); 2017. p. 2062–2065.

[8] Antoniou M and Cherniakov M. GNSS-based bistatic SAR: a signal processing view. EURASIP Journal on Advances in Signal Processing. 2013;2013(1):98.

[9] Ma H, Antoniou M, Cherniakov M, *et al.* Maritime target detection using GNSS-based radar: experimental proof of concept. In: 2017 IEEE Radar Conference (RadarConf); 2017. p. 0464–0469.

[10] Beltramonte T, Braca P, Bisceglie MD, *et al.* Simulation-based feasibility analysis of ship detection using GNSS-R delay-Doppler maps. IEEE Journal of Selected Topics in Applied Earth Observations and Remote Sensing. 2020;13:1385–1399.

[11] Willis NJ. Bistatic Radar. vol. 2. SciTech Publishing, London; 2005.

[12] Hiatt RE, Siegel KM, and Weil H. Forward scattering by coated objects illuminated by short wavelength radar. Proceedings of the IRE. 1960;48 (9):1630–1635.

[13] Glaser JI. Bistatic RCS of complex objects near forward scatter. IEEE Transactions on Aerospace and Electronic Systems. 1985;AES-21(1):70–78.

[14] Glaser JI. Forward scatter radar for future systems. WSTIAC Quarterly. 2011;10(3).

[15] Hu C, Sizov V, Antoniou M, *et al.* Optimal signal processing in ground-based forward scatter micro radars. IEEE Transactions on Aerospace and Electronic Systems. 2012;48(4):3006–3026.

[16] Sizov V, Cherniakov M, and Antoniou M. Forward scattering radar power budget analysis for ground targets. IET Radar, Sonar & Navigation. 2008;01 (1):437–446.

[17] Ustalli N, Lombardo P, and Pastina D. Detection performance of a forward scatter radar using a crystal video detector. IEEE Transactions on Aerospace and Electronic Systems. 2018;54(3):1093–1114.

[18] Ustalli N, Pastina D, and Lombardo P. Target motion parameters estimation in forward scatter radar. IEEE Transactions on Aerospace and Electronic Systems. 2020;56(1):226–248.

[19] Ustalli N, Lombardo P, and Pastina D. Target localization and velocity estimation in near forward scatter radar systems: preliminary results. In: 2019 International Radar Conference (RADAR); 2019. p. 1–6.

[20] Lombardo P, Pastina D, Daniel L, *et al.* Target motion estimation via multi-node forward scatter radar system. IET Radar, Sonar & Navigation. 2016;01:10.

[21] Rashid NEA, Antoniou M, Jancovic P, *et al.* Automatic target classification in a low frequency FSR network. In: 2008 European Radar Conference; 2008. p. 68–71.

[22] Cherniakov M, Abdullah RR, Jancovic P, *et al.* Automatic ground target classification using forward scattering radar. IEE Proceedings—Radar, Sonar and Navigation. 2006;153(5):427–437.

[23] Griffiths HD and Baker CJ. An Introduction to Passive Radar. Norwood, MA: Artech House; 2017.

[24] Suberviola I, Mayordomo I, and Mendizabal J. Experimental results of air target detection with a GPS forward-scattering radar. IEEE Geoscience and Remote Sensing Letters. 2012;9(1):47–51.

[25] Clemente C and Soraghan JJ. GNSS-based passive bistatic radar for micro-Doppler analysis of helicopter rotor blades. IEEE Transactions on Aerospace and Electronic Systems. 2014;50(1):491–500.

[26] Clemente C, Parry T, Galston G, *et al.* GNSS based passive bistatic radar for micro-Doppler based classification of helicopters: experimental validation. In: 2015 IEEE Radar Conference (RadarConf); 2015. p. 1104–1108.

[27] Martelli T, Colone F, and Lombardo P. First experimental results for a WiFi-based passive forward scatter radar. In: 2016 IEEE Radar Conference (RadarConf); 2016. p. 1–6.

[28] Losito A, Stentella M, Martelli T, *et al.* Automatic vehicles classification approaches for WiFi-based passive forward scatter radar. In: International Conference on Radar Systems (Radar 2017); 2017. p. 1–6.

[29] Colone F, Martelli T, and Lombardo P. Quasi-monostatic versus near forward scatter geometry in WiFi-based passive radar sensors. IEEE Sensors Journal. 2017;17(15):4757–4772.

[30] Krysik P, Kulpa K, and Samczyński P. GSM based passive receiver using forward scatter radar geometry. In: 2013 14th International Radar Symposium (IRS). vol. 2; 2013. p. 637–642.

[31] Contu M, De Luca A, Hristov S, *et al.* Passive multifrequency forward-scatter radar measurements of airborne targets using broadcasting signals. IEEE Transactions on Aerospace and Electronic Systems. 2017;53(3):1067–1087.

[32] Patel JS, Fioranelli F, and Anderson D. Review of radar classification and RCS characterisation techniques for small UAVs or drones. IET Radar, Sonar & Navigation. 2018;12(9):911–919.

[33] Rzewuski S, Kulpa K, Salski B, *et al.* Drone RCS estimation using simple experimental measurement in the WIFI bands. In: 2018 22nd International Microwave and Radar Conference (MIKON); 2018. p. 695–698.

[34] Gashinova M, Daniel L, Hoare E, *et al.* Signal characterisation and processing in the forward scatter mode of bistatic passive coherent location systems. EURASIP Journal on Advances in Signal Processing. 2013;2013 (1):36.

[35] Ilioudis C, Cao J, Theodorou I, *et al.* GNSS based passive radar for UAV monitoring. In: IEEE Radar Conference 2019; 2019.

[36] Skolnik MI. Radar Handbook, Third Edition. Electronics Electrical Engineering. New York, NY: McGraw-Hill Education; 2008.

[37] Persico AR, Kirkland P, Clemente C, *et al.* CubeSat-based passive bistatic radar for space situational awareness: a feasibility study. IEEE Transactions on Aerospace and Electronic Systems. 2019;55(1):476–485.

[38] Steigenberger P, Thoelert S, and Montenbruck O. GNSS satellite transmit power and its impact on orbit determination. Journal of Geodesy. 2018; 92(6):609–624.

[39] Rino C. Satellite Orbit Computation; 2010. uk.mathworks.com/ matlabcentral/ fileexchange/ 28888-satellite-orbit-computation.

[40] Woodward PM. Probability and Information Theory, With Applications to Radar: International Series of Monographs on Electronics and Instrumentation. vol. 3. Oxford: Elsevier; 2014.

[41] Davis ME. Advances in Bistatic Radar. vol. 2. SciTech Publishing, London; 2007.

[42] Malanowski M. Signal Processing for Passive Bistatic Radar. Norwood, MA: Artech House; 2019.

[43] Van Rossum WL and De Wit JJM. Forward scatter and bistatic radar tracking of people. In: 2018 IEEE Radar Conference (RadarConf18); 2018. p. 0435–0440.

[44] MathWorks. Variable Fractional Delay; 2019. uk.mathworks.com/help/dsp/ref/variablefractionaldelay.html.

[45] GoogleEarth. Caplaw model Flying Group; 2019. earth.google.com/web/search/Caplaw+model+Flying+Group.

Chapter 9

Radar UAV and bird signature comparisons with micro-Doppler

Matthew Ritchie[1], Colin Horne[1] and Nial Peters[1]

9.1 Introduction

This chapter reviews the similarities and differences between micro-Unmanned Aerial Vehicles (UAVs), also referred to as drones, and bird targets from the signals they present to radar sensors. The proliferation of small UAV platforms for commercial and personal use has increased significantly in recent years. All projections show a rapid increase in the utilisation of commercial civilian micro-UAV platforms across a number of use cases. In the United States alone, the consumer drone market was valued at $355M in 2015 and projected to be worth approximately $4BN by the year 2024 [1]. This broad range of applications includes capture of high-quality video imagery, remote surveying capabilities, delivery, agriculture and racing. The trend of increasing use has been highlighted in many journal articles and news reports, it seems that drones are here to stay in society and will only increase in their integration in everyday life.

With the significant increase in quantity of available platforms comes the increase in potential threat posed by them. Illegal activity that has been associated with the usage of drone platforms includes drug trafficking, use of surveillance by insurgents, attempted assassination, flying in the restrictive airspace of protected sites and disruptive political statements (e.g. displaying of controversial flag at a football game). This diverse set of illegal/unwanted activity has increased the demand for solutions that can first provide the ability to sense the presence of a drone and then also provide effectors to remove the drone from the situation. The first challenge of sensing a drone has generated solutions that provide a range of capabilities that include the basic ability to detect a drone, identify the intention of a UAV platform and even identify unusual payloads on a commercial drone which may prioritise it as a target. Both solutions do need to work in conjunction with each other in order to have a complete Counter UAV (C-UAV) system. The range of effectors that have been proposed include not only traditional military kinetic effectors but also falcons that can snatch a UAV from the air, other drones with nets

[1]Department of Electronic and Electrical Engineering, University College London, London, UK

to capture a UAV and kamikaze drone systems that purposely fly into the target UAV platform to destroy it.

The term UAV is generically used for all types of unmanned flying vehicles and that can cover a very broad range of targets. UAVs vary significantly in their size, range of flight, weight and cost. This range includes platforms such as the US Global Hawk system which is 45 m in length with a 35-m wingspan all the way down to tiny insect like platforms such as the Black Hornet nano-UAV, which is less than 20 g and only 10×25 cm in size. The majority of this chapter will focus on smaller platforms that are in the mini-, micro- and nano-taxonomy areas as defined by Table 9.1.

The detection and discrimination of drone targets are currently a high-priority area in the field of RF sensing. Alternative sensing options to RF do also exist and these include acoustic sensors that are able to detect and distinguish the motor sounds created by the UAV platforms [2,3]. The limitations of the distance these acoustic signals travel mean that typically the detection ranges for this category of sensor are reduced compared to an RF equivalent and they are very susceptible to background noises disturbing their probability of detection. Optical sensing solutions have also been extensively applied to this challenge and are effective in a broad range of scenarios. The advantage of a radar sensor over optical will be in situations of low visibility (smoke, fog, precipitation) and at night, when even with night vision sensors the capabilities of optical systems are reduced. It has also been proposed to use RF SIGINT or Electronic Surveillance (ES) devices to identify the control signals that are being sent between a base station and the UAV itself. This has been demonstrated to be successful and is a useful tool in the C-UAV challenge, but it is not feasible in all scenarios. In some situations, the UAV platform is autonomous and does not require control from a base station in order to perform its tasks or arrive at its planned destination. With the growing advent of autonomy being realised through examples such as swarm intelligence, it is easy to imagine that in the near future UAV platforms are capable of making their own decisions and therefore require far fewer communications signals.

The following report provides a list of C-UAV products that are categorised by sensing class (radar, Electro-Optical (EO), IR, acoustic, RF signals) [4]. This shows that >500 commercial products exist internationally that all look to address this growing and ever-changing challenge. This is a significant increase compared to

Table 9.1 Drone classifications

NATO UAS class	Maximum take off weight	Common taxonomy
I(a)	<200 g	Nano
I(b)	200 g to 2 kg	Micro
I(c)	2–20 kg	Mini
I(d)	20–150 kg	Small
II	150–600 kg	Tactical
III	>600 kg	MALE or HALE

the market survey completed by Sandia Labs years earlier [5]. The diversity in solutions demonstrates both the complex challenge that C-UAV systems have in detecting and mitigating these emerging threats and the variety of sensing modalities that can be employed individually or collectively in an attempt to solve the problem. It states that at least 95 countries now have drone capabilities which can at the very least provide much higher fidelity battlespace video feed intel than had been previously possible. Increasingly non-state actors are also using UAVs for both surveillance and actual attacks [6], this prioritises the ability to provide effective UAV surveillance systems to within the role of force protection.

The C-UAV systems broadly attempt to follow the following three key steps:

1. **Detection** – This is through the use of a variety of sensing modes aiming at identifying if a target is present or not. This step does not include the verification that the detected target is in fact a UAV platform or hostile.
2. **Classification** – Once detected, additional information gathered from sensors can be used to classify what the target is. This analysis may use image neural network recognition algorithms, acoustic pattern matching, or in the case of radar micro-Doppler, signature recognition.
3. **Action** – This may be the decision to take kinetic strike action, use electronic warfare methods to disable the drone or simply avoid it.

A key challenge to C-UAV systems is the discrimination between targets of interest and confuser targets that represent unwanted detections. Bird targets represent the most common type of confuser target. It is challenging to discriminate between them and UAV as they are a comparable target in size, speed and Radar Cross Section (RCS). They are present almost everywhere that radar systems have to operate and have been detected by radar since the early origin of radar engineering. Due to the similarity in radar signature, birds can cause common misclassification between them and the priority drone targets.

Within this chapter, the focus will be on the first two components of this process. The detection of drones with radar systems is very much feasible and comes down to the ability to transmit sufficient RF energy in the right direction such that enough is backscattered to a receiver. Of course, the practicalities of doing this in a compact form that is power efficient and fits the specific design constraints is a complex challenge but obtainable. The more elaborate step is the classification stage, a system that has a 100% probability of detection of UAV platforms but is never able to discriminate between a bird and person walking, and a drone would never be accepted as a viable solution. This emphasises the need for detailed investigations into the best means of discrimination of drone platforms from confuser targets if real-world solutions are to succeed.

Other areas of radar classification techniques such as High-Resolution Range Profiles (HRRPs), Synthetic Aperture (SAR) or Inverse Synthetic Aperture (ISAR) imaging can be used to classify moving targets. SAR is more suited for the imaging of static scenes and targets due to the way in which the image is formed using the phase information of a target. The issue with applying HRRP or ISAR processing techniques is that for small UAV platforms the bandwidths required to obtain sufficient detail would be very significant. This is not an impossible task, but due to the increasingly congested Electromagnetic Environment it is unlikely that a C-UAV RF sensor would be able to guarantee access to the required bandwidth to effectively classify these platform types. This is why micro-Doppler signatures are a well-suited feature for classification as it is possible to acquire the required temporal integration on a target using a steerable beam in order to generate Doppler with sufficient resolution and hence visualise a drone targets recognisable rotor blade signature.

The remainder of this chapter includes a review of UAV and bird radar signature research, followed by a description of target motion modelling. Next follows sections describing laboratory measurements of a small UAV using a fully polarimetric sensor and descriptions of the radar sensors we have used for bistatic and multistatic measurements of UAVs and birds. Acquisition of controlled radar measurements of bird and UAV targets using the NetRAD radar system are then described, followed by drone payload measurements, also employing NetRAD. Simultaneous X-band and L-band measurements of drone and bird targets using the NeXtRAD sensor are described along with details of the drone filter tested against these data. The final section of this chapter contains concluding remarks.

9.2 Review of UAV and bird radar signatures research

The detection of bird signatures using radar has been observed since the very early developments of modern radars in the 1940s. A Nature article from 1945 [7] describes how it was given permission for the first time by the Air Defence Research and Development Establishment to state that radar is capable of detecting birds. Experiments from 1941 detected gannets from a system based on Dover, which was initially published in a SECRET report and has since been declassified. These early observations of bird signatures were confirmed via simultaneous visual inspection. The phenomena that started the investigation into these detections were large flocks of starlings leaving their perches in the early morning creating propagating rings on a radar PPI display. These affects were dubbed as 'angels' by Lack and Varley [7] and Buss [8]. Ever since these early beginnings of radar systems, the detection of birds has been either a challenge or a focus of a wide range of research outputs and real-world systems.

Since these early developments there has been a constant strand of research that has been focused on the detection and observation of birds using radar for many decades. This domain is now growing significantly due to the desire to discriminate between these naturally appearing targets and small drones as birds represent the dominate confuser target in most scenarios. The challenge is not easily solvable due to the similar characteristics of these two classes of target. Both can fly low to the terrain, alternate between hovering and forward motion, represent low RCS signatures due to their size and materials. This is at odds with the traditional targets that large ground-based radar systems typically would detect and track. These traditional targets would be at very long ranges 10s–100s km, have a much larger RCS (particularly commercial airliners) and move at much faster speeds with reduced flight pattern changes. For all these reasons, the detection and tracking of small drone targets has required a significant amount of effort to adapt previous techniques that are no longer suitable.

The observation of birds only has been the sole focus of a number of publications in the area of biology and zoology journals. This includes analysis of bird migrations [9,10], monitoring of nesting seabirds [11] or identifying bird species using radar [12]. This zoology- and biology-focused research has mostly looked to leverage radar as a sensor able to detect, count biomass or migration numbers or detect areas birds travel to or through. An early review of radar use in detecting birds as well as insect targets can be found in [13], which includes a summary plot of RCSs from a number of sources for these targets plotted as a function target weight against RCS which is a useful review. Clearly radar has been successful in supporting this category of research but there is little work on highly detailed discrimination of bird targets from other targets, mostly because the UAV discrimination challenge is a new paradigm that has not been present for sufficient time to investigate extensively.

The radar signature from large sea birds was analysed within [14]. This work used a radar, a system in conjunction with a camera to gather data on sea birds and evaluate how the radar signature changed with the physical movement of the bird. Micro-Doppler signatures of the flying gannets showed that flapping movements were detectable within a micro-Doppler spectrogram. From these signatures, it was found that the ratio of the body to wing signature varied significantly from −30 to 5 dB across a range of azimuth angles of the target with respect to the radar. A challenge with this work is accurate estimation of the azimuth angle of the bird from the available video footage.

An excellent modern comparison of drone and bird signatures was reported within [15]. A total of 11 targets were included in the capture data set of which 10 were drones of different configurations (including helicopter, quadcopter and fixed wing classes). The micro-Doppler signatures of these platforms were then evaluated, and features extracted from them were based on eigenvalue decomposition. The data originated from an X-band Frequency-Modulated Continuous Wave (FMCW) 9.5 -GHz monostatic radar. The work did not look to fix aspect angle, distance or velocity within each recording; therefore, these variables change within each captured file analysed. The classification success achieved was high at 95% using a non-linear SVM classifier, which is impressive for an 11-class problem. These were measured in relatively controlled environments and with high SNR signatures but show promising signal processing methodologies to help support the challenge of drone and bird classification.

The use of polarimetric radar as a sensing method for discrimination between UAVs and birds was extensively reviewed by Torvik in his doctoral thesis [16]. Polarisation is widely used in radar remote sensing to provide additional information on the target or scattering phenomena that is occurring. In meteorological radar systems, polarisation can be used to differentiate between the type of precipitation and level of precipitation that is occurring by observing the ratio of HH/VV signatures and therefore calculating the 'roundness' of the raindrops and hence how heavily it is raining. In the context of UAV or bird platforms, polarisation can be used to evaluate the overall scattering ratio from a fixed target in order to help characterise it or differentiate it from another type of target. Even more effective is the analysis of how this polarisation ratio changes over a period of time. In the case of a bird, one can imagine how the superstructure of the bird's body and wings alter significantly as they move through a flapping motion. This change in shape provides a significant difference in how the EM wave interacts and therefore changes the resultant polarisation scattering [17]. In comparison, a drone does not undergo the same superstructure change over time and therefore would not be expected to present the same change in polarisation scattering as a bird. Analysis was performed to extract polarisation-based parameters which included values such as the Linear Depolarisation Ratio (LDR). This is defined as

$$\delta = \frac{2\langle |S_{HV}|^2 \rangle}{\langle |S_{HH}|^2 \rangle} \tag{9.1}$$

where δ is the LDR, S_{HV} and S_{HH} are the scattering cross and co-polar components of the target scattering matrix $\boldsymbol{S}$, respectively. It is assumed that components S_{HV} and S_{VH} are equal in this case, so they are interchangeable. By evaluating this change of this LDR over time, it is possible to discriminate between drone and non-drone (bird) targets using a fully polarimetric radar system.

This thesis used both modelling and empirical captured data to evaluate the difference in scattered returns from a variety of UAV and bird classes. The conclusions showed that the use of polarimetric method can enable improved discrimination between UAV and bird platforms. In terms of scattering signatures, it was found that the use of simple spheroid component parts in EM scattering modelling broadly represented the signature of bird targets in L and S frequency bands. The variation in RCS from broadside to front/end was found to be between 15 and 30 dB. At end/front of geometries the addition of the wings was found to be sufficient to require these to be incorporated into a model as they provided approximate equal scattering signatures as the body of the bird. At S- and L-band frequencies, the scattering from the wings of the bird is in the Rayleigh region. Therefore, the scattering from them is highly dependent on the polarisation of the illuminating signal.

The additional information that polarimetry provided for the classification of targets was found to be beneficial at both L and S bands. Without polarisation parameters being extracted, in general from the empirical data study, it has been shown that the classification success was found to be higher in horizontal polarisation compared to vertical. When polarisation is taken into account via the use of specially extracted polarisation features, it was found to increase classification success rates by 25% and 21% for L and S bands, respectively, which is certainly a significant benefit that would justify the required additional hardware capability [16]. This additional benefit is contradictory to previous comments from Skolnik on the benefits of polarimetry for target classification [18]. Here it was stated that the perceived lack of benefit for classification purposes was down to a few reasons which include interleaved pulses of different polarisation illuminating the target at a slightly different time step, lack of isolation between polarisations, multipath contributions and how the net polarisation of N scatters making up a target differs from the individual polarisation of each of these separately.

Recent measurements at the higher frequency bands of 24 and 94 GHz were performed by researchers at the University of St. Andrews [19]. These higher frequencies allow for wider bandwidths and greater sensitivity of measurement of the two target classes. This work compared the signatures of three small UAVs (two quadcopter platforms and one hexacopter) to the signature of four birds (Northern Hawk Owl, Harris Hawk, Indian Eagle Owl and Tawny Eagle). The average modal RCS values for the drone targets varied from −20 to −8 dBsm, while the bird signatures ranged from −30 to −19 dBsm at 24 GHz. The measurements at 94 GHz reconfirmed these modal RCS values to within ~1 dBsm. This does show that radar engineers should expcct larger reflections from drone platforms in comparison to bird signatures on average. Although in comparison the maximum RCS values from the bird signatures were found to range from −5 to 5 dBsm at 24 GHz which was much more comparable to the −6.2–8 dBsm values the drone exhibited. Perhaps this is linked to the bird signatures creating a larger flash in reflectivity when their wing structure is oriented such that a greater amount of RF energy is reflected back to the radar system.

9.3 Target motion models

This section provides details on the modelling of UAV rotor blades micro-Doppler signatures. A model was adapted from a simple helicopter-based micro-Doppler signature model which used a single-point scatterer to represent the body of the helicopter and four-point scatterers to represent the tips of the rotor blades. The adaptions made were used to create a model for UAV micro-Doppler signatures which has flexibility in the selection of the parameters given next. The model imposes no restrictions on the values of these parameters, leading to the possibility of practically unrealisable models:

- number of rotors,
- number of blades,
- blade length,
- body to rotor arm length,
- UAV radial motion,
- counter rotation of alternate rotors,
- body/rotor orientation,
- aligned or random initial blade angles,
- multiple scatterers per blade,
- body and blade tip RCS and
- additional blade scatterers RCS scale factor.

This degree of flexibility was introduced to allow for modelling a wide variety of scenarios and to isolate what are the fundamental components creating the observed micro-Doppler signatures. By tailoring these values, it is possible to model a specific UAV of interest, or to observe how any changes made to them induce changes in the observed micro-Doppler signatures. With the current model, all the parameters remain constant for the duration of a model run, no variation in RCS or number of scatterers occurs. The RCS of the UAV body and the RCS of the blade scatterers are set to constant values for each model run. As such there is no dependence in the model for these values on the body or rotor materials, or on the wavelength of the illuminating signal.

The geometry of the UAV platform and its rotor blades is initially setup in one of the configurations illustrated in Figure 9.1. This allows for the changing of the orientation and location of the rotors with respect to the sensor as well as the alignment of each of the rotor blades. The sensor and UAV are assumed to reside in the same horizontal plane.

Typical parameters used for the modelling of the micro-Doppler signatures are shown in Table 9.2. The simplest case which can be modelled is a single blade with a single-point scatterer at the blade tip, along with the UAV body, the signature of which is shown in Figure 9.2(a). The rotation rate in this example is slowed to only 5 rps to visualise the individual blade flash returns. The sinusoidal structure seen in Figure 9.2(a) is a direct result of the changing Doppler frequency induced by the blade tip scatterer on the radar return as the blade tip traverses its circular path. This

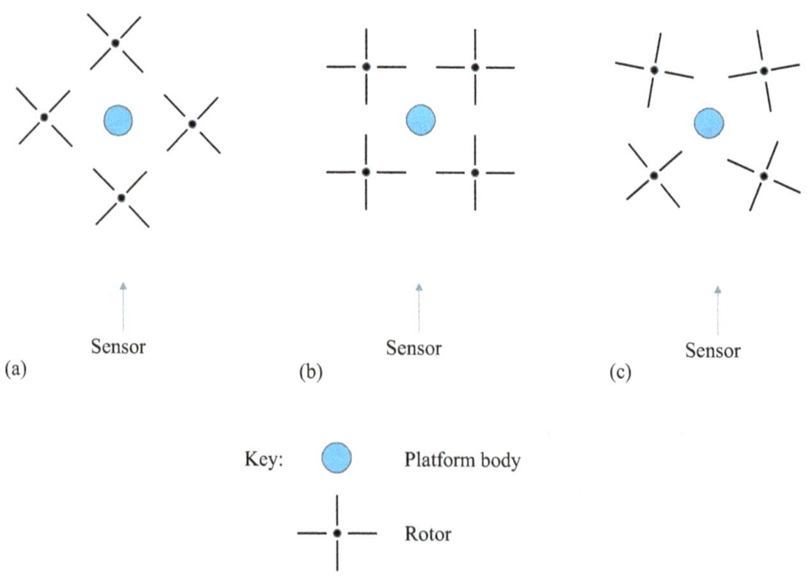

Figure 9.1 UAV and sensor configurations (a) aligned start with diamond orientation, (b) aligned start with square orientation, (c) random start with square orientation

Table 9.2 UAV micro-Doppler model parameters

Parameters	Value
Number of rotors	1–4
Rotor rotation	5–150 rps
Blades/rotor	2 blades per rotor
Blade length	0.13 m (typical – small UAV)
Scatterers/blade	Multiple
Radial velocity	20 m/s
Radar frequency	8.5 GHz
Body RCS	>1 m^2
Blade RCS	0.1 m^2

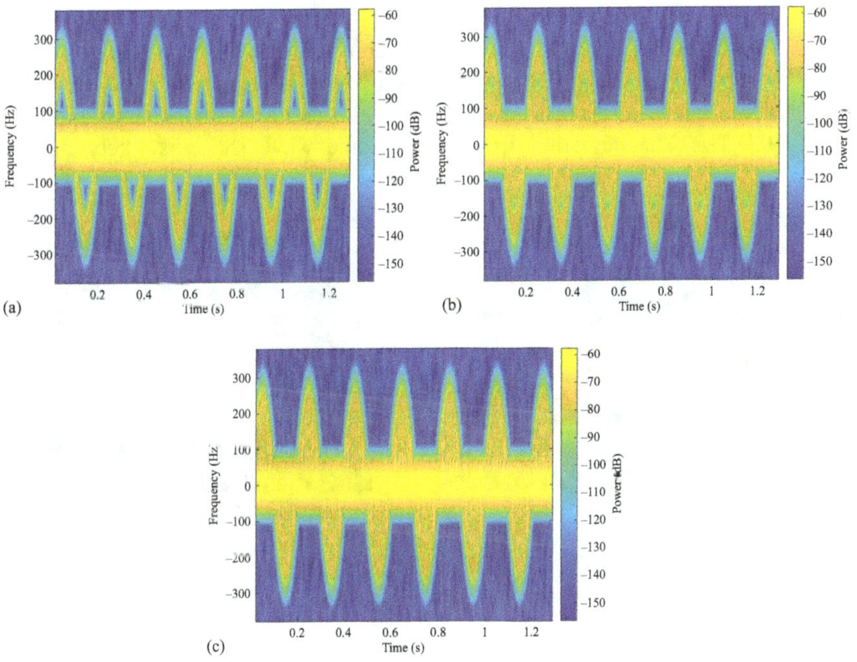

Figure 9.2 Micro-Doppler of UAV body and single blade with (a) a single scatterer per blade, (b) two scatterers per blade, (c) three scatterers per blade (carrier frequency is 8.5 GHz, PRF is 1 kHz, rotor rotation rate is 5 rps, spectrogram window is 50 ms, spectrogram overlap is 90%)

scenario is then extended to include two and three scatterers per rotor blade, see Figure 9.2(b) and (c). It can be seen that as more scatterers are added, the area under the sinusoidal blade flash outline becomes more occupied due to the range of Doppler frequencies created by the various scatterers. The results shown from the model used parameters that match either the NeXtRAD X-band system (8.5 GHz) or a Volcano monitoring FMCW radar system (10.25 GHz) developed by Dr Nial Peters at University College London (UCL) to allow validation of the modelling result with real experimental captures.

When moving from one blade per rotor to two and then three, the changes observed can be seen within Figure 9.3. This shows how the increase in blade numbers increases the observed frequency that blade flashes occur and eventually results in a smoothing of the signature such that overlapping blade flash signals are almost continuously present.

Model outputs were compared with controlled laboratory-based measurements carried out using an X-band FMCW radar with 400 MHz of bandwidth, for validation purposes. Measurements representing single scatters at the blade tips were undertaken by simulating each scatterer by a small metal ball, of approximately 12 mm radius, mounted on a cork disc driven by a low-speed motor. Additional balls are added between the tip and the disc hub to represent more scatterers on each blade.

The single scatterer per blade result is shown in Figure 9.4 for the balls being placed at a distance of 8 cm from the disc centre and a rotation rate of approximately 0.8 Hz. The deviation from a pure sine wave shape is likely due to

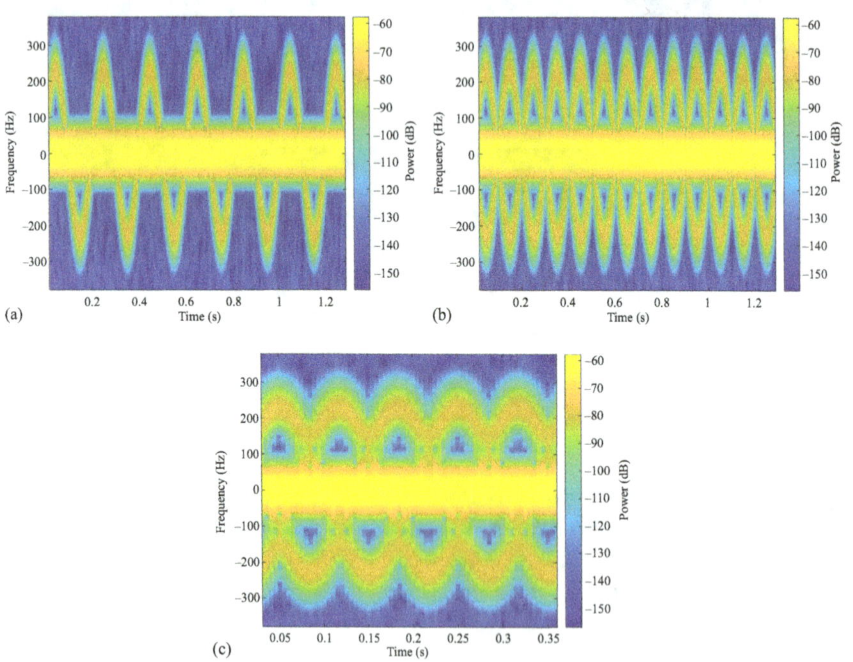

Figure 9.3 Micro-Doppler of UAV body and single scatter per blade with (a) a single blade, (b) two blades, (c) three blades (carrier frequency is 8.5 GHz, PRF is 1 kHz, rotor rotation rate is 5 rps, spectrogram window is 50 ms, spectrogram overlap is 90%)

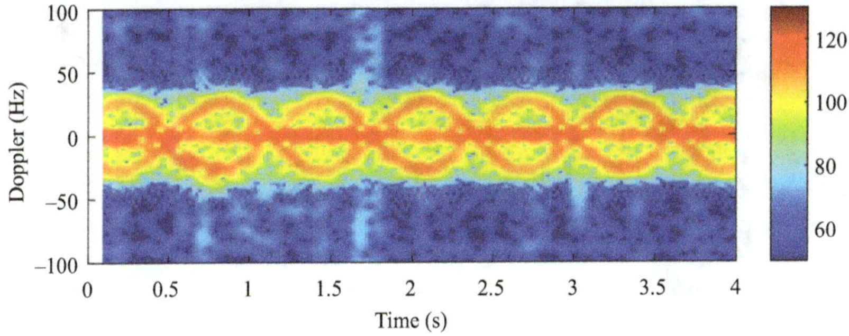

Figure 9.4 Laboratory measurement of two ball bearings placed at 8 cm from motor centre representing point scatterers at rotor tips (carrier frequency is 10.25 GHz, PRF is 1 kHz, rotor rotation rate is approximately 0.82 rps, spectrogram window is 250 ms, spectrogram overlap is 90%)

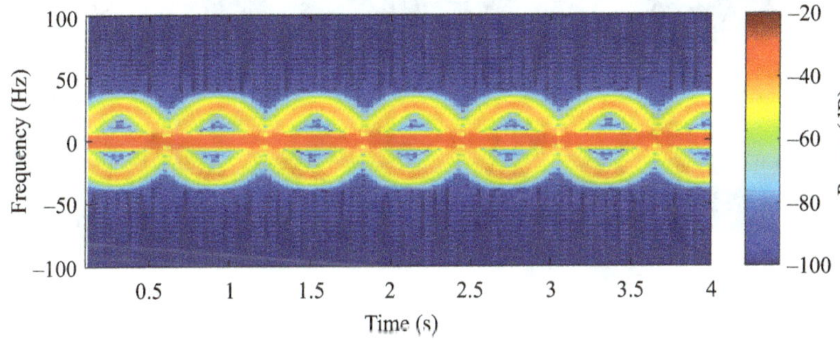

Figure 9.5 UAV radar model output for a single scatterer placed at each blade tip for a two-bladed rotor of radius 8 cm (carrier frequency is 10.25 GHz, PRF is 1 kHz, rotor rotation rate is 0.82 rps, spectrogram window is 250 ms, spectrogram overlap is 90%)

inaccuracies in the experimental setup, for example the disc not being perfectly flat with respect to the motor shaft and positional errors for the balls.

The equivalent output from the UAV radar model is shown in Figure 9.5 and confirms the same basic structure. The model output shows a cleaner, more symmetric structure as none of the experimental inaccuracies are included. The area under the sine wave structure in the experimental result is smeared, not seen in the model output, due to the spread of Doppler components caused by the experimental setup.

Replacing the disc and ball bearing with a 26-cm diameter plastic rotor from a UAV provides the result shown in Figure 9.6. The use of a real rotor results in the sine wave structure becoming less pronounced and instead a vertical structure

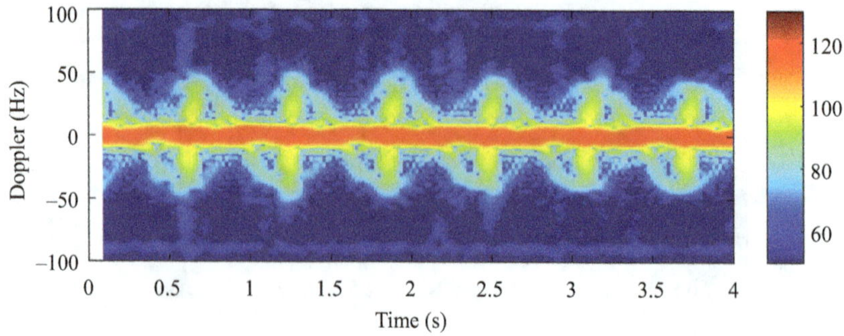

Figure 9.6 Laboratory measurement of 13-cm radius UAV plastic rotor (carrier frequency is 10.25 GHz, PRF is 1 kHz, rotor rotation rate is approximately 0.82 rps, spectrogram window is 250 ms, spectrogram overlap is 90%)

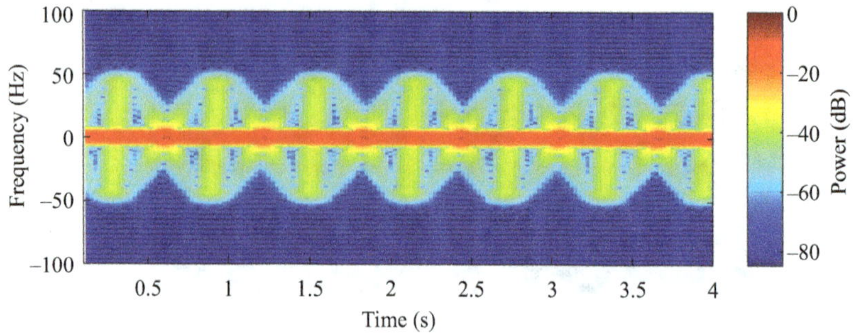

Figure 9.7 UAV radar model output for nine scatterers per blade (carrier frequency is 10.25 GHz, PRF is 1 kHz, rotor rotation rate is 0.82 rps, spectrogram window is 250 ms, spectrogram overlap is 90%)

becoming the dominant feature, caused by the interaction of multiple scatterers along the blade.

This experimental result in Figure 9.6 was replicated using the UAV radar model and showed good degree of similarity when employing nine, equally spaced, scatterers per blade, see Figure 9.7. It was found that when employing less than nine scatterers, additional vertical line segments appeared in the structure.

In comparison to the previous figures where the predominate signal that is observed is based on a blade flash, as the rotation rate of the rotor is increased to more realistic speeds and the Short-Time Fourier Transform (STFT) applied to the signals is long compared with the rotation time, the micro-Doppler phenomena change to become HElicopter Rotor Modulation (HERM) lines. This affect presents itself as horizontal lines in the micro-Doppler spectrograms compared to the vertical blade flash lines previously shown, see Figure 9.8.

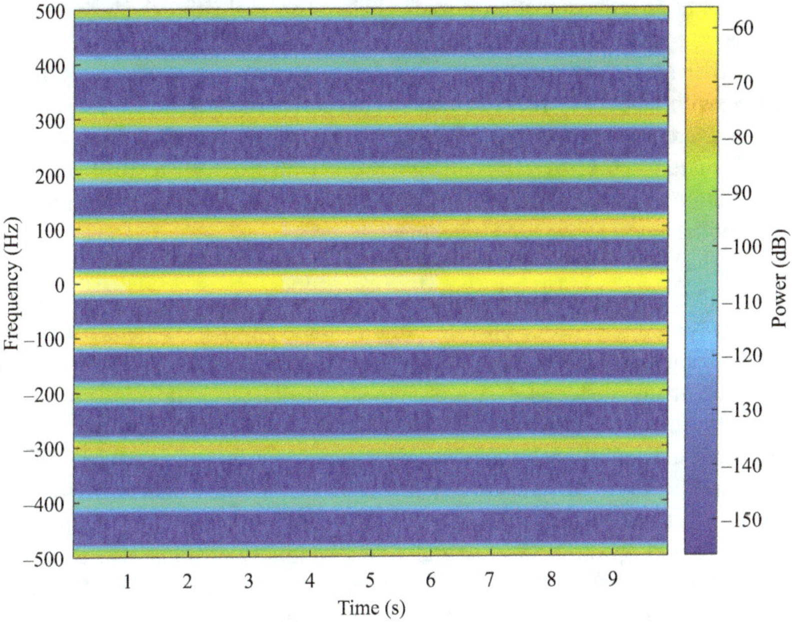

Figure 9.8 Micro-Doppler of modelled UAV with a single, two-bladed rotor, rotating at 150 rps with a single scatterer per blade (carrier frequency is 8.5 GHz, PRF is 1 kHz, rotor rotation rate is 150 rps, spectrogram window is 200 ms, spectrogram overlap is 90%)

The HERM line separation is defined as being

$$\Delta H = \Omega \times N \tag{9.2}$$

where the ΔH is a product of propeller rotational rate Ω, and number of blades N, per propeller. These parameters are independent of the radar RF in comparison to the maximum blade flash value which is directly related to RF via

$$\{f_d\}_{\max} = \left(4\pi L \Omega \frac{1}{\lambda}\right) \cos\theta \tag{9.3}$$

where L is the blade length, λ the wavelength, and θ is the angle between the sensor and the plane of the UAV rotor. The sample rate of the radar signal (the pulse repetition frequency (PRF)) and the STFT applied to the range profile data are jointly responsible for which phenomena are observed in the Doppler-time domain. If the PRF is too low and the integration time of the STFT is too long, the observed phenomena will be the HERM lines (as shown in Figure 9.8). On the other hand, if the PRF of the radar is sufficiently high along with short STFT window periods, the rotor blade signature is represented by the actual blade flash contributions (see Figure 9.3). Considerations should be taken when evaluating what phenomena are expected within a radar signal processing chain when evaluating the best means for both detecting and exploiting the signals that are present.

The HERM line presentation of the micro-Doppler signatures may also be compared with laboratory measurements. This comparison is shown in Figure 9.9 for the measured rotor (left) and the model (right), where the HERM lines are seen at approximately 1.6-Hz spacing, in agreement with (9.2) for a two-bladed rotor with a rotation rate of approximately 0.8 Hz. The measured results from the rotor and the model again show good agreement.

The other key target class of interest to this research is birds. It is an over-simplification to label all birds under one target class as there is a great deal of variation between species of bird in terms of their physical characteristics (size/weight/wingspan) as well as how they fly. Their flight movements are complex and can include many different types of movement. The general classes for their movements can be thought of as two key modes, flapping and gliding. The frequency and duration of each mode are very dependent on species, where they are flying, duration of flight and if they are flying in a fixed course or frequently changing direction. The biomechanics of bird flight has been studied in great detail for a broad range of birds and a useful recent summary example can be found in [20]. In addition to this, previous studies on bird RCS have been conducted to evaluate the expected average reflected energy at a given frequency [21]. The mean RCS for a pigeon target were quoted as being 15 cm^2 at X-Band and 80 cm^2 at S-band, whereas a smaller sparrow was found to have a mean RCS of only 1.3 cm^2 at X-band and 12 cm^2 at S-band.

Although flight movements have been studied from a biomechanics perspective, very little prior work exists on the linking of micro-Doppler signatures and bird flight dynamics from a modelling perspective. Prior applied literature in the area focuses more on the application of avoiding 'bird strikes' to aircraft [22,23].

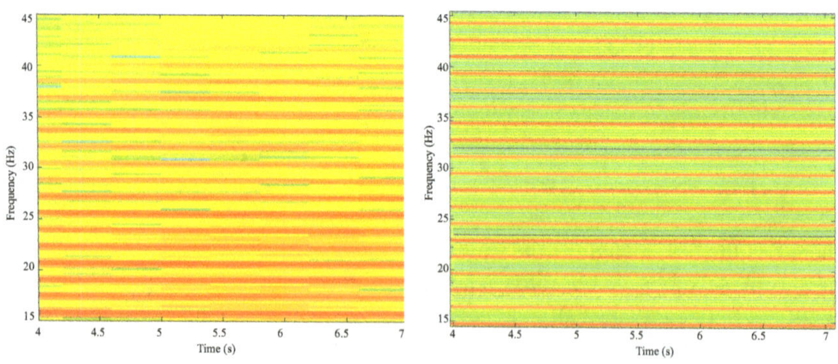

Figure 9.9 Laboratory measurement of 13 cm radius UAV plastic rotor (left) and UAV radar model output (right) (carrier frequency is 10.25 GHz, PRF is 1 kHz, rotor rotation rate is 0.82 rps, spectrogram window is 4 s, spectrogram overlap is 90%)

This is where a single bird or a flock would hit an aircraft and cause severe damage. This work focuses on the modelling of the detectability of a bird based on RCS calculation and radar parameters and not the modelling of micro-Doppler for classification. Torvik also presented some evaluation of RCS modelling of bird targets within [16], but again this did not replicate any micro-Doppler signature that a bird may produce. Therefore, a research gap clearly exists in high fidelity micro-Doppler motion models for bird targets to allow for the test and evaluation of both detection and classification algorithms.

9.4 Fully polarimetric, multiple observation angle laboratory measurements of UAV target

In this section, we describe the physical arrangement and measurements taken in an anechoic chamber environment of a small UAV (DJI Spark). Measurements were taken at various sensor-target observation angles for each polarimetric configuration, HH, HV, VH and VV. The data were captured using an X-band FMCW radar, further information can be found here on the system [24]. The operating parameters used in this measurement campaign are shown in Table 9.3.

The physical arrangement of the experimental measurements is based on a boom which is initially placed horizontally (0°) and can be raised such that it forms an angle with the horizontal plane up to a maximum angle of 90°. The radar antennas were mounted on a fixed platform at the lower end of the boom such that their pointing direction follows the angle of the boom and were directed towards the upper end of the boom. The target UAV is mounted on a small platform which pivots at the upper end of the boom such that the UAV can be operated in as close as possible to a normal horizontal hovering position, while allowing the observation angle between the sensor and the target to vary. Along the boom, a small metal ball was attached in order to act as a calibration target. The setup is illustrated in Figure 9.10.

Each data collection involves raising the boom to the required angle, ensuring that the UAV is horizontal on the pivot platform. The UAV motors are started and allowed to run at the default idle rate. In order to capture the polarimetric variation, the antennas were physically rotated to the desired configuration at each elevation angle.

Table 9.3 Summary of radar parameters used for laboratory-based polarimetric measurements of UAV

Radar type	**FMCW**
Centre frequency	10.25 GHz
Bandwidth	400 MHz
PRF	2 kHz
ADC sample rate	1.4 MHz
Chirp period	0.4 ms
Data collection period	20 s

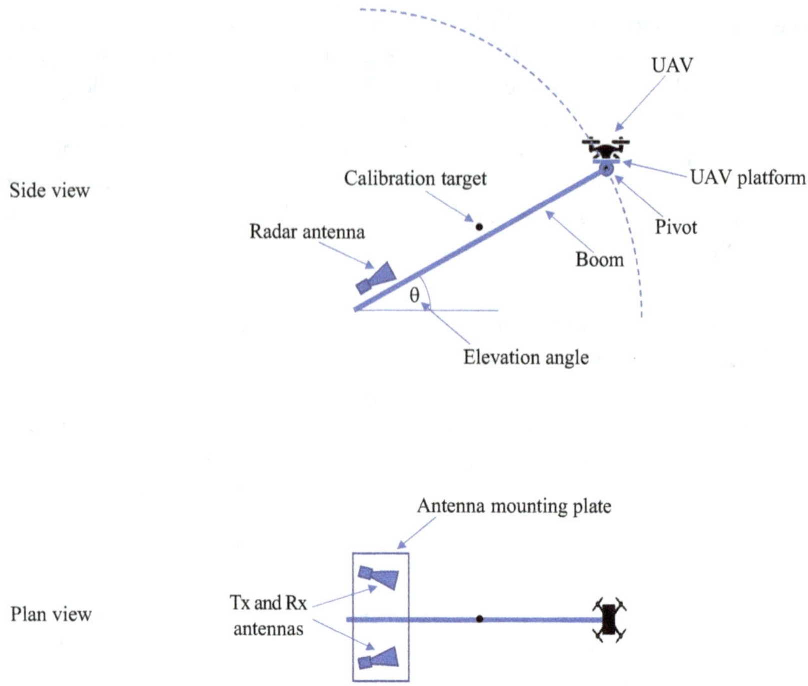

Figure 9.10 Illustration of polarimetric measurement setup

Measurements were taken at HH, HV, VH and VV polarisations and with observation angles at 20° steps between 0° and 80° and an additional measurement at 90°. The 0° angle observation represents the in-plane measurement where the radar sensor measures the UAV side-on. The 90° angle case is where the UAV is vertically above the radar sensor. The 90° observation might also be interpreted as a fixed wing, multi-propeller aircraft receding from the radar sensor.

For each polarisation, a calibration measurement was taken. This was taken at 0° inclination angle and without the drone rotors spinning. The mean intensity of the zero-Doppler frequency for the range cell containing the calibration target was taken as the calibration value for all the measurements taken for that polarisation. This is a preferable strategy to using the zero-Doppler intensity of the drone body as a calibration, since the RCS of the drone body will change with inclination angle. The mean HERM line intensities presented in this section are then computed as follows: for each radar time sample, the maximum intensity over a small window (typically around 100 Hz) of Doppler frequencies encompassing the HERM line of interest is computed. These values are then averaged over the full measurement time span (20 s) and converted to dB using the previously recorded calibration value as the 0-dB point. This calculation can be expressed as follows:

$$\overline{h} = \frac{\sum_{t_{idx}=0}^{t_{end}} \max\left(D_{\left[t_{idx},h_f-h_{lim}:h_f+h_{lim}\right]}\right)}{t_{end}+1} \tag{9.4}$$

$$\overline{H} = 20 \cdot \log_{10}\left(\frac{\overline{h}}{c}\right) \tag{9.5}$$

where $\overline{H}$ is mean HERM line intensity in dB, c is calibration value, t_{end} is the time index of the final measurement, D is Doppler–time–intensity array, h_f is the HERM line frequency of interest, and h_{lim} represents the range of frequencies considered.

The strongest SNR HERM line at ~175 Hz and its negative frequency counterpart were selected to allow a clear comparison between aspect angles to be evaluated.

Figures 9.11–9.14 show spectrograms of the range cells containing the drone at 0°, 40° and 90° inclination angles for HH, HV, VH and VV polarisations, respectively. These spectrograms were produced using time windows of 400 samples (corresponding to 0.2 s) with an overlap of 90%. The results shown in Figures 9.15 and 9.16 were derived from these spectrograms (and similar ones at different inclinations which are not shown here for brevity).

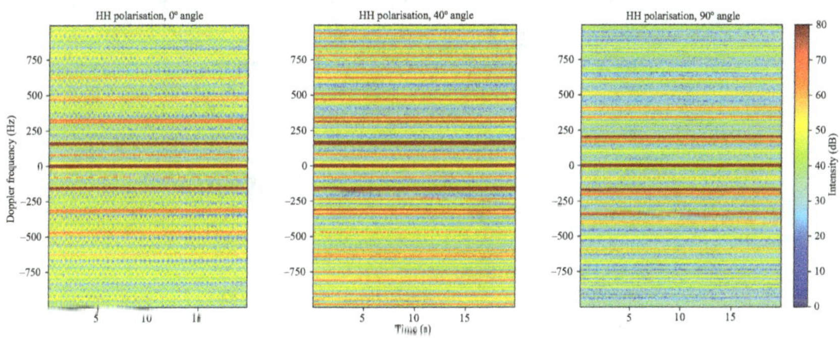

Figure 9.11 Spectrogram for HH polarisation at 0°, 40°, 90° inclination angles. Intensities are not calibrated to the calibration target

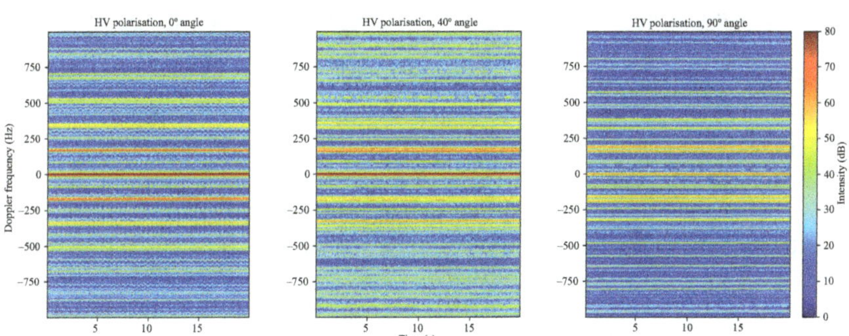

Figure 9.12 Spectrogram for HV polarisation at 0°, 40°, 90° inclination angles. Intensities are not calibrated to the calibration target

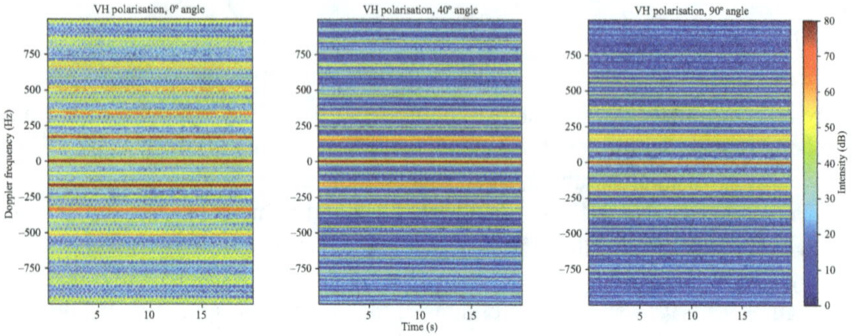

Figure 9.13 Spectrogram for VH polarisation at 0°, 40°, 90° inclination angles. Intensities are not calibrated to the calibration target

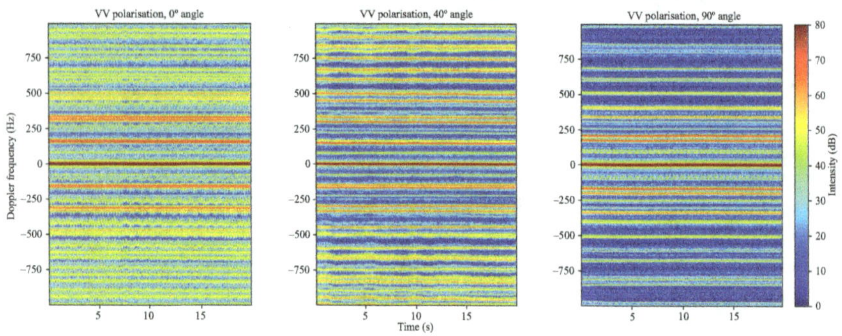

Figure 9.14 Spectrogram for VV polarisation at 0°, 40°, 90° inclination angles. Intensities are not calibrated to the calibration target

Figure 9.15 shows the variation of mean HERM line intensity (as defined earlier) with inclination angle of the boom (drone elevation angle) for each of the four polarisation configurations. Where we refer to a polarisation by a two-letter abbreviation, we list transmit polarisation first, e.g. HV refers to horizontal polarisation on transmit and vertical on receive. Figure 9.16 shows how the mean HERM line intensity at 0° inclination angle changes with polarisation configuration.

Figures 9.15 and 9.16 show that an HH polarisation configuration gives the highest intensity HERM lines across all viewing angles and is therefore the most suitable configuration for C-UAV systems. In general, HERM line intensity drops towards mid-range elevation angles then recovers somewhat towards high elevation angles (although VH is an exception to this).

Somewhat surprisingly, HV and VH configurations show markedly different behaviours to each other, with HERM lines in VH showing a consistent decrease in intensity with elevation angle, whereas HV shows some recovery of intensity towards high angles (similar in behaviour to HH and VV). Above 20° elevation

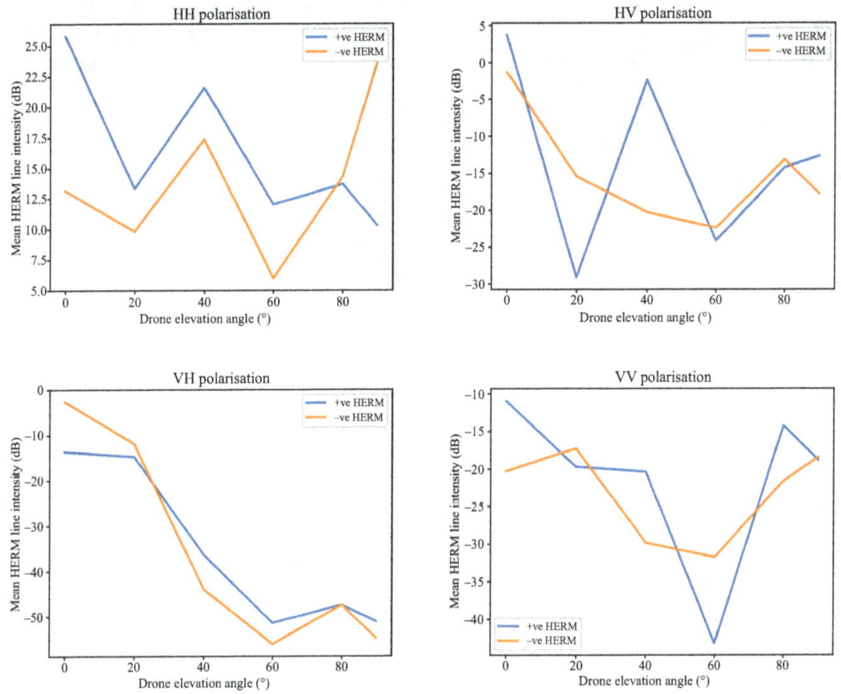

Figure 9.15 Mean HERM line intensity against drone elevation angle plots for each of the polarisation configurations. Results for both the positive Doppler HERM line and its negative Doppler counterpart are shown

angle, VH consistently provides the poorest HERM line intensity making it the least suitable configuration for C-UAV applications.

Figure 9.15 shows that both HH and HV polarisations show a peak in HERM line intensity at 40° elevation angle. We believe that this is caused by the rear (furthest from the radar) rotors of the drone being obscured from the radar's line of sight by the body of the drone when it is at 20° elevation angle. This drastically reduces the HERM line intensity at 20° elevation angle, resulting in a peak in intensity at 40° when the rotors are no longer obscured. Such an effect is not visible in the VH and VV data; this may be due to the blades being less obscured when they are parallel to boresight and more visible in the vertical polarisation.

HERM lines are clearly visible at 90° elevation angle for all polarisations. This is unexpected since there should be very little motion along boresight in this configuration. However, the results show not only that HERM lines are visible but also that they are of higher intensity than for mid-range elevation angles (with the exception of VH). A detailed explanation of this phenomenon is beyond the scope of this study; however, we hypothesise that there may be multiple contributing factors to HERM line creation. Although classical blade flash, as previously described in this chapter, is clearly one of them, by itself it does not explain the

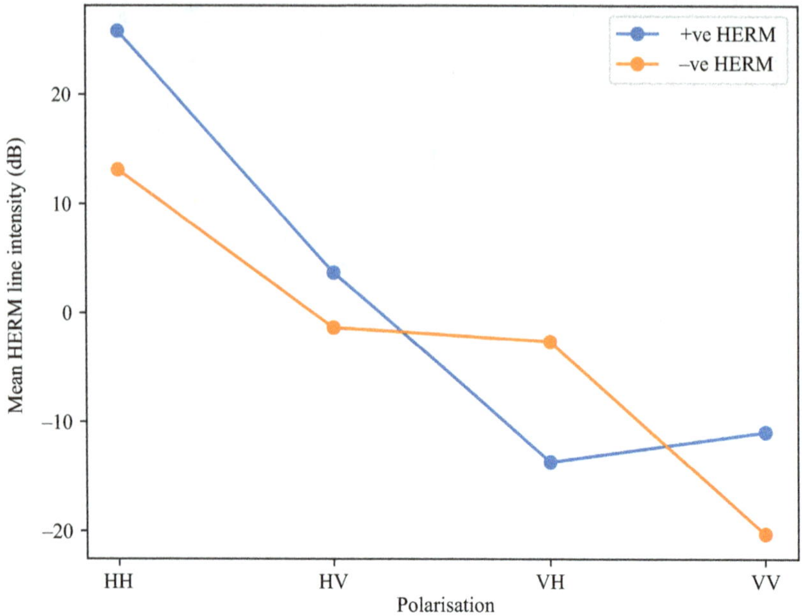

Figure 9.16 Mean HERM line intensity at 0° drone elevation angle for the different polarisation configurations. Results for both the positive Doppler HERM line and its negative Doppler counterpart are shown

visibility of HERM lines at 90° elevation angle. Other possible factors could be obstruction of the blades by the drone body at a particular rotation angle leading to a periodic variation in target RCS, or a periodic modulation of target return caused by blades moving in and out of alignment with the system polarisation.

The 90° elevation angle measurements are approximately equivalent to the fixed-wing UAV case, where the rotor is mounted in the vertical plane, although to be entirely analogous the radar would have to be mounted above the drone. However, the fact that HERM lines can still be observed in this orientation suggests that the techniques developed in this study for distinguishing drones from non-drone targets may also be applicable to fixed-wing drones.

As in the NeXtRAD data, see Section 9.8, there is a distinct asymmetry between the positive and negative Doppler frequency HERM lines. As shown in Figure 9.15, the level of asymmetry appears to be dependent on the elevation angle of the drone, and for the case of the HV polarisation positive and negative even show significantly different trends with angle. Specifically, the negative Doppler frequency line shows no peak in intensity at 40° elevation angle in sharp contrast to the positive Doppler frequency line. Positive Doppler frequency lines are caused by the leading edges of the blades, whereas their negative frequency counterparts are caused by the trailing edges of the blades. Since the two edges of the blades are physically different in shape, it seems likely that their RCSs will not only be different but will also have different responses with respect to viewing angle. This may explain the asymmetries in HERM line intensities that we observe.

9.5 Bistatic and multistatic radars used to gather bird and drone data

Two multistatic radar systems have been developed by both UCL and University of Cape Town (UCT) to gather data on both drones and bird signatures. These two radar systems are NetRAD and NeXtRAD which are both three-node pulsed Doppler radars. The first system developed was NetRAD, which started development in the early 2000s and supported a number of PhD projects over a period of almost 15 years.

9.5.1 NetRAD

NetRAD is a multistatic radar sensor system developed by UCL and UCT [25–27]. This radar system was developed initially as a three-node pulsed Doppler multistatic sensor network. Each node is identical and is capable of operating as a transceiver unit to both send and receive radar waveforms, see Figure 9.17. The sensor system was designed to operate in the S-band frequency range (2.4 GHz), this was because commercial components are readily available in this frequency band, as well as the fact that ISM band allows for low power transmissions without requiring a spectrum licence. The programmable parameters this system is capable of varying include its bandwidth, pulse length and PRF. Initially the system was capable of operating synchronously through the use of wired connections between the nodes. In collaboration with UCT, it was adapted so that synchronisation was provided via a GPS Disciplined Oscillator (GPSDO) solution [28]. This enabled wireless synchronisation across the multi-node network where baselines of up to 5 km were proven feasible. The specification for the system can be seen in Table 9.4.

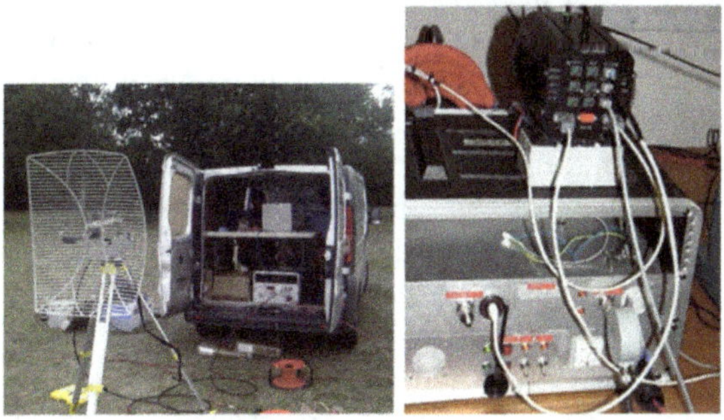

Figure 9.17 Image showing single node of a NetRAD radar system deployed in a van and on a bench top with GPSDO device injecting the required synchronisation signals

Table 9.4 NetRAD radar parameters list

Parameters	Value
Central frequency	2.4–2.5 GHz – S-band
Bandwidth	45 MHz
Output power without high power amplifier	+23 dBm
Output power with high power amplifier	+57.7 dBm
Radar waveform	Pulsed
PRF	1 Hz–10 kHz
Pulse length (typical)	0.1–20 μs
Antenna gain	27 dBi
E-plane/H-plane 3 dB beamwidth	11°/8°

The system has now been retired, but previous experimental campaigns using this system include measurements of individuals walking, carrying simulated rifle objects, as well as drone (i.e. UAV) and bird targets. Through capturing the movements of these targets in a variety of multistatic geometries, the advantages of a radar-sensing network have been quantified through the classification of the micro-Doppler signatures observed.

This system was used in a number of multistatic drone micro-Doppler radar measurements, some of which will be reviewed next. The key areas of investigation of interest to this chapter are distinguishing if a drone is carrying a payload and the analysis of bird and drone micro-Doppler using a multistatic system.

9.5.2 NeXtRAD

The NeXtRAD radar system is also a multistatic radar system developed jointly by UCL and UCT. This system is the successor to NetRAD and has a significant number of improvements over that system. It is also a coherent pulsed Doppler radar which is made up of three radar nodes. The key differences are that it is a multichannel system which enables polarimetric experimental captures; it is capable of dual frequency operation as well as having a much improved digital back-end and Command and Control architecture. One node operates as the transceiver and the other two as identical receivers. The system can operate at L-band with one receiving channel and with two at X-band. This allows for simultaneous co- and cross-polarised measurements at X-band to be performed. To provide the synchronous local oscillator signals, GPSDOs are used to input a common 100-MHz reference as well as a phase synchronous one pulse per second. The transmitted waveforms that have been used for all the data shown here are Linear Frequency-Modulated (LFM) signals which have a 50-MHz bandwidth. The system does have arbitrary waveform control and therefore is flexible in this design, but only LFM signals, with varying duration, have been utilised in the UAV measurement campaigns. Full specifications for the system can be seen in Table 9.5.

Table 9.5 NeXtRAD specifications

Parameters	Value
X-band peak output power	400 W
L-band peak output power	1.4 kW
No of Rx channels (X/L band)	2/1
PRF	1 Hz–10 kHz
Antenna gain	27 dBi
L-band antennas beamwidth	13.9° azimuth/12.4° elevation
X-band antenna beamwidth	9.1° azimuth/10.4° elevation

This system has been used to capture the measurement of sea clutter, drones and small boat targets while based in Simon's Town region in South Africa [29–34].

These two radar systems are almost unique in their position as coherent multistatic radars that are capable of networked sensing of targets using GPSDO synchronisation [28]. Very few comparable systems exist in open academic environments and therefore the data from these systems represent a rare opportunity for investigation of multistatic target signatures as well as clutter returns.

9.6 NetRAD bird and drone S-band measurements

The NetRAD sensor was deployed in 2016 for trials specifically focused on capturing both bird and drone signatures. This measurement campaign used 'cooperative' bird targets which were arranged via a falconry centre. This enabled the pre-planning of way point positions such that known ranges, orientations and flight paths could be configured to maximise the utility of the data and to arrange clear meta-data on the position of the bird which is typically impossible with 'non-cooperative' natural bird targets. The radar parameters used for the capture were a PRF of 5 kHz, transmit power of 23 dBm, LFM chirp with a bandwidth of 45 MHz and each capture was for 30 s, but only the sections which had the target in flight were selected for analysis. The measurements shown in this section were first published within [35]; readers are referred to this paper for further details.

The bird targets that were captured included a Hooded Vulture, Eurasian Eagle Owl and a Barn Owl. These varied significantly in their size, wingspan and weight, with the Vulture being the largest with a wingspan of approx. 1.7 m and weight of 1.8 kg, compared to the barn owl which had only an 80-cm wingspan and weighed 280 g, while the eagle owl represented a medium value between these two extremes. These three bird targets were then compared to a quadcopter drone target which was flown over the same waypoint positions once the birds were safely put away. This type of measurement is useful as it allows direct comparison of bird and drone signatures from a prearranged geometry which is rarely available.

The micro-Doppler signatures from the three bird targets can be seen in Figure 9.18. The images have been normalised to a peak of 0 dB in each case but

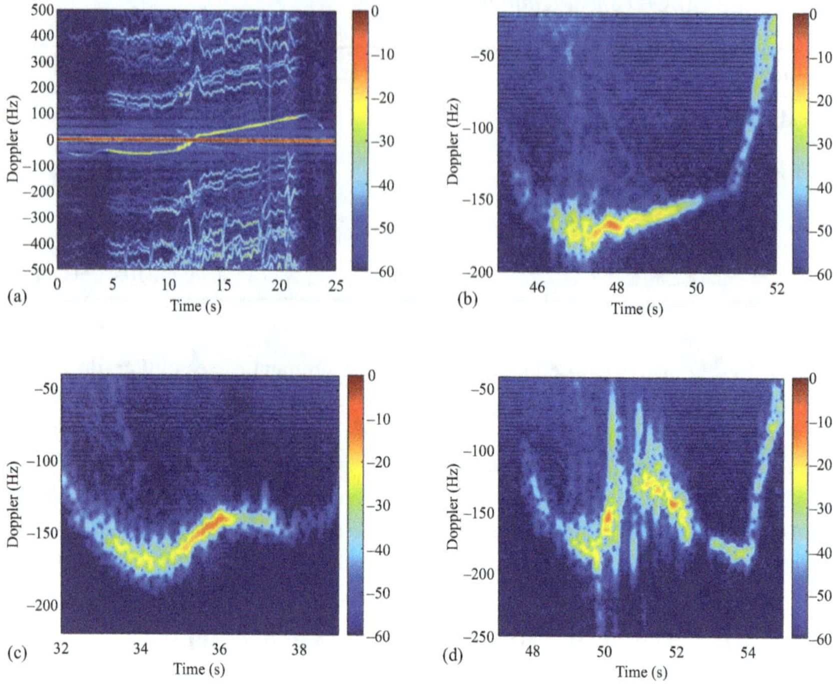

Figure 9.18 Drone and bird monostatic micro-Doppler signatures from HH polarisation NetRAD node: (a) DJI Phantom 2 Quadcopter, (b) Eurasian Eagle Owl, (c) Barn Owl, (d) Hooded Vulture

are not calibrated; therefore, you cannot compare a given SNR from one figure to another. The spectrogram window used to create the data shown was 0.3-s long, with an overlap of 95%. Clearly a significant difference can be observed between the captures shown for the drone target and all three bird targets. The drone target produced HERM line responses in the outer Doppler bins which were not present in any of the bird captures. The origin of these HERM lines was described in detail within Section 9.3. In comparison, the bird targets show two clear different motion types: the flapping of the wings and gliding movements. When the bird is taking off, landing or changing its direction, a flapping movement is required; this can be seen at different times within all three bird signatures in Figure 9.18. At other points within the capture, the birds are gliding which does not require any additional movement of their wings and hence does not create a micro-Doppler signature. When comparing the bird signatures, it was found that the smaller bird (the barn owl) produced a much more rapid flapping action which was clearly observable across multiple different micro-Doppler signatures. On the other hand, for the larger birds (the Vulture and Eagle Owl) the flapping that did occur was less frequent but produced higher micro-Doppler excursions from the main body, which must be due to the wingspan difference. These different movement mechanics are the key differentiator between the drone and the bird targets when evaluating the micro-Doppler. Otherwise, the birds are shown to be just as agile as the drone platform with the exception that they do not hover at any point. Some bird species do hover (particularly birds of prey surveying an area for prey) but this was not captured in the trials performed.

From this data, it was possible to extract features which could then be used to classify which target was present. The data were broken up into subsections to allow for N samples to be produced from the repeated flights that the birds took between their perch locations. The features extracted from these sections were taken from three different domains: the spectrogram domain (denoted as Time–Frequency (TF) and shown in Figure 9.18), the Singular Value Decomposition (SVD) domain and the Cadence Velocity Diagram (CVD) domain. The TF domain is the traditional space where micro-Doppler signals are analysed and hence makes it a first choice for creating features. The SVD domain looks to compress the information within the micro-Doppler signals into ordered eigenvalue contributions and therefore is well suited when trying to select features that represent the key contributors to the overall signal. Finally, the CVD domain represents the rate with which each frequency contained in the TF spectrogram repeats, hence is suitable for creating features for periodic movements. Features from all three of these domains have been successfully used previously for micro-Doppler classification challenges and it was felt that they would also be applicable to this situation [36–39].

From within each domain, the selected features were the mean, variance and entropy of the complex samples. In the SVD domain, the diagonal S output matrix was selected and the mean, variance and entropy of this were used as features. These same features were extracted from the full CVD domain matrix to make a total of nine features used to classify the targets. The classification problem was setup as a four-class problem with the aim to identify exactly which target it was from the three different birds and one drone possibilities. The classifier used was a simple linear discriminant analysis machine learning model in all cases and the results from this can be seen in Table 9.6. The classifier was either fed data using features from a single domain (TF, SVD or CVD) or the total set of nine features that represents all three domains. The individual domain features were able to achieve a classification success ranging from 42% to 69% with the TF domain getting the best results and the SVD the lowest. Using features from all three

Table 9.6 Classification results for bird vs. drone data set

Classification challenge	Classifier success rate (%)
Bird vs. drone (4 classes) – only TD domain features	69.75
Bird vs. drone (4 classes) – only SVD domain features	42.5
Bird vs. drone (4 classes) – only CVD domain features	50.75
Bird vs. drone (4 classes) – all 3 domains features	87

domains, it was possible to increase this success rate to 87% for the four-class problem presented. Further improvements will likely be possible through a careful selection of features used and increasing the quantity of empirical data available to both train and test classifiers.

9.7 NetRAD drone payload experiments

The second key challenge that was addressed using the NetRAD radar system was discriminating whether a drone has been modified to carry a payload or not. This was perceived to be an important challenge that represents a sensing capability that is over and above the mere detection phase. If a payload can be inferred using information from the returned radar signal, it will be possible to prioritise this target over other platforms that do not have an additional payload. The research that is summarised in this section was published in [40–42]; review those articles for full information on the experiments and their results as well as within Chapter 11 within this book.

The measurements were taken in HH polarisation which maximises the returns from the quadcopter rotor blades. For the moving captures, the drone proceeded approximately 15 m forwards and backwards in a radial direction relative the central radar node. Measurements were repeated with the drone carrying a payload of 0, 200, 300, 400 and 500 g. For these measurements, the drone that was used was a DJI Phantom 2 quadcopter. At the maximum payload of 500 g, the UAV was visibly more restricted in terms of speed and manoeuvrability.

The radar signals from the drone captured by each radar node were processed in order to extract the micro-Doppler information. The full capture of 30 s for each payload weight was then divided into subsets each with 5,000 pulses representing 1 s of data at the 5-kHz PRF used. An example of the micro-Doppler signatures detected for no payload and for 500 g can be seen in Figure 9.19. Key features were then extracted in order to classify the payload the drone was carrying. The features

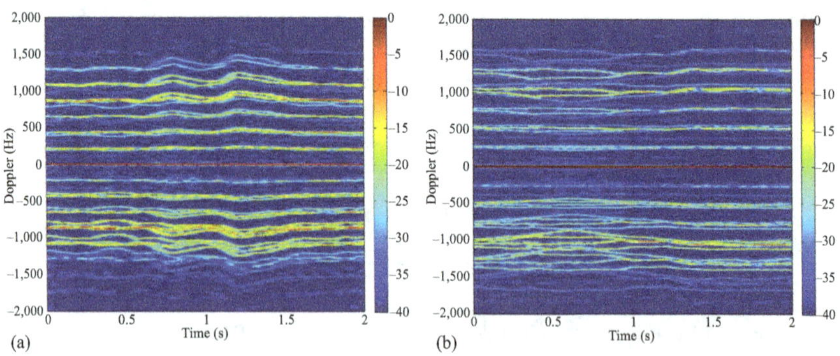

Figure 9.19 Drone micro-Doppler signature for (a) no payload monostatic and (b) 500-g payload monostatic

used were extracted from both the Range–Time–Intensity (RTI) domain and the TF domain in order to compare the success of payload classification using only range domain data compared to when the micro-Doppler information is also available. For the time domain data only, the mean reflected energy as well as the standard deviation from the range bin the drone was present within were used as features. For comparison, the features extracted from the micro-Doppler data were either derived from the SVD domain generated directly from the micro-Doppler signatures or were created by evaluating the 'centroid' of the spectrogram signatures.

The SVD features selected for classification were the mean and standard deviation of the $\boldsymbol{U}$ output matrix. It was decided that only two features were to be used to make a fair comparison with the range domain classifier features and upon inspection the $\boldsymbol{U}$ matrix was found to be the most effective in differentiating different payload classes. It was noted that the largest separation within the feature space was achieved within the monostatic node data, compared to the two bistatic nodes.

Table 9.7 shows the results of payload classification produced for centroid and SVD features. This was an average value over 100 repeats after selecting 50% of data for training and 50% for use randomly as 1-s sections of data from which to produce the features. Overall, it was found that the highest successful classification results were achieved at a level of 96%–97% for this five-class problem using the centroid features. This high level of classification was achieved through a fused classification process using the data from all three radar nodes using a classifier per radar node and fusing the decision, including a confidence threshold within the decision (the bottom row of Table 9.7). When data from only the monostatic node was used, the classification rate was limited to a maximum of 85% and 93% for SVD and centroid features, respectively. It is important to note that simply using all three-radar nodes data in a single classifier produced a worse result than the monostatic only case. It was found that using independent classifier for each radar nodes data was key in order to obtain an improvement over the monostatic success rates.

The multistatic network classification level demonstrates the potential for such a system to identify drones with varying payloads attached, but it must be noted that the empirical data captured were in ideal conditions with a hovering drone of the same type and as short ranges to maximise SNR. In real world challenges with unknown drone model that is erratically flying to evade detection, this same problem will likely have a much reduced classification success rate. A comprehensive review of the payload classification research is provided within Chapter 11 of this book.

Table 9.7 Table of classification success rate using centroid and SVD features

Classifier data used	Classification success rate (%) centroid features	Classification success rate (%) SVD features
Monostatic only	93.60	85.20
3 nodes single classifier	81.70	71.00
3 nodes voting	96.90	94.20
3 nodes vote+confidence threshold	96.90	94.70

9.8 NeXtRAD L- and X-band drone and birds measurements

The NeXtRAD radar system is currently housed in the South African Institute of Maritime Technology (IMT) on the coast of Simons Town, South Africa. In December 2018 and December 2019, NeXtRAD trials focused on bird and drone measurements were conducted, predominantly stationed from the IMT building. The NeXtRAD system consists of three nodes, one Transceiver (Node 0), which is typically based on the IMT building roof, and two portable receivers (Node 1 and Node 2) which were located at different places along the coast of Simons Town Bay area to enable the desired multistatic geometry required. Node 2 suffered technical problems during the trials; therefore, data from this node are not presented here. However, the different positions used for Node 1 during the trials provide data at bistatic angles of ~10° and ~88° with baselines of ~150 m and ~2 km, respectively. During both trials, measurements were taken in a marine environment containing a variety of non-cooperative targets, including birds, flocks of birds, a marine buoy, a lighthouse and in some cases passing boats. A DJI Matrice hexacopter UAV was used as a cooperative target for both trials. The field setup is summarised in Figure 9.20. Ground truth data of the UAV's flightpath is provided by its onboard GPS. These ground-truth GPS data have been aligned with the radar data, allowing us to automatically extract range cells that contain a drone target (e.g. for micro-Doppler analysis).

Figure 9.21 shows a spectrogram created across range cells containing a drone target in a NeXtRAD L-band data set. A time window of 200 samples was used to create the spectrogram, which, given NeXtRAD's PRF of 1 kHz, corresponds to 0.2 s. The bulk-Doppler signal can be seen varying around 0 Hz as the drone's flight path takes it towards and away from the radar node. Either side of the bulk-Doppler signature, HERM line structures (characteristic of a UAV's micro-Doppler signature), can be observed approximately 20 dB below the level of the bulk-Doppler. In contrast to the model data shown in Section 9.3, the combination of HERM line contributions from each of the six rotors and wrapping of HERM lines due to the low PRF of the NeXtRAD system has led to a rather confused pattern of HERM lines. In such a case, reliable measurements of HERM line spacing to determine drone characteristics (such as rotor rotation rate) are challenging. Despite this, the distinctive HERM line structure is still a powerful tool in distinguishing UAVs from non-UAV targets as will be demonstrated in the following section.

NeXtRAD's dual-polarisation receive capability allows HH and HV data to be acquired simultaneously at X-band. Figure 9.22 shows spectrograms of range cells containing a drone target from one such acquisition. It should be noted that NeXtRAD's X-band transmit channel is significantly lower power than the L-band one, resulting in much lower target SNR and subsequently less definition in micro-Doppler signatures at X-band. Despite this, HERM line structures are still evident in both co- and cross-polarised measurements. Due to the higher frequency, bulk-Doppler variations are far more pronounced than at L-band. In both

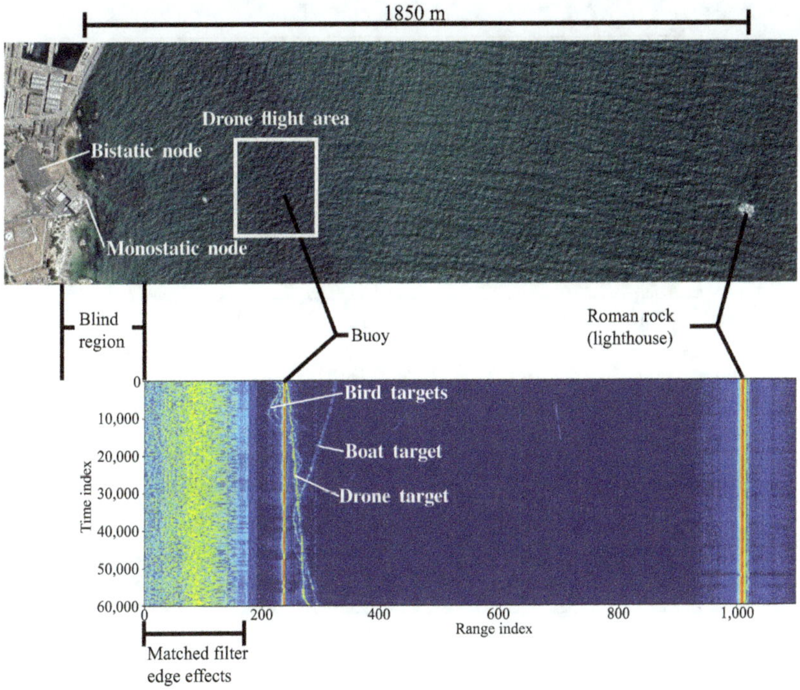

Figure 9.20 Google Earth image (top) of the NeXtRAD trial site and corresponding L-band monostatic RTI plot recorded on 13 December 2018 (bottom). Key features of the data which appear throughout the data set used in this study are labelled

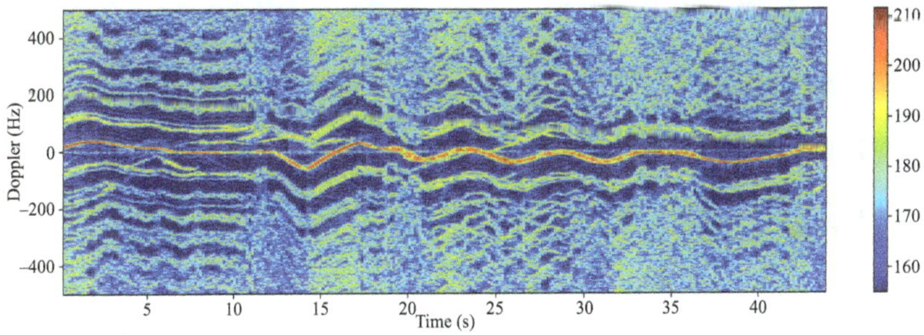

Figure 9.21 Spectrogram of UAV target at L-band using a time window of 0.2 s

the co- and cross-pol cases, the HERM lines again appear at an intensity of ~20 dB below that of the bulk-Doppler. The noteworthy feature of Figure 9.22 is that neither co- nor cross-pol data consistently provide the best clarity of HERM lines. For example, at around 20–30 s, the co-pol data provide the most distinct HERM lines at positive Doppler frequencies; however, at negative Doppler frequencies

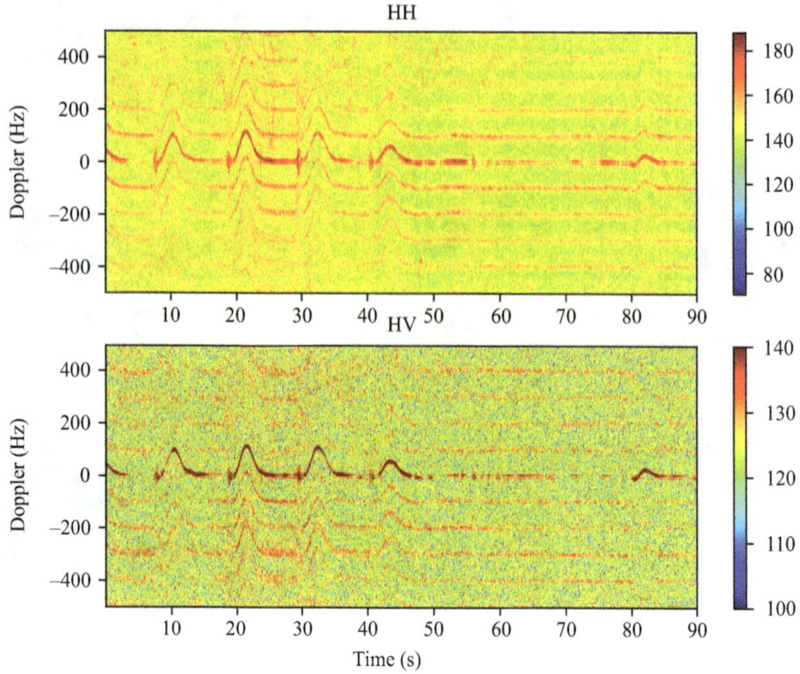

Figure 9.22 Spectrograms of a drone target at X-band using both co-polarised (upper) and cross-polarised (lower) transmit and receive. A time window of 0.2 s was used to produce the spectrogram

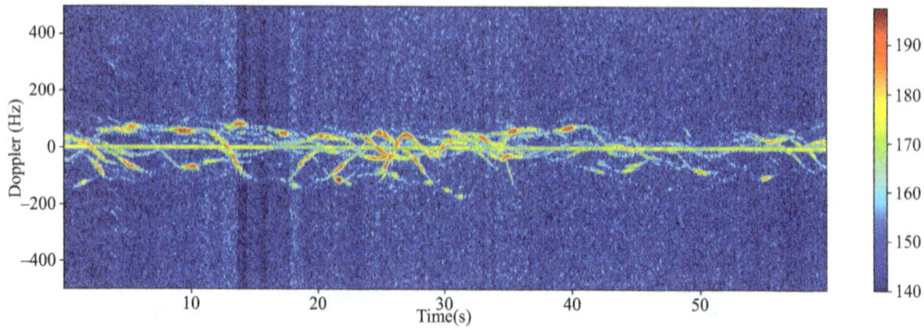

Figure 9.23 Spectrogram of several bird targets at L-band. The micro-Doppler signature is clearly distinct from that of a drone

the cross-pol data have greater clarity. This observation is typical across the full NeXtRAD data set and strongly suggests that UAV detection approaches may benefit from making the use of polarimetric data.

Figure 9.23 shows a spectrogram of single range cell containing several bird targets in an L-band data set. As these are non-cooperative targets, their specific behaviour during the radar acquisition is not known (e.g. it is not known if they were gliding or flapping their wings). However, it seems reasonable to assume that over the full 60 s acquisition some mixture of behaviours will have been captured. The spectrogram was created using a time window of 0.2 s. The contrast of the bird micro-Doppler with the UAV micro-Doppler signatures shown in Figure 9.23 and Figure 9.21 is clear, and we show in the next section how this may be used to classify UAV targets.

9.8.1 Drone filtering

The stark contrast between the micro-Doppler signature of a UAV and a non-UAV target can be used as a simple means to discriminate between the two. As a demonstration, we present an algorithm which takes RTI data as its input and returns a likelihood map (in range and time) of a drone target being present. We refer to this as 'drone filtering'. The drone filter algorithm exploits the symmetry of the HERM lines about the bulk-Doppler for a UAV target compared to the asymmetry of the micro-Doppler signatures from other targets (especially birds). For a typical scene, outputs from the filter span many orders of magnitude and are best presented on a logarithmic (dB) scale. Our preferred normalisation for this scale is to normalise against a static non-drone target in the scene (hence drone targets will be indicated by high, positive dB values). For this purpose, we are using the Roman Rock lighthouse which is ~1.9 km from the monostatic node (see Figure 9.20).

Figure 9.24 shows results of applying the algorithm to L-band monostatic data. As is clear from the figure, the algorithm is extremely effective at removing the buoy target and multiple bird targets from the scene. There are occasional false positives, which typically occur when there are multiple bird targets appearing at

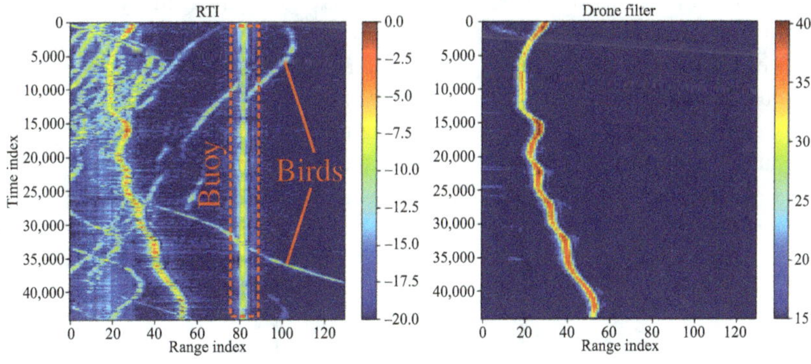

Figure 9.24 RTI plot (left) of L-band monostatic acquisition containing a drone target, birds and the marine buoy. RTI plot after applying the drone filter algorithm to the data (right). The filter is shown to effectively remove bird targets and the buoy target

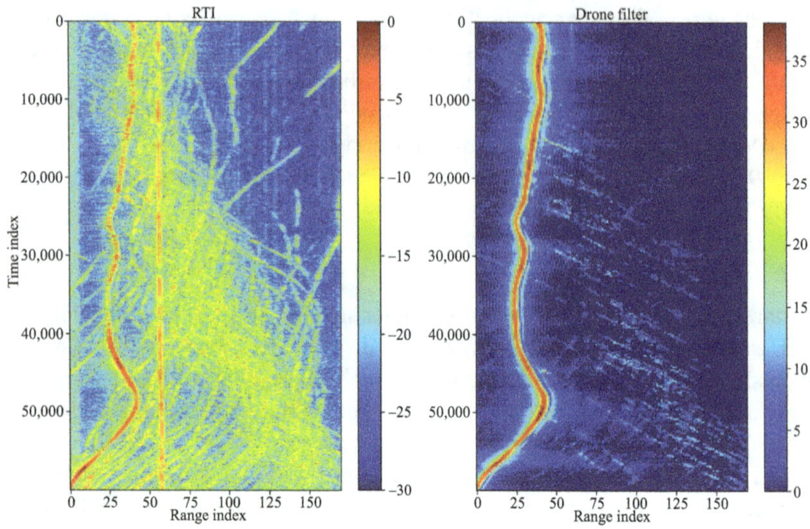

Figure 9.25 RTI plot (left) of L-band monostatic acquisition containing both a drone target and a flock of birds. RTI plot after applying the drone filter algorithm to the data (right)

the same range. However, these are transient and easily distinguishable from the drone track.

A more challenging scene than that shown in Figure 9.24 is one containing a flock of birds. The 2018 NeXtRAD data set contains one such acquisition (recorded at L-band). Figure 9.25 shows this data set and the results of applying the drone filter algorithm to it. Within the recording, a static target can be observed at range bin 60 which is thought to be caused by reflections from a small buoy on the sea surface. The vast majority of returns otherwise within the data were generated by a significant flock of bird targets.

The results shown in Figure 9.25 illustrate that the drone filter algorithm is still very effective at picking out the drone within the flock of birds while suppressing non-drone targets. However, it is also worth noting that multiple birds coinciding in the same range cells give rise to some 'speckle' in the algorithm output (seen in particular in the lower right region of the drone filter data in Figure 9.25). The speckle is ~20 dB below the amplitude of the drone target and therefore easily distinguished as non-drone. It may be further suppressed by median filtering the output from the drone filter.

9.9 Concluding remarks

This chapter has reviewed both modelled and real empirical data sets of bird and drone signals as perceived by monostatic and multistatic radar systems with a strong focus on the micro-Doppler signature of these targets. Analysis of the data has been presented, and key points have been summarised within each section. This area of research is in its relatively infancy due to the ever-changing problem that drone detection represents, but the research shown is some of the current cutting-edge outputs. The use of drones in both military and civilian applications has continued to show strong growth for a number of years. With this current trend, they are clearly going to be part of current and future military sensing challenges. Research is required to best understand how to sense this paradigm shift from large fast moving expensive airborne platforms to significant numbers of low cost, slow and small RCS platforms.

It is clear from the material presented that radar micro-Doppler has been proven to be a vital method in differentiating between these two target categories. If only time domain signatures are used to differentiate bird and drone targets, it is a very challenging problem. Techniques based on extracting information from the micro-Doppler signatures will no doubt be leveraged in any real-world system tasked with this objective. It is important to note that the micro-Doppler contributions from rotor blades of a drone and the flapping wings of a bird can be significantly below the main body return, but if they are detectable, exploiting them should be a priority for any radar system tasked with differentiating these two classes of target.

Assistance may be found when using fully polarimetric systems in order to differentiate a drone and bird target but without processing other signal characteristics, this high-priority challenge will remain unsolved. The use of micro-Doppler comparisons between target classes has been found to be very effective in both differentiating drone and bird targets, using the explained filtering techniques, as well as even differentiating different drone payloads. The authors believe that if Automatic Target Recognition techniques are going to be applied to identify threat drone systems compared to benign birds, micro-Doppler has to be one of the sensing methodologies. Overall a total solution will require data fusion from a number of sensing domains in order to maximise its potential, including EO, ES and potentially even acoustic sensing methods.

Future currently unaddressed challenges in the area of drone micro-Doppler include the fully quantitative characterisation of how bistatic angle affects the perceived drone micro-Doppler signatures and hence how best can a multistatic system be deployed to protect a given know high-value area, e.g. a port, airport or military base. As well as this research into flight dynamics of drone and birds from a tracking perspective should be connected with the micro-Doppler research outputs such that one can inform the other. In that, if a target is found to have given micro-Doppler signature this could help subcategorise its potential class and therefore inform the flight dynamics model used to improve the track fidelity achieved. Of course, it is also of great interest to validate the detection and discrimination concepts developed in real-world scenarios and at ranges which allow for effective responses to situations to be made. If some very effective techniques are limited only to short ranges, then their utility in the real world may be restricted compared to other methods that have lower detection or classification rates but can achieve much greater ranges.

Acknowledgements

The authors are very grateful to the many exceptional researchers who have contributed to the development of the NetRAD system Chris Baker, Karl Woodbridge,

Shaun Doughty, Tom Derham, Waddah Al-Ashwal and Stephan Sandenbergh. As well as those who helped to develop the NeXtRAD system, particularly Hugh Griffiths, Francesco Fioranelli, Riccardo Palama, Simon Lewis and Brad Kahn.

The authors would like to acknowledge the support of the many supporters and funding agencies that have enabled the research conducted at UCL and UCT, including the EPSRC, IET for the A. F. Harvey Award, FFI Norway, Royal Academy of Engineering, the US ONR(G) and the IMT. In particular, the authors would like to thank both Dstl and the Defence and Security Accelerator (DASA) which funded some of the most recent analysis of UAV signatures presented within this chapter.

References

[1] "Consumer Drone Market Analysis By Product (Multi-Rotor, Nano) By Application And Segment Forecasts To 2024," Report ID: 978-1-68038-831-2, Grand View Research; 2016. https://www.grandviewresearch.com/industry-analysis/consumer-drone-market

[2] M. Z. Anwar, Z. Kaleem, and A. Jamalipour, "Machine learning inspired sound-based amateur drone detection for public safety applications," *IEEE Trans. Veh. Technol.*, vol. 68, no. 3, pp. 2526–2534, 2019.

[3] Z. Shi, X. Chang, C. Yang, Z. Wu, and J. Wu, "An acoustic-based surveillance system for amateur drones detection and localization," *IEEE Trans. Veh. Technol.*, vol. 69, no. 3, pp. 2731–2739, 2020.

[4] A. H. Michel, Counter-Drone Systems, 2nd Edition. Center for the Study of the Drone at Bard College; 2019.

[5] G. C. Birch, J. C. Griffin, and M. K. Erdman, "UAS detection classification and neutralization: Market survey 2015," *Sandia Rep.*, 2015, p. 74.

[6] S. Balkan, *Daesh's Drone Strategy Technology and the Rise of Innovative Terrorism*. SETA Foundation for Political, Economic and Social Research, 2017.

[7] D. Lack and G. C. Varley, "Detection of birds by radar," *Nature*, vol. 156, no. 3969, p. 629, 1945.

[8] I. O. Buss, "Bird detection by radar," *Auk*, vol. 63, no. 3, pp. 315–318, 1946.

[9] B. Bruderer, "The radar window to bird migration BT," *Avian Migration*, Berlin, Heidelberg: Springer; 2003, pp. 347–358.

[10] H. Alfiya, "Surveillance radar data on nocturnal bird migration over Israel, 1989–1993," *Isr. J. Ecol. Evol.*, vol. 41, no. 3, pp. 517–522.

[11] D. F. Bertram, "Use of radar for monitoring colonial burrow-nesting seabirds," *J. Field Ornithol.*, 1999.

[12] G. W. Schaeffer, "Bird recognition by radar: A study in quantitative radar ornithology," in *The Problems of Birds as Pests*, R. K. Murton and E. N. Wright, Eds. Academic Press, London, 1968, pp. 53–86.

[13] C. R. Vaughn, "Birds and insects as radar targets: A review," *Proc. IEEE*, vol. 73, no. 2, pp. 205–227, 1985.

[14] B. Torvik, K. E. Olsen, and H. D. Griffiths, "X-band measurements of radar signatures of large sea birds," in *2014 Int. Radar Conf. Radar 2014*, 2014, pp. 1–6.

[15] P. Molchanov, R. I. A. Harmanny, J. J. M. De Wit, K. Egiazarian, and J. Astola, "Classification of small UAVs and birds by micro-Doppler signatures," *Int. J. Microwave Wireless Technol.*, vol. 6, no. 3–4, pp. 435–444, 2014.

[16] B. Torvik, *Investigation of Non-cooperative Target Recognition of Small and Slow Moving Air Targets in Modern Air Defence Surveillance Radar*, University College London, London, 2016.

[17] B. Torvik, K. E. Olsen, and H. Griffiths, "Classification of birds and UAVs based on radar polarimetry," *IEEE Geosci. Remote Sens. Lett.*, vol. 13, no. 9, pp. 1305–1309, 2016.

[18] M. Skolnik, *Introduction to Radar Systems.* New York, NY: McGraw Hill, 2001.

[19] S. Rahman and D. A. Robertson, "In-flight RCS measurements of drones and birds at K-band and W-band," *IET Radar Sonar Navig.*, vol. 13, no. 2, pp. 300–309, 2019.

[20] B. W. Tobalske, "Biomechanics of bird flight," *J. Exp. Biol.*, vol. 210, no. 18, pp. 3135–3146, 2007.

[21] G. Pollon, "Distributions of radar angels," *IEEE Trans. AESS*, vol. AES-8, no. 6, pp. 721–726, 1972.

[22] N. Huansheng, C. Weishi, M. Xia, and L. Jing, "Bird-aircraft strike avoidance radar," *IEEE Aerosp. Electron. Syst. Mag.*, vol. 25, no. 1, pp. 19–28, 2010.

[23] J. Gong, J. Yan, D. Li, and R. Chen, "Comparison of radar signatures based on flight morphology for large birds and small birds," *IET Radar Sonar Navig.*, vol. 14, no. 9, pp. 1365–1369, 2020.

[24] N. J. Peters, C. Oppenheimer, P. Brennan, L. B. Lok, M. Ash, and P. Kyle, "Radar altimetry as a robust tool for monitoring the active lava lake at Erebus Volcano, Antarctica," *Geophys. Res. Lett.*, vol. 45, no. 17, pp. 8897–8904, 2018.

[25] S. Doughty, K. Woodbridge, and C. J. Baker, "Characterisation of a multi-static radar system," in *Proc. 3rd Eur. Radar Conf. EuRAD 2006*, 2006, pp. 5–8.

[26] T. Derham, S. Doughty, K. Woodbridge, and C. J. Baker, "Realisation and evaluation of a low cost netted radar system," in *CIE International Conference of Radar Proceedings*, 2006.

[27] M. Ritchie, A. Stove, K. Woodbridge, and H. Griffiths, "NetRAD: Monostatic and bistatic sea clutter texture and Doppler spectra characterization at S-band," *IEEE Trans. Geosci. Remote Sens.*, vol. 54, no. 9, pp. 5533–5543, 2016.

[28] J. S. Sandenbergh and M. R. Inggs, "A common view GPSDO to synchronize netted radar," in *IET Int. Conf. Radar Syst. 2007*, 2007, pp. 24–24.

[29] P. Beasley, M. Ritchie, H. Griffiths, *et al.*, "Multistatic radar measurements of UAVs at X-band and L-band," in *IEEE Radar Conference*, Florence, Italy, 2020.

[30] R. Palama, F. Fioranelli, M. Ritchie, M. R. Inggs, S. Lewis, and H. Griffiths, "Measurements of multistatic XL band radar signatures of UAVS," in *2019 Int. Radar Conf. RADAR 2019*, 2019, pp. 15–19.

[31] M. R. Inggs, S. Lewis, R. Palama, M. Ritchie, and H. Griffiths, "Report on the 2018 trials of the multistatic NeXtRAD dual band polarimetric radar," in *2019 IEEE Radar Conf.*, 2019, pp. 1–6.

[32] M. Inggs, S. Coetzee, H. Griffiths, F. Fioranelli, M. Ritchie, and K. Woodbridge, "Database design for an experimental, dual band, polarimetric radar," in *2015 IEEE Radar Conference – Proceedings*, 2015.

[33] S. Alhuwaimel, S. Coetzee, P. Cheng, *et al.*, "First measurements with NeXtRAD, a polarimetric X/L Band radar network," in *2017 IEEE Radar Conference, RadarConf 2017*, 2017.

[34] M. Inggs, H. Griffiths, F. Fioranelli, M. Ritchie, and K. Woodbridge, "Multistatic radar: System requirements and experimental validation," in *2014 International Radar Conference, Radar 2014*, 2014.

[35] M. Ritchie, F. Fioranelli, H. Griffiths, and B. Torvik, "Monostatic and bistatic radar measurements of birds and micro-drone," in *2016 IEEE Radar Conference, RadarConf 2016*, 2016.

[36] B. K. Kim, H.-S. Kang, S. Lee, and S.-O. Park, "Improved drone classification using polarimetric merged-Doppler images," *IEEE Geosci. Remote Sens. Lett.*, pp. 1–5, 2020.

[37] S. Zhang, G. Li, M. Ritchie, F. Fioranelli, and H. Griffiths, "Dynamic hand gesture classification based on radar micro-Doppler signatures," in *2016 CIE International Conference on Radar, RADAR 2016*, 2017.

[38] G. Li, R. Zhang, M. Ritchie, and H. Griffiths, "Sparsity-based dynamic hand gesture recognition using micro-Doppler signatures," in *2017 IEEE Radar Conference, RadarConf 2017*, 2017, pp. 0928–0931.

[39] M. Ritchie and A. M. Jones, "Micro-Doppler gesture recognition using Doppler, time and range based features," in *IEEE Radar Conference*, Boston, 2019.

[40] M. Ritchie, F. Fioranelli, and H. Borrion, "Micro UAV crime prevention: Can we help Princess Leia?" in *Crime Prevention in the 21st Century: Insightful Approaches for Crime Prevention Initiatives*, 2017, pp. 359–376.

[41] F. Fioranelli, M. Ritchie, H. Griffiths, and H. Borrion, "Classification of loaded/unloaded microdrones using multistatic radar," *Electron. Lett.*, vol. 51, no. 22, pp. 1813–1815, 2015.

[42] M. Ritchie, F. Fioranelli, H. Borrion, and H. Griffiths, "Multistatic micro-Doppler radar feature extraction for classification of unloaded/loaded micro-drones," *IET Radar Sonar Navig.*, vol. 11, no. 1, pp. 116–124, 2017.

Chapter 10

Radar recognition of multiple UAVs

Gang Li[1] *and Wenyu Zhang*[1]

10.1 Introduction

Radar detection and recognition of small unmanned aerial vehicle (UAV) have attracted significant attention, as a misuse of UAVs has led to a series of illegal events such as filming of restricted areas, disruption of public events and even transport of illicit materials [1]. It has been demonstrated that radar micro-Doppler signatures (MDSs) are helpful for the detection and recognition of small UAVs [2–8]. Especially when the radar bandwidth is limited, it is difficult to find the small UAVs from the radar images with poor resolutions. In this case, the micro-Doppler analysis provides a reliable solution for small UAV recognition. In [2], the feasibility of distinguishing between loaded and unloaded small UAVs based on MDS observed by the multistatic radar system is investigated. In [3,4], three kinds of small UAVs are classified by using a dual-band radar system and exploiting the multipath components, respectively. In [5], empirical mode decomposition is applied to extract features for small UAV classification. The problem of small UAV detection is investigated in [6] from the perspective of their extremely low radar cross section (RCS). More studies on detection and classification based on micro-Doppler analysis are summarised in [7]. Most of existing studies on the radar classification of small UAVs assume that the radar signal is reflected from a single UAV. However, in recent years, the civilian and military applications for UAV swarms keep growing [9,10]. When there are multiple UAVs in the observed scene, the recognition of multiple UAVs becomes more complicated because of the superimposition of a number of micro-Doppler components. In particular, it is challenging to determine if a specific kind of target exists when there are multiple targets in the observed scene. The spatial separation of multiple targets is partially solved if the radar range resolution is high enough, but this requires large bandwidths that also increases the hardware cost. For radar systems with lower range resolution and azimuth resolution, there may exist multiple UAVs in a single range-azimuth resolution cell. Under this situation, it is necessary to investigate how to detect and classify multiple UAVs from their MDSs.

In this chapter, we present two algorithms for radar recognition of multiple UAVs. In the first algorithm, the cadence frequency spectrum (CFS) is presented as

[1]Department of Electronic Engineering, Tsinghua University, Beijing, China

a kind of low-dimensional feature of small UAVs [11]. The CFS of each class of small UAVs is obtained by accumulating cadence-velocity diagram (CVD) over the Doppler frequency axis. Experimental results on measured data have demonstrated that the combination of CFS and the *K*-means classifier is capable of providing a satisfactory accuracy of recognition of multiple UAVs. The second algorithm is based on dictionary learning [12]. The dictionary is learnt by performing the *K*-SVD algorithm on the CVDs of training samples from each class of small UAVs. Then, the learnt dictionaries of all the classes are combined together to generate a recognition dictionary. In this way, the problem of recognition of multiple UAVs is converted to a problem of signal decomposition. Subsequently, the sparse representation of the CVD from multiple UAVs is solved by the orthogonal matching pursuit (OMP) with the learnt recognition dictionary and recognition result is obtained by evaluating the magnitudes of the sparse representation coefficients. Experimental results on measured radar data have demonstrated the effectiveness of the dictionary learning-based algorithm for the recognition of multiple UAVs and the superiority of the dictionary learnt from CVD over that learnt from a time–frequency spectrogram. It is worth emphasizing that the training data sets used in these two algorithms only consist of radar echoes reflected from every single class of small UAVs.

10.2 Recognition of multiple UAVs based on CFS

In this section, an algorithm of the recognition of multiple small UAVs based on the CFS features is presented.

10.2.1 Signal model

Assume that there are C different classes in total and $\mathbf{s}_{il} \in \mathbb{C}^{n_s \times 1}$ is the lth training signal collected from the ith target class and n_s is the number of samples in $\mathbf{s}_{il}$, for $i = 1, 2, \ldots, C$, $l = 1, 2, \ldots, N_i$, and N_i is the number of training signals in the ith target class. The time–frequency spectrogram, denoted by $\mathbf{Z}_i$, is obtained by performing STFT on the radar echo:

$$\mathbf{Z}_{il}(t,f) = \int_{-\infty}^{+\infty} \mathbf{s}_{il}(\tau)h(\tau - t)e^{-j2\pi f\tau}d\tau, \tag{10.1}$$

where t is the sample instant, f is the Doppler frequency, and $h(t)$ is the window function. The repetition frequency of the time–frequency spectrogram describes the periodic characteristics of the micro-motions of the small UAVs, which can be visualised via CVD. The CVD, denoted by $\mathbf{F}_{il}$, can be obtained by performing the Fourier transform along the time axis on the time–frequency spectrogram $\mathbf{Z}_{il}$, i.e.,

$$\mathbf{F}_{il}(\nu,f) = \int_{-\infty}^{+\infty} \mathbf{Z}_{il}(\tau,f)e^{-j2\pi\nu\tau}d\tau, \tag{10.2}$$

where ν is the cadence frequency.

In what follows, a visualised example on the time–frequency spectrogram and CVD is provided. The small UAVs to be recognised, including a helicopter, a quadcopter and a hexacopter with almost the same size, are shown in Figure 10.1. A single-channel continuous wave radar system is used for collecting the data reflected from the flying UAVs. The carrier frequency of the radar is 25 GHz and the sampling frequency is 8 kHz. The details about the radar system and experimental settings are shown in Table 10.1. The spectrograms of the radar data reflected from these three small UAVs are shown in Figure 10.2, which are

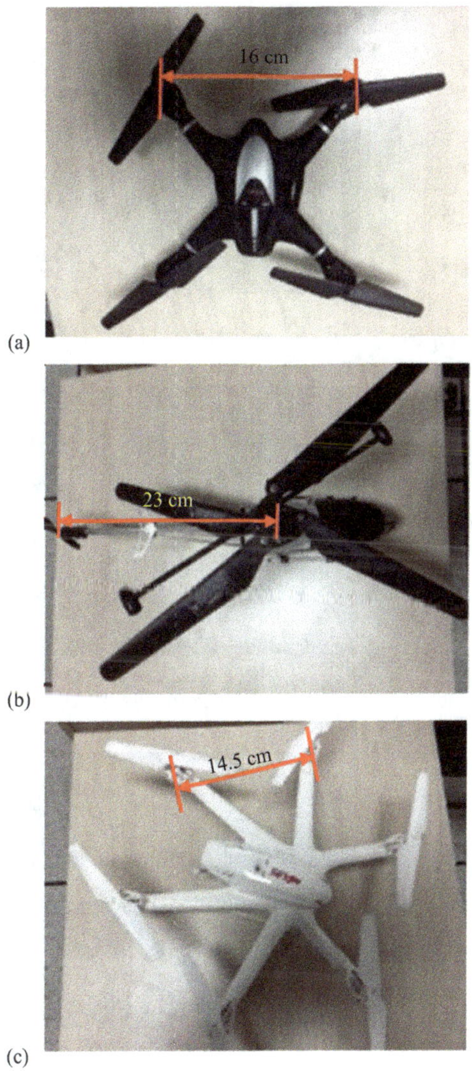

Figure 10.1 Appearance of the UAVs: (a) quadcopter, (b) helicopter and (c) hexacopter

Table 10.1 Experimental settings

Experimental settings	Details
Radar	Ancortek SDR-KIT 2500B
Waveforms	Continuous wave
Carrier frequency	25 GHz
Sweep time	0.5 ms
Number of samplers per sweep	512
Sampling frequency	8 kHz
Range	8 m
Room size	6 m × 10 m

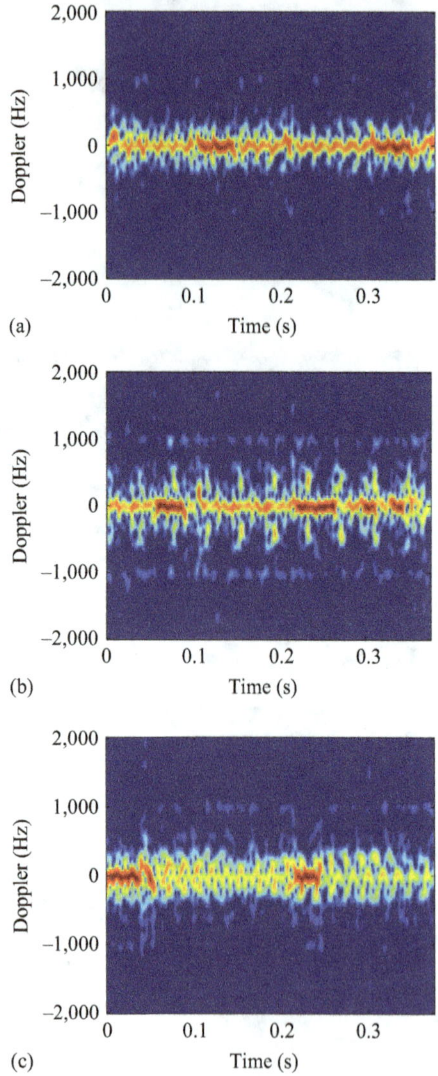

Figure 10.2 The time–frequency spectrograms of the small UAVs: (a) quadcopter, (b) helicopter and (c) hexacopter

obtained by performing STFT on the received signals as expressed in (10.1). As suggested in [13], the translational motions of the UAVs have been compensated, which turns the central Doppler frequency in Figure 10.2 to zero. Further, the CVDs of these three small UAVs are presented in Figure 10.3, which are obtained according to (10.2). From the comparison between Figures 10.2 and 10.3, one can see that the CVD gives a measure on how often each Doppler frequency component

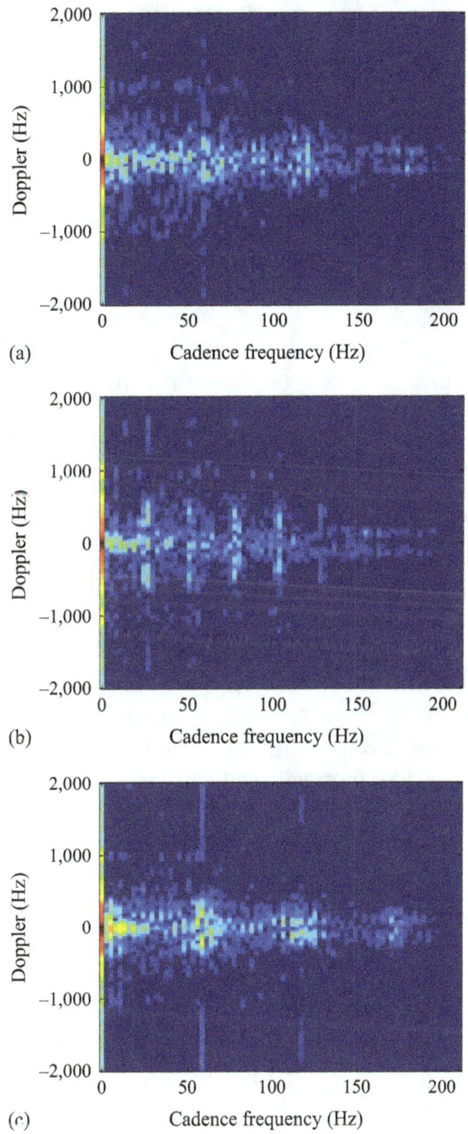

Figure 10.3 The CVDs of the small UAVs: (a) quadcopter, (b) helicopter and (c) hexacopter

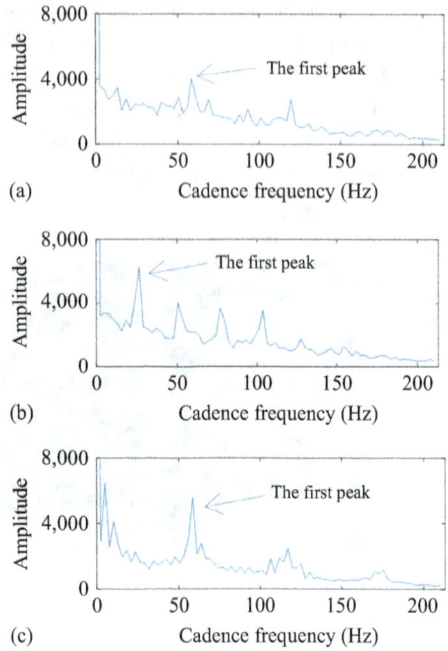

Figure 10.4 The CFSs of the small UAVs: (a) quadcopter, (b) helicopter and (c) hexacopter

repeats. Moreover, it is evident that the CVD is independent of the time shift of the radar echo, which may make the recognition of small UAVs more robust.

The CFS, which is defined as the sum of all the rows of the two-dimensional CVD, can be regarded as a compressed expression of the CVD. The CFSs of the three kinds of UAVs are shown in Figure 10.4, in which the positions of the first peaks in nonzero cadence-frequency areas represent the repetition frequencies of the corresponding time–frequency spectrograms. One can see that the peak amplitudes and peak positions of these CFSs are significantly different due to the diverse numbers of blades and RCSs of the three kinds of UAVs. In comparison with CVD, the CFS decreases the dimension of the features and therefore makes the computation easier.

10.2.2 Classification method and experimental results

The measured data are collected by using the radar and the UAVs introduced in Section 10.2.1. The time duration for each data recording is 6 s, and each measurement data is divided into 20 segments of 0.3 s. Such a segment is long enough to contain several periods of the radar signals reflected from the UAVs. To ensure the stability of data, the first and the last segments are abandoned, so there are 18 segments available per data record. The same data collection process is repeated ten times. Thus, the total number of the segments corresponding to each class of UAVs is (18 segments)$\times$(10 repeats) $=$ 180 (Figure 10.5).

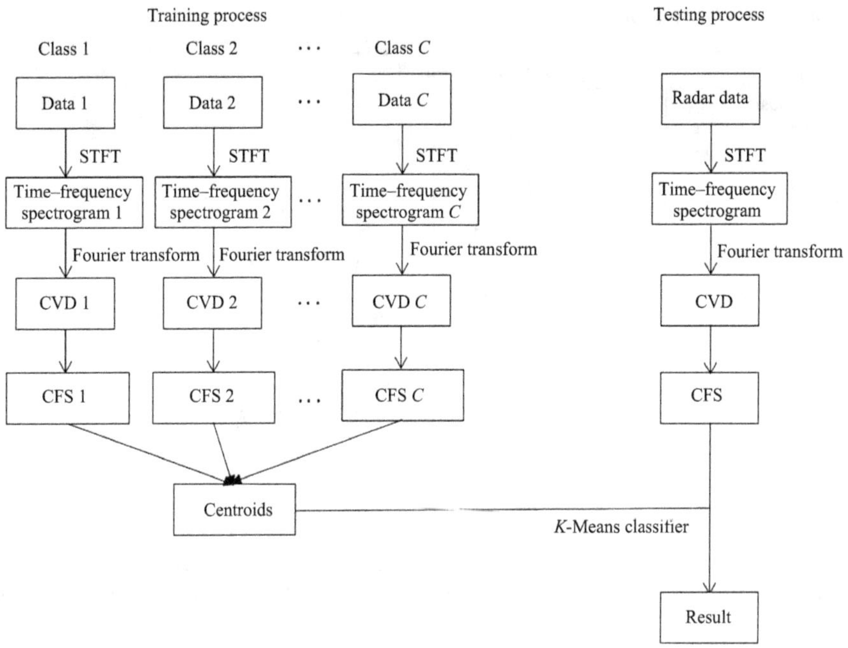

Figure 10.5 The scheme of the CFS-based method, where C is the number of classes of UAVs

Table 10.2 Possible scenarios

Case	Target in scene
Scenario A	A helicopter
Scenario B	A hexacopter
Scenario C	A quadcopter
Scenario D	A helicopter and a quadcopter
Scenario E	A hexacopter and a quadcopter
Scenario F	A helicopter and a hexacopter
Scenario G	A helicopter, a hexacopter and a quadcopter

In the training phase, the CFSs of those training data corresponding to each single class of UAV are extracted as described in Section 10.2.1 and fed into the K-means classifier [11] to obtain the clustered centroid in the feature space. Since three kinds of UAVs are used in the experiments, there are seven possible scenarios totally, as shown in Table 10.2. The feature centroids of the scenarios which contain more than one UAV are obtained by adding the fusion of the corresponding feature centroids of the single UAV scenarios.

In the testing phase, the CFS of the testing signal is extracted and compared with all the feature centroids to determine the classes of the UAVs existing in the

Table 10.3 Confusion matrix

True scenarios	Recognition results (%)						
	A	B	C	D	E	F	G
A	100	0	0	0	0	0	0
B	0	88.02	11.98	0	0	0	0
C	0.15	9.56	90.28	0	0	0	0
D	0	0	0	84.86	2.36	12.78	0
E	0	0	0	0.68	96.30	3.02	0
F	0	0	0	12.00	2.86	85.14	0
G	0	0	0	0	0	0	100

Table 10.4 UAV recognition accuracy (%)

Training ratios	**20% training**	**30% training**	**40% training**	**50% training**
Min	88.89	89.34	89.29	88.73
Max	94.25	94.21	94.71	95.24
Average	**91.85**	**91.92**	**92.00**	**92.23**

observed scene. Here, 50% of the single-target data sets are used for training and the remaining data are used for testing. The recognition process is repeated 500 times with a random selection of the training/testing data and the average over all the trials is calculated to assess the recognition accuracy. The confusion matrix is given in Table 10.3.

It is clear from Table 10.3 that (1) the average recognition accuracy for single UAVs (i.e., Scenarios A–C) is 92.77%, (2) the average recognition accuracy for two-UAV scenarios (i.e., Scenarios D–F) is 89.10%, (3) the recognition accuracy for three-UAV scenario (i.e., Scenario G) is 100% and (4) the average classification accuracy of all the scenarios is 92.23%.

From Table 10.3, one can observe a block diagonal pattern in the confusion matrix, while the obvious misclassification results are marked in red. The block pattern is caused by the difficulty in distinguishing UAVs within the same pattern (i.e., single- and dual-UAV scenarios). Both the quadcopter and the hexacopter have at least four rotors, which are the main components that produce MDSs. The superimposition and interaction of the various micro-Doppler scattering from the multiple rotors of multiple UAVs increase the difficulty of classification.

To further investigate the performance of the presented method, experiments under different proportions of training data, i.e., 10%, 20%, 30% and 50%, are carried out. Table 10.4 shows the minimum, maximum and average accuracy over 500 trials with randomly selected training data. It is observed that the classification accuracy of the presented method is high enough with even 20% training data, which indicates that the presented method is suitable for classification in the case of small data sets.

It is worth mentioning that the CFS-based method has low computational complexity. The average running time of the method, including training and testing, is only 0.128 s. The hardware platform is a laptop with an Intel® Core™ i5-3230M CPU inside, and the CPU clock frequency and the memory size are 2.6 GHz and 7.89 GB, respectively. The software platform is MATLAB® 2017a and the operation system is Windows 7. Thus, it is promising to achieve real-time processing of the CFS-based method on embedded computing platforms in practical applications.

10.3 Recognition of multiple UAVs via dictionary learning

In this section, an algorithm of the recognition of multiple small UAVs based on dictionary learning is presented. Being different from the CFS-based method, the aim of the dictionary learning-based method is to determine whether a specific kind of small UAV exists when there are multiple small UAVs in the observed air space.

10.3.1 Recognition method

The scheme of the method based on dictionary learning is illustrated in Figure 10.6. The presented method contains two sub-processes: (1) training and (2) testing. The training process is composed of two steps. First, the CVD of each training signal is computed for each UAV as formulated in (10.2). Second, the *K*-SVD algorithm [14] is applied to the CVDs to learn the dictionary of each class of small UAVs. Next, the learnt dictionaries of all classes are concatenated to form the overall recognition dictionary. In the testing process, the sparse representation of the CVD of the test signal is computed by using the OMP algorithm [15] with the learnt recognition dictionary. Subsequently, the l_2-norm of several segments of the sparse solution is compared with a threshold to determine the presence or absence of each class of UAVs.

The data set is divided into three parts: training, testing and validating. The validating set is used for determining the threshold in the testing process. It is worth emphasizing that the training set only consists of radar echoes reflected from every single class of small UAVs.

10.3.1.1 Training phase

In the training phase, the CVDs of the training data are extracted as described in Section 10.2.1. Then, the two-dimensional CVD matrices $\mathbf{F}_{il}$ are reshaped into one-dimensional vectors $\mathbf{y}_{il}$ by stacking the columns of $\mathbf{F}_{il}$ one underneath the other in sequence, i.e.,

$$\mathbf{F}_{il} = [\mathbf{m}_1, \mathbf{m}_2, \ldots, \mathbf{m}_n] \rightarrow \begin{bmatrix} \mathbf{m}_1 \\ \mathbf{m}_2 \\ \vdots \\ \mathbf{m}_n \end{bmatrix} = \mathbf{y}_{il}, \tag{10.3}$$

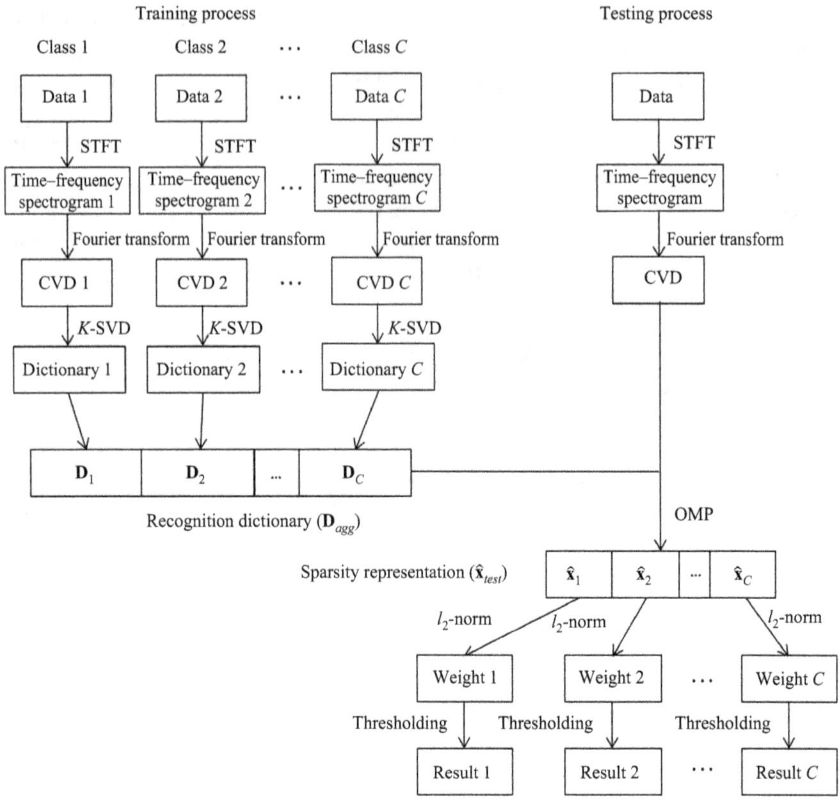

Figure 10.6 The scheme of the dictionary learning-based method, where C is the number of classes of UAVs

where $\mathbf{m}_i$ is the ith column vector of the $\mathbf{F}_{il}$. Assume that $\mathbf{F}_{il}$ is an $L_1 \times L_2$ matrix. As a result, $\mathbf{y}_{il}$ is an $(L_1 \cdot L_2) \times 1$ vector.

Next, the features are extracted from the CVDs via dictionary learning. Compared with classical feature extraction methods such as empirical feature selection and principle component analysis, sparse representations are more representative and more discriminative with lower dimensionality [16]. Moreover, existing studies [17–19] have shown that compared with traditional sparsity coding methods with predesigned dictionaries, dictionary learning can dramatically improve the performance of sparse representation thanks to its capability of adaptively adjusting the dictionary based on the training data set. Dictionary learning aims to learn a set of atoms so that a given signal can be accurately approximated by a sparse linear combination of atoms and has produced promising results in the field of image processing [20–22]. In what follows, a typical algorithm of dictionary learning, *K*-SVD, [14] is applied to CVDs for feature extraction.

The linear decomposition of CVD of the ith target class $\mathbf{y}_i$ can be expressed as

$$\mathbf{y}_{il} = \mathbf{D}_i \mathbf{x}_{il}, \tag{10.4}$$

where $\mathbf{y}_{il}$ is an $(L_1 \cdot L_2) \times 1$ measurement vector, $\mathbf{D}_i \in \mathbb{C}^{(L_1 \cdot L_2) \times n_d}$ is the dictionary matrix with normalised columns, and $\mathbf{x}_{il} \in \mathbb{C}^{n_d \times 1}$ is a sparse vector. Assume the sparsity of $\mathbf{x}_{il}$ is K, i.e., the number of the nonzero coefficients in $\mathbf{x}_{il}$ is not larger than K. The dictionary $\mathbf{D}_i$ is expected to be capable of representing $\mathbf{y}_{il}$ as the sum of its K columns. Here we have concatenated all the samples $\mathbf{y}_{il}$ as columns of the matrix $\mathbf{Y}_i \in \mathbb{C}^{(L_1 \cdot L_2) \times N_i}$ and similarly stacked the representations coefficient vector $\mathbf{x}_{il}$ to build the matrix $\mathbf{X}_i \in \mathbb{C}^{n_d \times N_i}$ where N_i is the number of training signals in the ith target class. The sparse representation of $\mathbf{Y}_i$ can be achieved by solving the following problem:

$$\langle \hat{\mathbf{D}}_i, \hat{\mathbf{X}}_i \rangle = \arg\min_{\mathbf{D}_i, \mathbf{X}_i} \| \mathbf{Y}_i - \mathbf{D}_i \mathbf{X}_i \|_2^2 \quad \text{s.t.} \ \| \mathbf{x}_{il} \|_0 \leq K, \tag{10.5}$$

where $\| \cdot \|_0$ and $\| \cdot \|_2$ denote l_0-norm and l_2-norm, respectively. The K-SVD algorithm [14] introduces an iterative approach to efficiently solve the optimisation problem of (10.5). The K-SVD algorithm divides the numerical treatment of the optimisation problem into two stages, i.e., a sparse-coding stage that optimises $\hat{\mathbf{X}}_i$ assuming knowledge of $\hat{\mathbf{D}}_i$ and a dictionary update stage that updates $\hat{\mathbf{D}}_i$ using the known representation $\hat{\mathbf{X}}_i$. In the sparse-coding stage, the optimal sparse representation $\hat{\mathbf{X}}_i$ can be solved by the OMP algorithm [15]. In the dictionary update stage, the dictionary $\hat{\mathbf{D}}_i$ is updated by computing the SVD decomposition of the representation residual. Readers are referred to [14] for a more detailed description of the K-SVD algorithm.

After all the dictionaries are learnt via the K-SVD algorithm from the samples of each target class, the dictionaries from all the classes are combined together to form the recognition dictionary $\mathbf{D}_{agg} \in \mathbb{C}^{(L_1 \cdot L_2) \times (C \cdot n_d)}$:

$$\mathbf{D}_{agg} = \left[\hat{\mathbf{D}}_1, \hat{\mathbf{D}}_2, \ldots, \hat{\mathbf{D}}_C \right]. \tag{10.6}$$

The recognition dictionary $\mathbf{D}_{agg}$ is utilized to realize signal decomposition in the testing process.

10.3.1.2 Testing phase

Assume that $\mathbf{y}_{test} \in \mathbb{C}^{(L_1 \cdot L_2) \times 1}$ is the CVD vector of the testing data, generated from the two-dimensional CVD matrix after stacking the columns one underneath the other in sequence.

Since the CVD is a linear transform of the radar echo, the CVD vector $\mathbf{y}_{test}$ can be viewed as the sum of the CVD vectors corresponding to all the small UAV. In this way, the problem of recognition of multiple small UAVs is converted into the problem of signal decomposition. Assume that each target signal lies in its own subspace class and all the subspace classes are non-overlapping. As a result, the dictionaries learnt from different classes are discriminative. Since $\mathbf{D}_i$ is the dictionary trained from the ith target class, the signal from ith target class may have higher correlation with the columns of $\mathbf{D}_i$ (also called atoms) than with the atoms of other dictionaries. In other words, $\|\mathbf{y}_i - \mathbf{D}_j\mathbf{x}_j\|_2^2$ reaches the minimum value when $j = i$. The sparse representation of the aggregation signal can be obtained by

$$\hat{\mathbf{x}}_{test} = \arg\min_{\mathbf{x}_{test}} \|\mathbf{y}_{test} - \mathbf{D}_{agg}\mathbf{x}_{test}\|_2^2 \quad \text{s.t.} \ \|\mathbf{x}_{test}\|_0 \leq C \times K. \tag{10.7}$$

In this chapter, the OMP algorithm [15] is used to solve (10.7).

As the recognition dictionary $\mathbf{D}_{agg}$ is composed of C dictionaries learnt from different target classes, the sparse solution $\hat{\mathbf{x}}_{test} \in \mathbb{C}^{(C \cdot n_d)\times 1}$ can be separated into C parts, i.e., $\hat{\mathbf{x}}_1, \hat{\mathbf{x}}_2, \ldots, \hat{\mathbf{x}}_C$, corresponding to the different target classes:

$$\hat{\mathbf{x}}_{test} = \left[\hat{\mathbf{x}}_1{}^T, \hat{\mathbf{x}}_2{}^T, \ldots, \hat{\mathbf{x}}_C{}^T\right]^T. \tag{10.8}$$

Next, thresholding is applied to the sparse solution $\hat{\mathbf{x}}_{test}$. That is, the target from the ith class is declared as present, if the strength $\|\hat{\mathbf{x}}_i\|_2$ exceeds the learnt threshold (τ_i). The threshold is designed as follows.

The testing data set can be divided into two parts, i.e., data set I that contains the ith type of UAV and data set II in which the ith type of UAV does not exist. We first define the rate of correct recognition for the ith target class $\mathcal{CR}_i(\tau_i)$ on data set I and the rate of incorrect recognition with respect to the ith target class $\mathcal{ER}_i(\tau_i)$ on data set II as

$$\mathcal{CR}_i(\tau_i) = \frac{\sum_{j=1}^{N_{\mathrm{I}}} \mathrm{I}\left[\|\hat{\mathbf{x}}_{ij}\|_2^2 \geq \tau_i\right]}{N_{\mathrm{I}}}, \tag{10.9}$$

$$\mathcal{ER}_i(\tau_i) = \frac{\sum_{j=1}^{N_{\mathrm{II}}} \mathrm{I}\left[\|\hat{\mathbf{x}}_{ij}\|_2^2 \geq \tau_i\right]}{N_{\mathrm{II}}}, \tag{10.10}$$

where

$$\mathrm{I}\left[P\right] = \begin{cases} 1 \text{ if } P \text{ is true} \\ 0 \text{ if } P \text{ is false} \end{cases}, \tag{10.11}$$

where N_{I} and N_{II} are the number of samples in data set I and data set II, respectively, $\hat{\mathbf{x}}_{ij}$ is the ith signal part in the jth testing signal segments, $i = 1, 2, 3, \ldots, C$ and $j = 1, 2, 3, \ldots, N_{\mathrm{I}}$ or N_{II}.

The threshold is determined by solving the following problem with the validating data set:

$$\tau_i = \arg\min_{\tau_i} \left(\mathcal{ER}_i(\tau_i) + (1 - \mathcal{CR}_i(\tau_i))\right). \tag{10.12}$$

The problem in (10.12) can be solved by searching algorithms [23,24].

10.3.2 Experimental results

The measured data are collected by using the radar and the UAVs introduced in Section 10.2.1. The measurement for each target class is repeated ten times with a duration of 6 s. Each measurement data is divided into 20 segments with the length of 0.3 s, which is long enough to contain several periods of oscillation of the reflected echo. Therefore, the number of signal segments of each scenario for training and testing is 10×6 s/0.3 s = 200.

In order to formulate the best sparse representation that is suitable for CVDs of small UAVs, the *K*-SVD algorithm is utilised to obtain the recognition dictionary during the experiment. Regarding the initialization of dictionary, there are three kinds of approaches [25], i.e., assigning the dictionary randomly, setting it to some randomly chosen training data points or running an initial clustering on the training data (such as discrete cosine transformation (DCT) and discrete wavelet transformation). In practice, a large quantity of training data are required for the first two kinds of methods, which is difficult to be satisfied in radar data sets of UAVs. The DCT dictionary has less limitation on the data set size and has produced satisfactory results in many existing studies [14,26]. Therefore, here we adopt the DCT dictionary for initializing the dictionary.

10.3.2.1 Recognition performance against multiple small UAVs

The objective of the dictionary learning-based method is to determine if there is a specific kind of target exists in the test scenario. Reference [27] proposes an approach for the recognition of different human behaviour types, in which the training data set is single-target data and testing data set contains multiple-target data. Here, we adopt a similar data set structure in this chapter.

In the experiments, the training data are composed of half of the measured data for each of the single-target case and 5% of the data are used as the validation data set, which is used for the determination of the threshold by solving (10.12). All the remaining single-target data and multi-target data are used for testing. The sparsity *K* is set to be 10, while the window size of STFT is set to be 4 ms. The recognition accuracy is obtained by averaging over 50 trials with randomly selected training data and validation data. The recognition accuracies for the scenarios of single, two and three small UAVs are presented in Table 10.5.

Here, we take Scenario D in Table 10.5 (i.e., the fourth row in Table 10.5) as an example, where the testing scene contains a helicopter and a quadcopter. The value in the fourth row and the first column (91.24%) is calculated according to (10.9), which represents the correct recognition rate of the helicopter. It means that the accuracy of the presented method in determining whether a helicopter exists in Scenario D is 91.24%. Similarly, the value in the fourth row and the third column (99.64%), which is also calculated according to (10.9), represents the accuracy of the presented method in determining whether a quadcopter exists. On the contrast, the value in the fourth row and the second column (7.20%), which is calculated by following (10.10), stands for the false alarm rate of hexacopter in Scenario D. This

Table 10.5 Recognition result yielded by the presented method

Scenario	Helicopter (%)	Hexacopter (%)	Quadcopter (%)
A: Single-UAV helicopter	95.47	10.69	2.53
B: Single-UAV hexacopter	2.44	99.11	0.31
C: Single-UAV quadcopter	0.11	0	99.98
D: Two-UAV helicopter+quadcopter	91.24	7.20	99.64
E: Two-UAV hexacopter+quadcopter	16.51	90.44	98.33
F: Two-UAV helicopter+hexacopter	88.91	95.20	5.82
G: Three-UAV helicopter+hexacopter+quadcopter	91.00	88.53	97.49
Rate of correct on average (%)	93.38 (the average of the green background rate)		
Rate of incorrect on average (%)	5.37 (the average of the white background rate)		

indicates that a small part of the sparse coefficients corresponding to the hexacopter also exceeds the decision threshold.

It is clear from Table 10.5 that (1) the rate of correct recognition on average for single-UAV scenarios (i.e., Scenarios A–C) achieves 98.19%, (2) the rate of correct recognition on average for two-UAV scenarios (i.e., Scenarios D–F) is 90.20%,

(3) the rate of correct recognition on average for three-UAV scenario (i.e., Scenario G) is 88.35%, and (4) the average rate of correct recognition for all the scenarios is 93.38%.

The recognition performance of the method proposed in [27] is also provided for comparison. The method proposed in [27], called the spectrogram-based method, recognises target classes by distinguishing the components of the testing signal using the dictionaries learnt directly from the time–frequency spectrogram. The proportions of training data are also set to be half the measured data for each single small UAV. The resulting recognition accuracies are shown in Table 10.6. The rate of correct recognition of the spectrogram-based method is 86.25% on average with the rate of incorrect recognition of 5.32%. It is clear from Tables 10.5 and 10.6 that the presented method obtains higher recognition accuracy.

10.3.2.2 Recognition performance with different sizes of the training data sets

In this experiment, the performance of the dictionary learning-based method is analysed with different sizes of the training data sets. The presented method is compared with the spectrogram-based method in terms of the recognition performance. The proportions of training data vary from 10% to 70% with a step size of 20%, and 5% of the data set is used for the determination of threshold. The sparsity is set to be 10 for each dictionary. The recognition performances are obtained by averaging over 50 trials with randomly selected training data. The boxplot and averaged value of resulting rates of correct and incorrect recognition are shown in Figure 10.7.

The presented method and the spectrogram-based method have comparable incorrect recognition rates as shown in Figure 10.7(b). Moreover, it is clear from Figure 10.7(a) that the presented method consistently outperforms the spectrogram-based method in terms of the correct recognition for different sizes of training data sets with higher mean, median, first quartile and third quartile values. In addition, the dispersion levels provided by the presented method are lower than that of the spectrogram-based method, which means the presented method is more stable for different data sets. When the proportion of the training data is 10%, the presented method still achieves satisfactory recognition performance with a correct recognition rate of 85%. When the proportion of training data reaches 70%, the rate of correct recognition of the presented method achieves 94% on average.

10.3.2.3 Recognition performance with different dictionary scales

In this subsection, we evaluate the performance of the presented method with different dictionary scales. The scale of the dictionary varies from 500 to 2,000 with a step size of 500. The sparsity is set to be 10, while the window size of STFT is set to be 4 ms. The proportion of the training set is 50%. The recognition performances are averaged over 50 trials with randomly selected training data. The experimental results are demonstrated in Table 10.7.

Table 10.6 Recognition result yielded by the method in [27]

Scenario	Helicopter (%)	Hexacopter (%)	Quadcopter (%)
` A: Single-UAV helicopter	84.88	18.60	0.58
B: Single-UAV hexacopter	16.50	98.68	0
C: Single-UAV quadcopter	0	0	100
D: Two-UAV helicopter+quadcopter	66.62	3.74	100
E: Two-UAV hexacopter+quadcopter	15.02	81.18	99.74
F: Two-UAV helicopter+hexacopter	91.10	99.14	0.64
G: Three-UAV helicopter+hexacopter+quadcopter	81.30	93.64	99.02
Rate of correct on average (%)	86.25		
Rate of incorrect on average (%)	5.32		

It can be seen from Table 10.7 that the recognition performance of the presented method changes slightly when the scale of the dictionary is larger than 1,000. This means the presented method is not sensitive to the dictionary scale when it is large enough. Based on the trade-off between the recognition accuracy

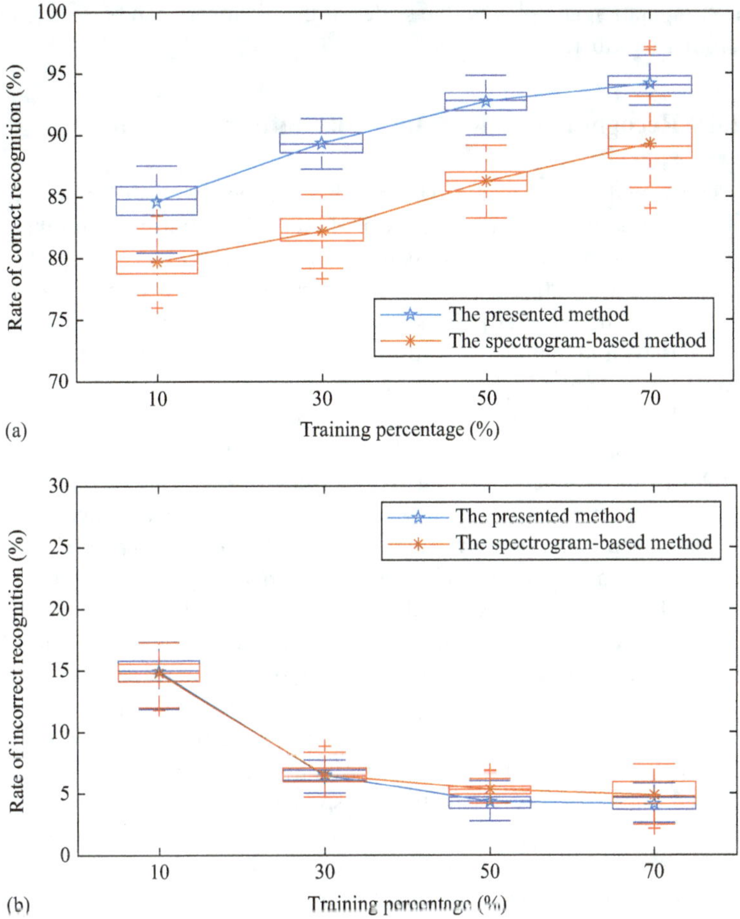

Figure 10.7 Recognition performance yielded by the presented method and the spectrogram-based method under different sizes of training sets: (a) rate of correct recognition and (b) rate of incorrect recognition

Table 10.7 Rate of correct recognition yielded by the presented method under different scales of the dictionary

Dictionary scale	500	1,000	1,500	2,000
Rate of correct recognition (%)	92.15	93.50	93.38	93.85

and the computation complexity, the scales of the dictionary can be selected in the range from 1,000 to 1,500.

10.3.2.4 Recognition performance with different sizes of sparsity

In this experiment, we evaluate the presented method with different values of sparsity K. The performance of the presented method is compared with that of the spectrogram-based method. The sparsity K varies from 1 to 19. The scale of the dictionary is set to 1,500, while the window size of STFT is set to be 4 ms. The proportion of the training set is 50%. The recognition performances are averaged over 50 trials with randomly selected training data. The resulting rates of correct recognition are shown in Figure 10.8.

Figure 10.8(a) shows the recognition performance of the presented method and the spectrogram-based method with different sparsity values while Figure 10.8(b) is the boxplot of the results. It is clear from Figure 10.8(a) that the rate of correct recognition of the presented method increases as the sparsity K increases when $K<10$. The reason is that more features are extracted as the sparsity K increases. The rate of correct recognition tends to be stable when $K \geq 10$. This means that no more useful features can be extracted by setting larger value of K. Meanwhile, a larger value of the sparsity K will lead to more computational burden. Considering the trade-off between the computation complexity and the recognition accuracy, the sparsity K can be selected in the range from 10 to 13. In order to more exactly determine the value of sparsity K, the training scheme method proposed in [28] can be used. That is, the proposed method with all the possible sparsity candidates is evaluated and the proper sparsity corresponding to the best performance is selected.

In addition, it is obvious from Figure 10.8(b) that the presented method consistently outperforms the spectrogram-based method for different sparsity values with higher mean, median, first quartile and third quartile values. Furthermore, the box sizes of the presented method are smaller than that of the spectrogram-based method, which means the dispersion levels of the presented method are lower than those of the spectrogram-based method.

10.3.2.5 Recognition performance with different window sizes

In this subsection, the performance of the presented method versus the window size of the STFT is investigated. The window size varies from 2 to 8 ms with a step size of 2 ms. The sparsity K is set to 10. The recognition accuracies obtained by the presented method are illustrated in Figure 10.9, where 50% and 5% of the data set are used for training and validating, respectively.

It can be seen from Figure 10.9 that the presented method achieves better accuracy when the window size is in the interval of 4 and 6 ms. The performance of the presented method declines when the window size is less than 4 ms or larger than 6 ms, this is because the frequency resolution or the time resolution of the

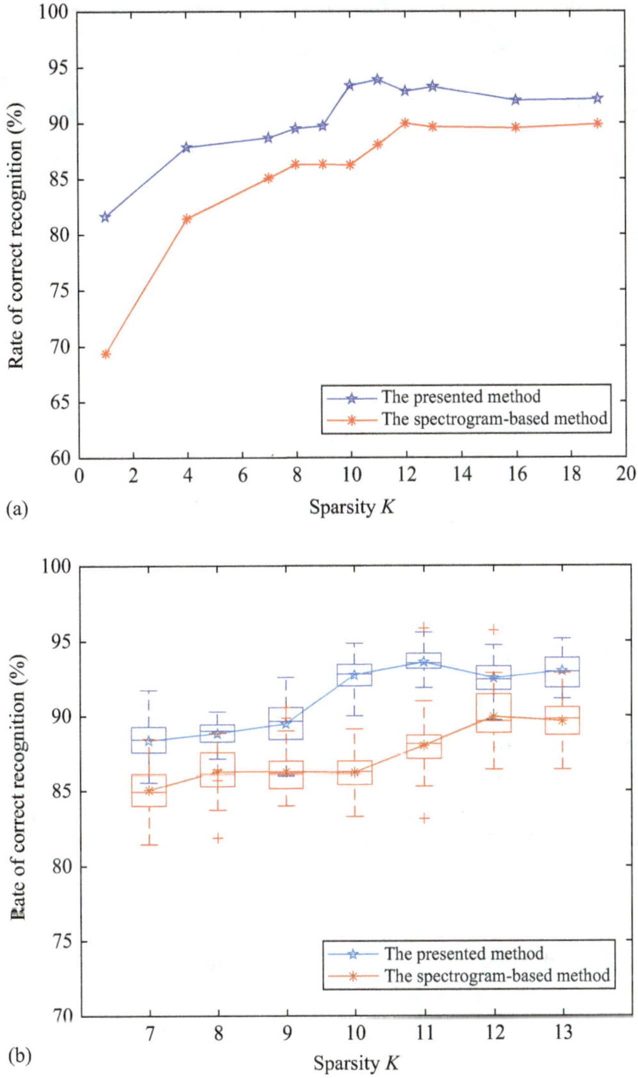

Figure 10.8 Rate of correct recognition yielded by the dictionary learning-based method and the method in [27] versus different values of sparsity K (a) on average and (b) on boxplot

time–frequency spectrogram are poor when the window size is too small or too large, respectively [21]. In addition, it is clear that the presented method consistently outperforms the spectrogram-based method for different window sizes with lower dispersion levels.

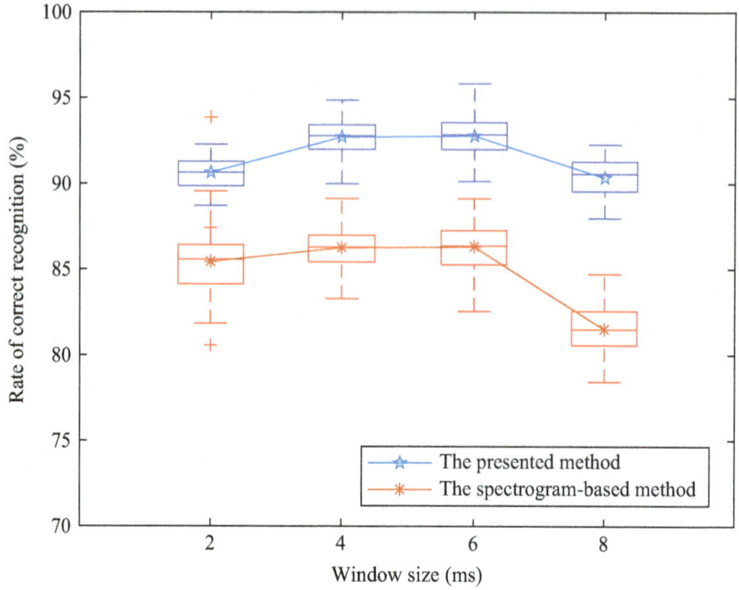

Figure 10.9 Rate of correct recognition yielded by the presented method and the spectrogram-based method in [27] versus different window sizes

10.3.2.6 Recognition performance with different noise levels

In this subsection, the presented method is evaluated under different levels of additive white Gaussian noise (AWGN). The signals received by the radar are mixed with simulated AWGN according to the following expression:

$$\mathbf{s}=\frac{\mathbf{y}}{\|\mathbf{y}\|_2}+10^{-\frac{SNR}{20}}\boldsymbol{\varepsilon}, \tag{10.13}$$

where $\mathbf{y}$ represents the received signal, $\|\mathbf{y}\|_2$ represents the l_2-norm of $\mathbf{y}$, SNR is the approximate signal-to-noise ratio of the AWGN and the received signal, and $\boldsymbol{\varepsilon}$ is the AWGN with zero mean and unit variance. It is worth emphasizing that there may be a small gap between SNR and the real SNR value of the mixed signal $\mathbf{s}$, because the received signal $\mathbf{y}$ also contains noise components.

In the experiments, the value of SNR varies from 0 to 20 dB with a step of 5 dB. Under each value of SNR, the recognition accuracies obtained by each method are measured by averaging over 50 trials. Here, the sparsity K is set to be 10, and the length of the window is set to be 4 ms. The proportion of the training set is 50%. It can be seen that the recognition accuracy obtained by the presented method is higher than 80% when the value of SNR is greater than 0 dB, as shown in Figure 10.10. Moreover, the presented method outperforms the spectrogram-based method under moderate noise with higher mean, median, first quartile and third quartile values.

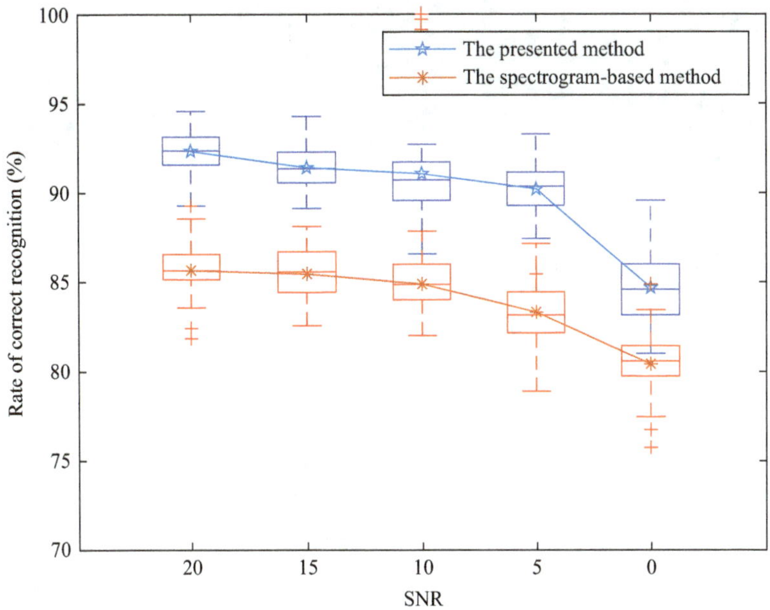

Figure 10.10 *Rate of correct recognition yielded by the presented method and the spectrogram-based method in [27] under different levels of AWGN*

10.3.2.7 Analysis about the time consumption

In this subsection, the computational time of the dictionary learning-based method is evaluated.

The hardware platform is a server with an Intel Xeon® E5-2640 CPU inside, and the CPU clock frequency and the memory size are 2.40 GHz and 128 GB, respectively. The software platform is MATLAB 2018b and the operation system is Windows Server 2016 Standard. In realistic applications, the recognition of UAVs needs to be real-time processing. Considering that the training process can be accomplished offline, the bottleneck of real-time processing is the time consumption for testing. For each method, the running time for testing is measured by averaging over 100 trials. The results of running time are presented in Figure 10.11, where the presented method has lower time consumption under different sparsity levels. It is obvious that the time consumption increases linearly as the sparsity rises. The reason is that the number of inner iterations of OMP algorithm performs linearly with the sparsity level. Since the running time for recognizing one sample by the presented method with MATLAB is only 0.28 s when the sparsity is set to be 10, it is promising to achieve real-time processing with optimised code on embedded computing platforms in practical applications.

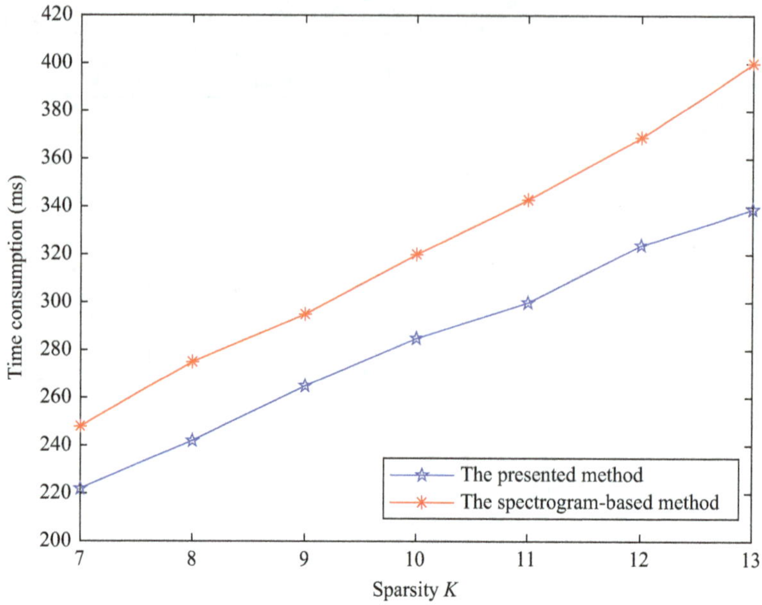

Figure 10.11 The time consumption of the presented method and the spectrogram-based method as a function of different sizes of sparsity

10.4 Conclusion

In this chapter, the problem of recognition of multiple UAVs with a radar sensor is studied. To some extent, this problem can be solved by spatial separation with high range resolution, which requires a large radar bandwidth and accordingly high hardware cost.

In order to recognise multiple UAVs with lower hardware cost, two algorithms for radar recognition of multiple UAVs with CW radar system are presented. In the first algorithm, the CFS is presented as a kind of low-dimensional features to realise the detection of multiple UAVs. CFS gives a general characteristic about the shape, size and variation tendency of CVD, which is independent of the time shift of radar echo and makes the recognition of small UAVs more robust. The experimental results on measured data demonstrate that the presented method is capable of dealing with multiple UAVs with high classification accuracy and low computational complexity. The second algorithm is based on dictionary learning, which aims to determine the presence of a specific type of small UAV. The dictionaries learnt by the K-SVD algorithm are combined together to determine the existence of the specific type of UAVs and accordingly the problem of recognition of multiple small UAVs is converted into the problem of signal decomposition. Experiments on real radar data have demonstrated the effectiveness of the dictionary learning-based algorithm.

The contents reported in this chapter are heuristic with several kinds of small UAVs. It is worth performing more experiments with more types of UAVs and a larger number of UAVs and evaluating the presented methods. When the data set is large enough, the features suggested in this chapter may be combined with artificial intelligence algorithms to further improve the recognition accuracy. Furthermore, the information of the UAVs in the time-range-Doppler domain could be exploited to improve the performance of recognition of UAV swarms.

References

[1] Ritchie, M., Fioranelli, F., and Borrion, H.: 'Micro UAV crime prevention: can we help Princess Leia?', Crime Prevention in the 21st Century: Insightful Approaches for Crime Prevention Initiatives, B. LeClerc and E.U. Savona, Eds. New York, NY, USA: Springer; 2017, pp. 359–376.

[2] Ritchie, M., Fioranelli, F., and Borrion, H.: 'Multistatic micro-Doppler radar feature extraction for classification of unloaded/loaded micro-drones', IET Radar Sonar Navig, 2017, 11, (1), pp. 116–124.

[3] Zhang, P., Yang, L., Chen, G., *et al.*: 'Classification of drones based on micro-Doppler signatures with dual-band radar sensors', 2017 Progress in Electromagnetics Research Symp., Singapore, Singapore, November 2017.

[4] Zhang, P., Li, G., Huo, C., and Yin, H.: 'Exploitation of multipath micro-Doppler signatures for drone classification', IET Radar Sonar Navigation, 2020, 14, (4), pp. 586–592.

[5] Oh, B.-S., Guo, X., Wan, F., *et al.*: 'Micro-Doppler Mini-UAV classification using empirical-mode decomposition features', IEEE Geosci Remote Sens Lett, 2018, 15, (2), pp. 227–231.

[6] Ritchie, M., Fioranelli, F., Griffiths, H., *et al.*: 'Micro-drone RCS analysis', IEEE Radar Conf., Johannesburg, South Africa, October 2015.

[7] Clemente, C., Balleri, A., Woodbridge, K., *et al.*: 'Developments in target micro-Doppler signatures analysis: radar imaging, ultrasound and through-the-wall-radar', EURASIP J Adv Signal Process, 2013, 47, (1), pp. 1–18.

[8] Fioranelli, F., Ritchie, M., and Griffiths, H.: 'Classification of unarmed/armed personnel using the NetRAD multistatic radar for micro-Doppler and singular value decomposition features', IEEE Geosci Remote Sens Lett, 2016, 13, (1), pp. 8–12.

[9] Zhou, Z., Zhang, C., Xu, C., *et al.*: 'Energy-efficient industrial internet of UAVs for power line inspection in smart grid', IEEE Trans Ind Inf, 2018, 14, (6), pp. 2705–2714.

[10] Shi, X., Yang, C., Xie, W., *et al.*: 'Anti-drone system with multiple surveillance technologies: architecture, implementation, and challenges', IEEE Commun Mag, 2018, 56, (4), pp. 68–74.

[11] Zhang, W. and Li, G.: 'Detection of multiple micro-drones via cadence velocity diagram analysis', Electron Lett, 2018, 54, (7), pp. 441–443.

[12] Zhang, W., Li, G., and Baker, C.: 'Radar recognition of multiple micro-drones based on their micro-Doppler signatures via dictionary learning', IET Radar Sonar Navig, 2020, 14, (9), pp. 1310–1318.

[13] Wang, Y., Feng, C., and Xu, D.: 'Micro-Doppler separation for target with rotating parts based on wavelet decomposition,' CIE International Conference on Radar, October 2016.

[14] Aharon, M., Elad, M., and Bruckstein, A.: 'K-SVD: an algorithm for designing overcomplete dictionaries for sparse representation', IEEE Trans Signal Process, 2006, 54, (11), pp. 4311–4322.

[15] Tropp, J.A. and Gilbert, A.C.: 'Signal recovery from random measurements via orthogonal matching pursuit', IEEE Trans Inf Theory, 2007, 53, (12), pp. 4655–4666.

[16] Wu, L., Wang, Y., and Pan, S.: 'Exploiting attribute correlations: a novel trace lasso-based weakly supervised dictionary learning method', IEEE Trans Cybern, 2017, 47, (12), pp. 4497–4508.

[17] Liu, L., Shao, L., Li, X., and Lu, K.: 'Learning spatio-temporal representations for action recognition: a genetic programming approach', IEEE Trans Cybern, 2016, 46, (1), pp. 158–170.

[18] Liu, L., Yu, M., and Shao, L.: 'Unsupervised local feature hashing for image similarity search', IEEE Trans Cybern, 2016, 46, (11), pp. 2548–2558.

[19] Patel, V.M., Qiu, Q., and Chellappa, R.: 'Dictionaries for image-based recognition', Information Theory and Applications Workshop, San Diego, CA, USA, February 2013.

[20] Wang, J., Yang, J., Yu, K., *et al.*: 'Locality-constrained linear coding for image classification', IEEE Computer Society Conference on Computer Vision and Pattern Recognition, San Francisco, CA, USA, June 2010.

[21] Li, G., Zhang, R., Ritchie, M., and Griffiths, H.: 'Sparsity-driven micro-Doppler feature extraction for dynamic hand gesture recognition', IEEE Trans Aerosp Electron Syst, 2018, 54, (2), pp. 655–665.

[22] Yang, M., Zhang, L., Feng, X., and Zhang, D.: 'Fisher discrimination dictionary learning for sparse representation', International Conference on Computer Vision, Barcelona, Spain, November 2011.

[23] LaValle, S.M., Branicky, M.S., and Lindemann, S.R.: 'On the relationship between classical grid search and probabilistic roadmaps', Int J Rob Res, 2004, 23, (7–8), pp. 673–692.

[24] Wu, S. and Manber, U.: 'A fast algorithm for multi-pattern searching', Technical Report TR-94-17, Department of Computer Science, University of Arizona, 1994.

[25] Zhang, Q. and Li, B.: Dictionary Learning in Visual Computing, San Rafael, CA, USA: Morgan & Claypool, 2015.

[26] Song, L. and Peng, J.: 'Dictionary learning research based on sparse representation', IEEE Conference on Computer Science and Service System, Nanjing, China, August 2012.

[27] Vishwakarma, S. and Ram, S.S.: 'Detection of multiple movers based on single channel source separation of their micro-Dopplers', IEEE Trans Aerosp Electron Syst, 2018, 54, (1), pp. 159–169.

[28] Molchanov, P.: Radar Target Classification by Micro-Doppler Contributions, Tampere, Finland: Tampere University of Technology, 2014.

Chapter 11

Advanced classification techniques for drone payloads

Francesco Fioranelli[1] and Julien Le Kernec[2]

11.1 Introduction

This chapter presents a summary of radar-based classification approaches developed for small drones carrying payloads. Specific focus is given to three types of techniques that were validated on the same multistatic radar data set collected using the University College London (UCL)-netted radar NetRAD. These techniques used, respectively, features extracted from the centre of mass and bandwidth of the micro-Doppler signatures; different radar data domains generated from the micro-Doppler data to be processed by pretrained Convolutional Neural Networks (CNNs) and spectral kurtosis analysis on the micro-Doppler.

All these approaches make the assumption that differences in the kinematics of the drone and rotation patterns of its blades induced by the presence of the payload can be captured by the analysis of the micro-Doppler signatures. While this might not always be the most effective approach for all operational scenarios and all payloads, some interesting results are presented.

Recently there has been a significant 'buzz' about small drones, due to the proliferation of these platforms becoming cheaper to buy off the shelf, or even to assemble in-house. For example, it is estimated that there will be a global market of £102bn for drone-based activities by 2025 [1], and over 2,000 licenses for commercial drone operators have already been granted in the United Kingdom.

Many novel leisure and commercial activities have arisen from this proliferation of small drones: from professional filming to autonomous inspections in critical infrastructures and environments, from support to agriculture and parcel delivery, to search and rescue, and support to Internet connectivity in remote areas [2].

Similarly, attention from the media and concerns from law enforcement agencies and air traffic controllers have been raised for many potential misuses of these platforms, either accidentally or voluntarily. These may include illegal filming, privacy violation, collision hazard with other aircraft (very sensitive for larger

[1]Department of Microelectronics, Microwave Sensing Signals and Systems (MS3) Group, TU Delft, Delft, The Netherlands
[2]James Watt School of Engineering, University of Glasgow, Glasgow, United Kingdom

aircraft taking off and landing) and people/assets on the ground, transport of illicit material (drugs into prisons, e.g., an issue in the United Kingdom), and terrorist attacks and weaponised drones (ISIS has been reported to try and use them) [2]. An example of the disruptive potential of these platforms is the closure of London Gatwick airport in December 2018, which was estimated to disrupt 120,000 passengers and block the departure of 700 flights scheduled in that period [3].

From the radar engineering perspective, small drones are an emerging class of targets posing challenging monitoring problems with traditional approaches and systems, often designed for monitoring larger aircraft that fly at higher altitude. Drones are on the contrary small in size, mostly made of plastic or materials that have low Radar Cross Section (RCS), and highly manoeuvrable, making them difficult to sense with radar, especially in complex and clutter-rich urban environments [4,5].

As a result of this challenge, research effort in various aspects of radar signatures of small drones has drastically increased in recent years. For example, the review paper in [4] published in 2018 counted more than 40 publications on the subject in 2017, whereas in previous years, only approximately 10 papers were published on this. A special issue on the subject of 'Innovative Radar Detection, Tracking and Classification for Small UAVs as an Emerging Class of Targets' was organised by IET Radar, Sonar & Navigation and published in 2020 [6]. The book of which this chapter is part is also an example of the current interest of the radar research community in this subject.

More specific research topics within the general 'radar signatures of small drones' theme include, for example the characterisation of RCS and the polarimetric response of these platforms [7–14]; algorithms for classification of different models of drones and drones vs birds [15–21]; algorithms for tracking drones [22,23]; systems employing passive radar techniques exploiting a variety of illuminator of opportunities, among others [24–28].

While performing an extensive literature review of all papers is a task well beyond the scope of this chapter, its specific focus is on the classification of payloads that drones could be carrying. Detecting the presence of payloads and if/where possible also additional information on their size, volume, material composition can be critical in situations where the drone could be carrying illicit or dangerous materials (e.g. drug smuggling, modified drones to carry weapons or explosives). On this specific focus, there is, to the best of our knowledge, not a large amount of work published in the open academic literature.

This chapter aims to bring together in one place some of the recent results published (Sections 11.2 and 11.3) and provide some considerations on open challenges and possible research trends (Section 11.4).

11.2 Radar system and experimental setup

The data discussed and analysed in this chapter were collected using the multistatic radar system NetRAD [29–31], which was developed at the UCL over a number of years and used for multiple experiments on multistatic targets and clutter signatures. NetRAD stands for 'Netted Radar' and is a coherent pulse radar composed of three identical but separated nodes that operate at 2.4-GHz carrier frequency (in the S-band).

In this specific experiment, the pulse waveforms used linear chirp modulation with 0.6-μs duration and 45-MHz signal bandwidth. The chosen Pulse Repetition Frequency (PRF) was 5 kHz to ensure that the whole micro-Doppler signatures of the small drone and its rotor blades were included in the unambiguous Doppler region. The radar transmitted relatively low power (+23 dBm) and was equipped with horizontally polarised antennas with approximately $10^\circ \times 10^\circ$ beamwidth, and 24 dBi gain. The choice of horizontal polarisation was determined by studies showing that this is more suitable to capture and characterise micro-Doppler patterns of the rotor blades of small drones [7,22].

The experiment was performed in July 2015 in a football field belonging to the UCL Sports Ground to the North of London. Figure 11.1 shows a sketch of the experimental setup, with the three NetRAD radar nodes deployed along a linear baseline with approximately 50-m internodal separation. The central node, Node 1, acted as the monostatic transceiver, whereas Nodes 2 and 3 were used as bistatic receiver-only nodes.

The model of small drone used in the experiment was a DJI Phantom Vision 2+, from which the original camera was removed to accommodate payloads of different weights. The weight of the drone prior to adding any payload was approximately 1.2 kg. The payloads used in this experiment included a small plastic tray mounted below the drone, containing small metallic discs. Multiple identical discs were stacked together to generate different weights for the payloads, namely 200, 300, 400, 500, and 600 g; these in addition to the case of free flight, i.e. no payload added. It should be noted that the 600-g payload appeared to be the limit for the drone to be able to take off and fly reasonably; hence, heavier payloads were not attempted with this drone. In this experiment, the volume, shape, and position of the payloads with respect to the body of the drone were not investigated; the only variable of interest was the overall weight of the payloads, which was simply increased by stacking more identical discs on the tray.

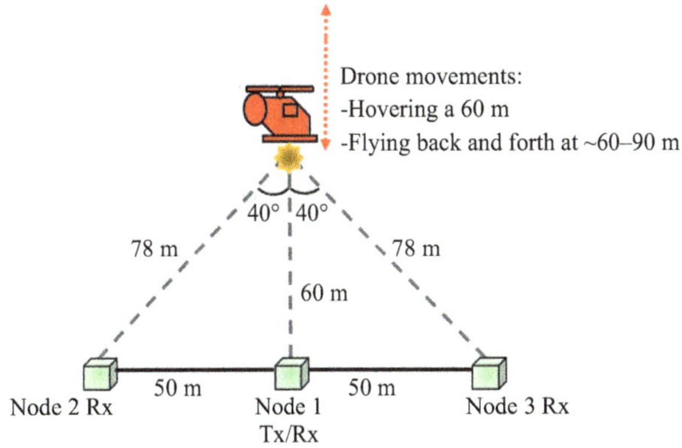

Figure 11.1 Sketch of the experimental setup for data collection

For each payload case, two types of movements were recorded. Initially, the drone was hovering at approximately 60-m distance from the baseline, resulting in a bistatic angle of 40° for both bistatic nodes. Then, simple back and forth linear trajectories were recorded for the drone flying from approximately 60 to 90 m, along the radial line of sight of Node 1. Each recording of radar data was 30-s long, corresponding to 150,000 recorded pulses. It should be noted that NetRAD operates as a staring radar, without the ability to rotate the antennas and track the target. For this reason, the antennas were aligned to the point of hovering for the drone, approximately 60 m away from the baseline.

11.3 Classification approaches for drones and payloads

In this section, details of the classification techniques used to identify whether the drone was carrying a payload or not based on radar signatures are described. Common to these techniques is the usage of micro-Doppler spectrograms, which result from taking the absolute value (intensity) or power (squared intensity) of the Short-Time Fourier Transform (STFT). STFT is a common choice as time–frequency distribution for the characterisation of radar micro-Doppler signatures and has been applied to many radar-based classification problems, including human movements and gait, automotive setting, human gestures for interaction with smart devices, animal locomotion, and ballistic targets, among others [32–36]. STFT and micro-Doppler spectrograms have also been extensively used for the classification of small drones, for example, to differentiate between drones and birds and to differentiate different models of drones [16,17].

Equation (11.1) shows the formula of the STFT, where the signal $x[n]$ is multiplied by the window $w[n]$ and transformed in the Doppler frequency domain. Note that k/N in (11.1) indicates the discretised values of the frequency, f.

$$STFT[f, m] = \sum_{n=-\infty}^{\infty} x[n] * w[n-m] e^{-j2\pi nk/N} \tag{11.1}$$

The window slides along the time domain signal with a given overlap between consecutive steps, and the result of this operation is a matrix as a function of time and Doppler frequency. An important parameter is the duration of the sliding window, as this influences the estimation of the frequency components in the micro-Doppler signature. If the window is sufficiently short with respect to the rotation of the blades of the drones, individual peaks (often called 'blade flashes') can be identified in the spectrograms. If the window is of longer duration, the contribution of the blades' signatures appear smeared over time onto horizontal lines in the spectrogram (often called 'HERM lines', Helicopter Rotor Modulation lines [14,37]).

11.3.1 SVD and micro-Doppler centroid features

Before resorting to the usage of micro-Doppler, an attempt was made to derive features directly from the Range–Time–Intensity (RTI) plots, which are provided by the NetRAD radar. This approach assumed to relate the presence of the payload to variations in the received power reflected by the drone, in case the payload modified the overall RCS or more in general reflectivity of the 'drone+payload' target.

The original 30-s recordings were divided into blocks of 1-s duration each, equal to 5,000 radar pulses at 5-kHz PRF. Then the standard deviation and mean value of the received signal were extracted, specifically looking at the signal in the range bin where the signature of the drone was contained. Figure 11.2 shows examples of these feature samples for five classes of interest (from no payload to 500-g payload) for data from the three radar nodes; feature 1 is, in this case, the mean value and feature 2 the standard deviation. Visual inspections of these plots revealed a poor separation between the classes of interest and led to conclude that classification algorithms based on these features only would perform poorly. While fluctuations of the reflectivity/RCS value were visible, it was not possible to establish a clear relationship between such changes and the presence of payloads of given weights. Hence, micro-Doppler features were considered more attractive for

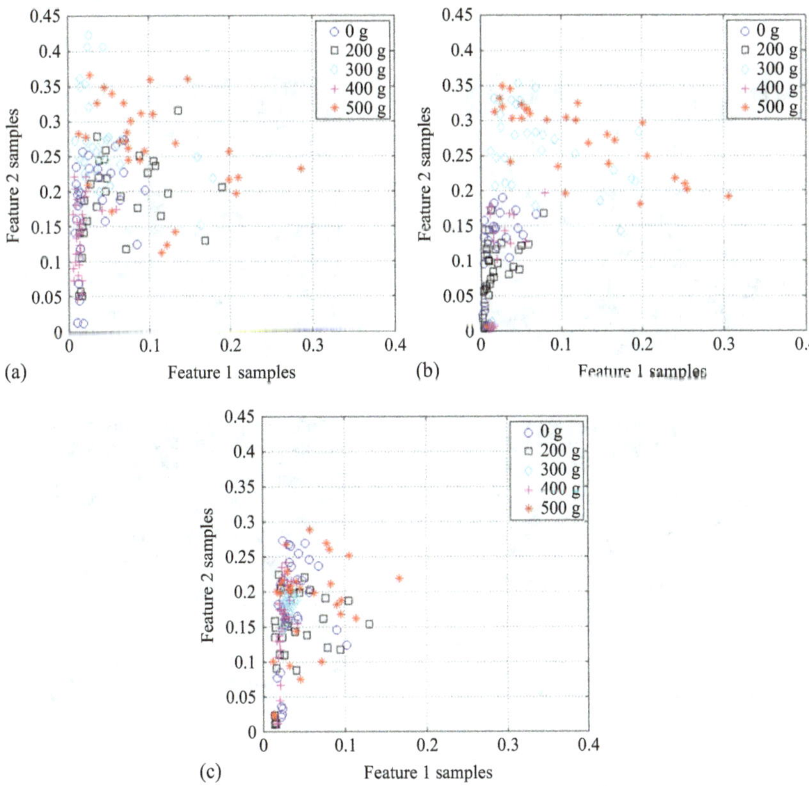

Figure 11.2 Feature samples extracted from the range–time–intensity images for the three radar nodes and five classes of payloads: (a) Node 1, (b) Node 2, and (c) Node 3

the classification problem. This does not imply that reflectivity values are always necessarily not informative for different data sets with different payloads size, volume, and material, and different models of small drones.

STFTs were applied on each recorded data set using a 0.1-s Hamming window and an overlap of 95% between adjacent windows. Each 30-s spectrogram was then divided into shorter segments to extract features. Different durations of each segment were considered to investigate the effect of this parameter, comparable to dwell time on target, on the classification. The five durations that were considered were 1, 1.5, 2, 2.5, and 3 s.

An example of 2-s long micro-Doppler spectrograms is shown in Figure 11.3, where monostatic and bistatic data of the two same recordings are compared, for the drone hovering without payload and with a 500-g payload. By visual inspection, one can see a difference in the patterns of HERM lines, with them assuming a more uniform and smoother trend over time for the heavy payload case, as well as stretching to reach higher positive and negative Doppler frequency values. This can be interpreted as an effect of the higher rotation speed of the rotor blades in the case of heavy payload presence, in order to provide higher lift force to deal with the weight.

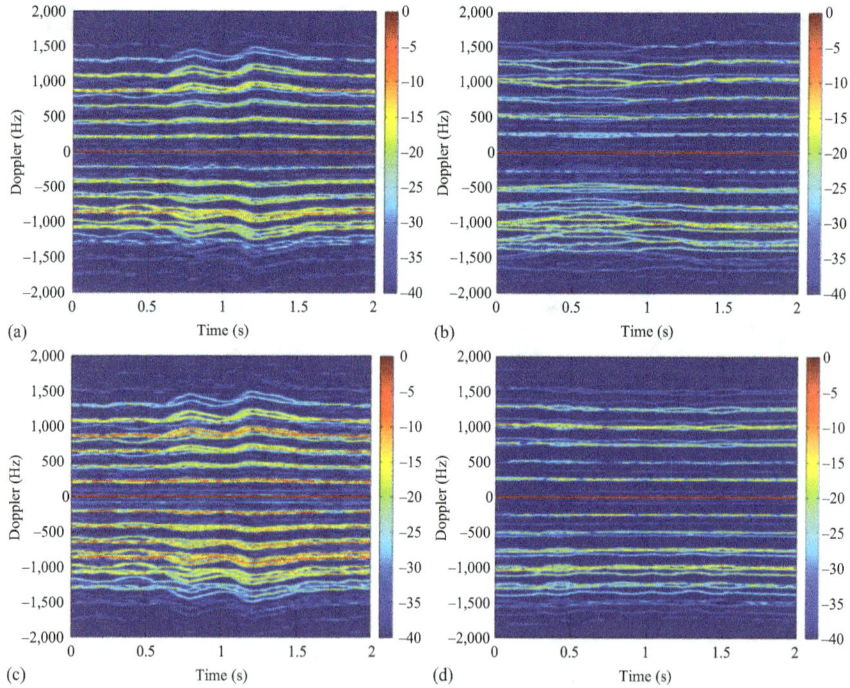

Figure 11.3 Micro-Doppler spectrograms for a small drone hovering: (a) monostatic data with no payload, (b) monostatic data with 500-g payload, (c) bistatic data with no payload, and (d) bistatic data with 500-g payload

In order to capture these empirical differences in quantitative metrics and features, the Singular Value Decomposition (SVD) of the spectrogram was first applied [38]. If the original spectrogram is defined as a matrix $\boldsymbol{M}$, its SVD decomposition is reported in the next equation:

$$\boldsymbol{M} = \boldsymbol{U}\boldsymbol{S}\boldsymbol{V}^T \tag{11.2}$$

where $\boldsymbol{S}$ is a diagonal matrix containing the singular values of $\boldsymbol{M}$, whereas $\boldsymbol{U}$ and $\boldsymbol{V}$ are matrices containing the right and left singular vectors of the original spectrogram matrix. Features extracted from the SVD decomposition of spectrograms have been demonstrated for the problem of classifying different models of drones [18], specifically capturing kinematic characteristics such as the velocity and periodicity of rotor blades that are specific to given models of drones.

Rather than focusing on a limited number of singular vectors as in [18], one can assume that the relevant information for the classification of payloads is spread across multiple vectors in the whole matrices $\boldsymbol{U}$ and $\boldsymbol{V}$. Empirical analysis on matrices $\boldsymbol{V}$ for the available data showed that there were no significant changes for the different classes of interest. On the contrary, the values on the diagonal elements of the $\boldsymbol{U}$ matrices show differences for different weights of the payloads. Therefore, the mean and standard deviation values of the diagonal elements were considered as features. Figure 11.4 shows feature samples for five different classes, from no payload to a heavy 500-g payload, where features 1 and 2 are the mean and the standard deviation values of the diagonals of $\boldsymbol{U}$ matrices resulting from SVD applied to the spectrograms. It is noticeable that feature samples for the 'no payload' case are well separable from the other classes for both monostatic and bistatic data; the best separation among all classes' samples can be seen for the case of monostatic data (Node 1).

Besides SVD-based features, other simple features were extracted, looking at the centroid and bandwidth of the micro-Doppler signatures [39]. The micro-Doppler centroid can be considered an estimate of the centre of mass of the signature, containing the main velocity component across the different Doppler bins. The micro-Doppler bandwidth can be considered an estimate of the signature spread and intensity around the centroid. Equations (11.3) and (11.4) show the formulae to calculate the centroid f_c and bandwidth B_c, where $M(i,j)$ is the value of the spectrogram $\boldsymbol{M}$ at the ith Doppler frequency bin and jth time bin.

$$f_c(j) = \frac{\sum_i f(i)M(i,j)}{\sum_i M(i,j)} \tag{11.3}$$

$$B_c(j) = \sqrt{\frac{\sum_i (f(i) - f_c(j))^2 M(i,j)}{\sum_i M(i,j)}} \tag{11.4}$$

The mean value of the centroid and the bandwidth are considered as features for the payload classification problem. Figure 11.5 presents feature samples extracted at the three radar nodes for the five classes of interest, from no payload to heavy 500-g payload. With these features, clear separation can be seen for all classes at all three radar nodes; hence, an improvement of classification performance compared to using the SVD-based features is expected.

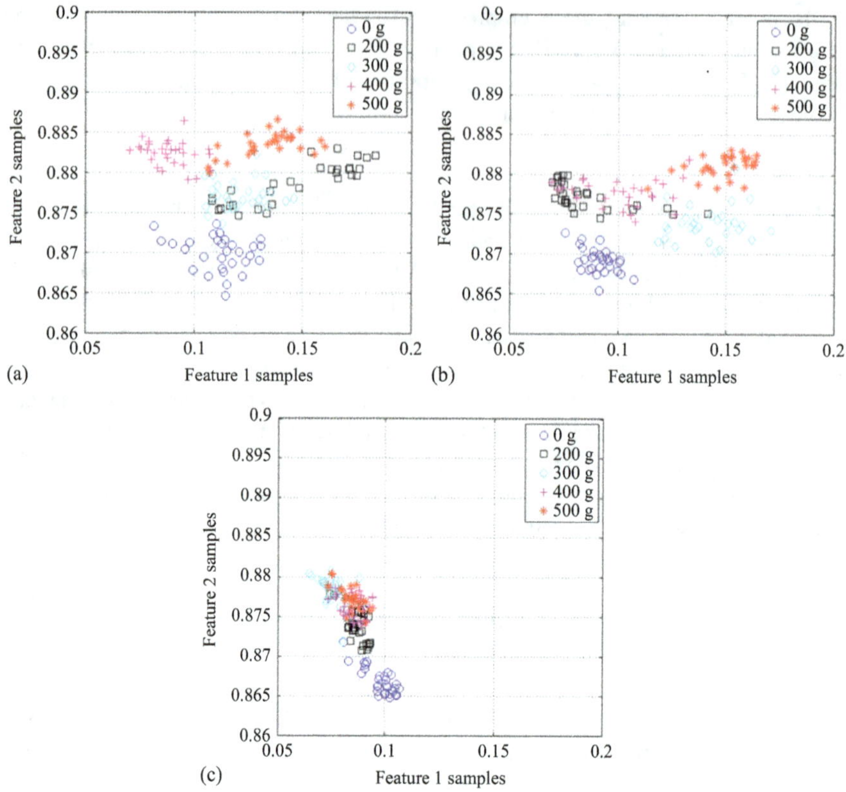

Figure 11.4 Feature samples extracted from the SVD of the micro-Doppler for the three radar nodes and five classes of payloads: (a) Node 1, (b) Node 2, and (c) Node 3

The SVD and centroid features were used as input to simple classifiers based on supervised learning framework, specifically naïve Bayes classifier and random forest classifier. The data set is randomly split into a training and testing data set for the classifiers and performance metrics such as testing classification accuracy are calculated. This cross-validation process is repeated 100 times to remove possible biases in the choice of training data and to assess the consistency of the classifiers' behaviour.

Three different approaches are tested when using multistatic radar data for this classification problem, and these can be compared with the usage of simple monostatic data.

- All feature samples from all the three radar nodes are combined into a single vector (feature fusion) and used to train and test a single classifier.
- Three separate classifiers are used to process the data of each individual radar node, and their partial decisions are fused in a simple majority-voting mechanism across the three nodes.

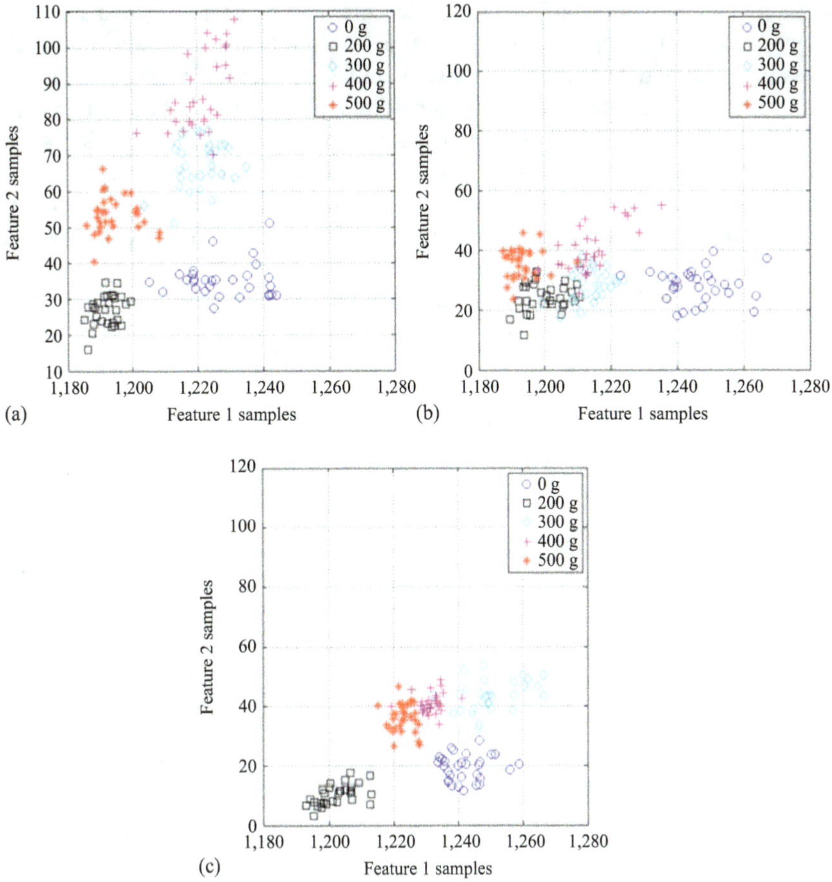

Figure 11.5 Feature samples extracted from centroid and bandwidth of the micro-Doppler for the three radar nodes and five classes of payloads: (a) Node 1, (b) Node 2, and (c) Node 3

- Similar to the decision fusion of the previous approach, but in this case, the confidence level in the partial decisions of each node is considered via a threshold when generating a final decision. If two nodes provide the same label and both have higher confidence than the given threshold, their label becomes the final label. If one of these two nodes has lower confidence than the threshold and the remaining third node has higher confidence than the two other nodes, then the final label is provided by the third node. The rationale of this approach is to prevent those nodes with lower confidence leading to wrong classifications. A value of the threshold was set at 65% through empirical investigation with different values in the range 55%–75% with the available data.

More complex decision level fusion schemes can be devised with a system of weights among the different radar nodes or exploiting the diversity of features across different nodes [40], but these were not tested with the available data.

Figure 11.6 presents the classification accuracy as a function of dwell time, i.e. the different duration of the spectrogram segments used to extract features, for data related to the hovering drone scenario. The sub-figures compare different classifiers, the naïve Bayes and the random forest, and each trace in the plots shows a different approach to use and fuse multistatic radar data. The training data set was in this case, 40% of all available data.

Regardless of the classifier and features used, specific trends can be observed with respect to the strategy for multistatic data fusion. The decision fusion approach outperforms both the feature fusion approach and the usage of only monostatic data; however, the usage of the threshold compared to the simpler voting approach in decision level fusion does not appear to bring a noticeable improvement in classification accuracy. The notable decrease in performance when

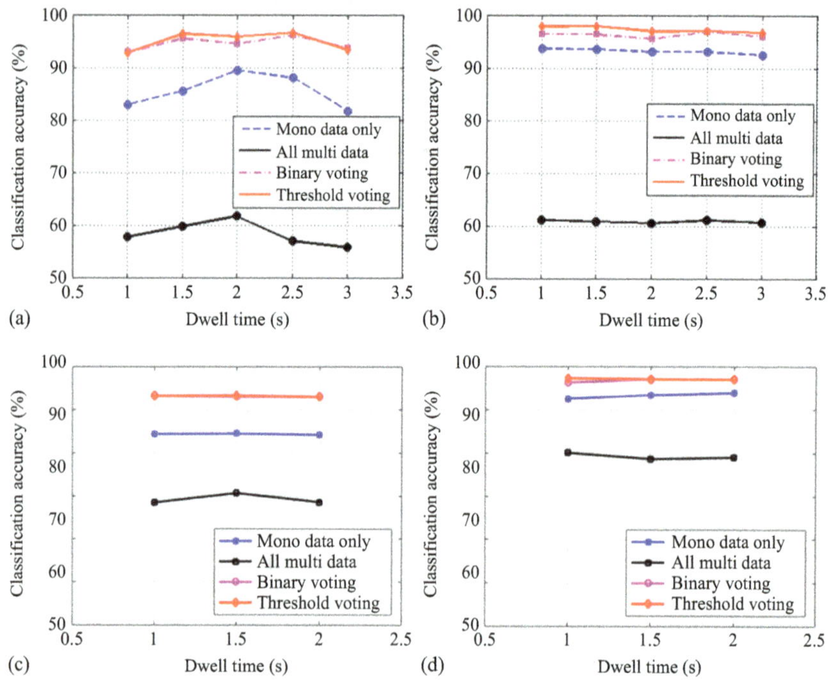

Figure 11.6 Classification accuracy as a function of dwell time (duration of spectrogram segment used to extract features) for hovering drone: (a) naïve Bayes classifier and SVD features, (b) naïve Bayes classifier and centroid features, (c) random forest classifier and SVD features, and (d) random forest classifier and centroid features

feature fusion at a single classifier is used as opposed to decision level fusion is interesting. This can be related to the differences in micro-Doppler signatures and feature samples seen by the different radar nodes because of the differences in spatial locations with respect to the target. Leveraging this spatial diversity into various feature extraction schemes and decision fusion can be a successful approach to boost classification in multistatic radar systems. Besides the problem at hand, similar results were also reported for classification problems on human micro-Doppler signatures [38–40]. For a given classifier, set of features, and multistatic data fusion strategy, there is no notable variation in classification accuracy as a function of the dwell time. This appears to suggest that acceptable performances can also be achieved with relatively short observation times, such as 1 and 1.5 s, at least for the data analysed in this case. Comparing SVD and centroid features, the latter appears to yield better classification performances with accuracy close to 95%–99% despite using only two relatively simple and easily explainable features.

Some results are also presented looking at the data recorded when the drone was flying back and forth, as shown in Figure 11.1. In performing these flying tests, the flight trajectory and velocity were kept as regular and constant as possible, compatibly with the wind conditions and with the different inertial response of the drone controls as the weight of the payloads was increased. In this case, the classification problem was slightly simplified considering three cases, no payload, the medium payload of 300 g (comparable to the weight of the default camera normally attached to the drone and removed for this test), and the heavy payload of 500 g.

The same features based on SVD and the centroid of the micro-Doppler signatures were considered. Figure 11.7 shows the classification accuracy as a function of the dwell time as the duration of the spectrogram segments used to extract features. As for the case of hovering drones, 40% of the available data were used to train the classifiers, and the decision fusion with threshold was used as a strategy to

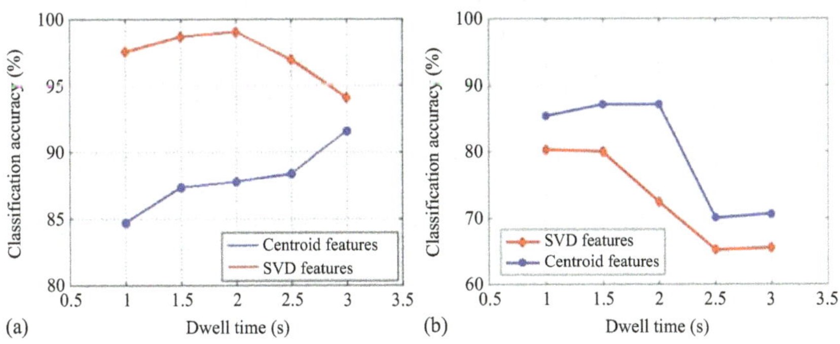

Figure 11.7 Classification accuracy as a function of dwell time (duration of the spectrogram segment used to extract features) for a flying drone: (a) naïve Bayes classifier and (b) random forest classifier

combine multistatic radar data. This was chosen as the previous results indicated that it outperformed other methods.

Results with the naïve Bayes classifier show that the classification accuracy monotonically improves with longer dwell time if centroid features are used, whereas this is no longer true for SVD-based features, for which a peak of accuracy is reported for dwell time equal to 2 s. When the random forest classifier is used, a different trend is observed in the general classification accuracy that is significantly decreased for longer dwell times. This difference might be related to the different mathematical formulations of the classifiers, with the Bayesian one assuming Gaussian distributions for the feature samples in each class. This could, in turn, generate different performances for different sizes of the feature space as a function of the dwell time. More investigation with a more extensive set of data with more diversity in terms of drone model, flying trajectories, types of payloads would be needed to validate the earlier hypothesis. However, the classification results reported for this three-class problem are promising, considering that these are achieved with relatively simple classifiers and features.

11.3.2 Pretrained convolutional neural networks

In this section, the classification of payloads carried by the drone is performed using deep learning techniques, specifically a pretrained CNN [41]. The network treats the radar input data as images and processes them in a supervised learning framework to identify the regions of interest in the images that contain relevant information for the classification. This is equivalent to the feature extraction step discussed in the previous section but is in this case implicitly performed within the network itself. It has been shown that some of the features highlighted by the network are related to patterns in the radar data, such as lines or contours, but in other cases, it is more difficult to explain what features or patterns are considered the most informative by the network.

Three different radar data formats, or representation domains of radar data, are presented here. The first is the spectrogram resulting from the STFT, as discussed in the previous section. The second uses cepstrograms. To generate this representation, the natural logarithm of the absolute value of the STFT is calculated. Then an inverse Discrete Fourier Transform (DFT) is applied over each short window similar to what is done for the STFT, followed by an absolute value operator. The formula for the cepstrogram is provided in (11.5), where k/N indicates the discretised values of the frequency, f. For drone signatures, cepstrogram can reveal periodic details in the original spectrogram that can be used as relevant features for the classification of drone targets [41].

$$Ceps[q,n] = \left| \frac{1}{N} \sum_{n=0}^{N-1} \{\log(|STFT[f,n]|^2 + 1)\} * e^{j2\pi kn/N} \right|^2 \tag{11.5}$$

The third representation is the Cadence Velocity Diagram (CVD), which is calculated by performing a DFT across the time dimension of the spectrogram. This results in a pattern of Doppler frequencies as a function of cadence frequency and enables to represent the periodicity of the Doppler and micro-Doppler modulations. Equation (11.6) provides the formula for the CVD calculation, where k_{cad}/N indicates the discretised values of the cadence frequency.

$$CVD[f, f_{cad}] = \left| \sum_{n=0}^{N-1} STFT[f, n] * e^{-j2\pi n k_{cad}/N} \right| \tag{11.6}$$

As an example, Figure 11.8 shows the different data domains for the same data set of hovering drones. In Figure 11.8(a), the RTI plot is shown, where the drone can be seen as a vertical line (i.e. more or less stable position over time) between range cells 34 and 36. The other vertical lines present in the plot are the direct signal from the monostatic transceiver (around range cell 10) and the reflection from a line of trees at the end of the experimental area in the field (around range cell 105).

Figure 11.8(b) shows the spectrogram resulting from STFT, with patterns similar to those shown in Figure 11.3. The rotor blades of the drone produce HERM lines, with values compatible with the operating specifications of the DJI Phantom drone. The maximum rotational rate is quoted at approximately 7,000 rpm in hovering conditions for this drone [20]. Equation (11.7) can then be used to calculate the expected rotation rate resulting from the strongest component in the micro-Doppler spectrogram at about 875 Hz: this yields approximately 5,220 rpm. Note that v is velocity, r is the length of a

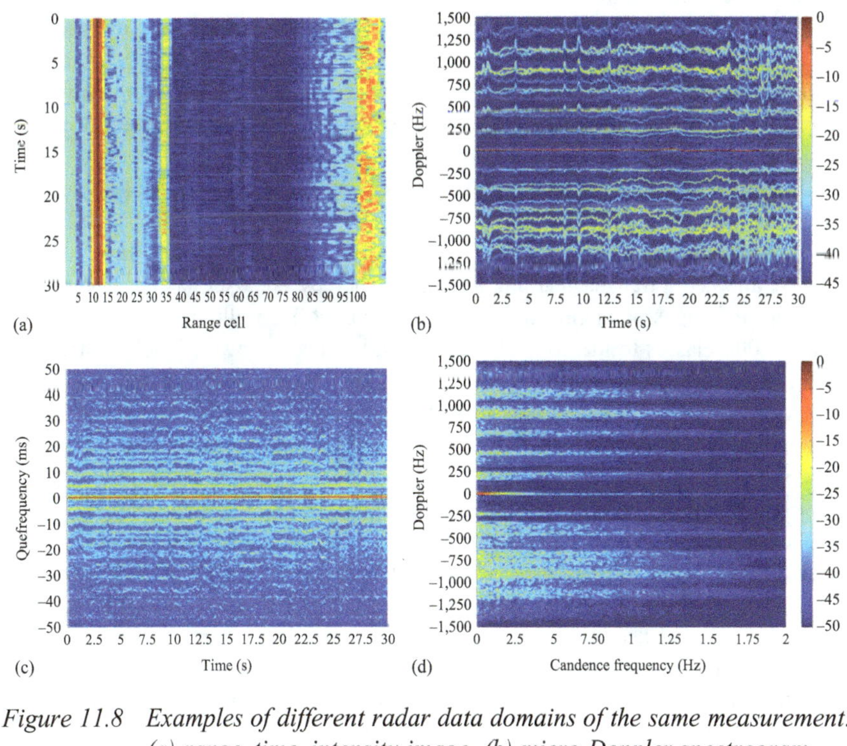

Figure 11.8 Examples of different radar data domains of the same measurement: (a) range–time–intensity image, (b) micro-Doppler spectrogram, (c) cepstrogram, and (d) Cadence Velocity Diagram (CVD)

single blade, ω the rotation velocity, and θ the aspect angle between the radar line of sight and the drone heading. Different scatter points on the blades rotating at different speed generate the spread of the signature over the Doppler axis, with an approximate space between the replicas of approximately 220 Hz [42]. It is believed that the component visible at 1,350 Hz is an intermodulation artefact, as its corresponding rotation speed of 8,050 rpm would be too high for this model of drone, and its signal strength appears to be ~15-dB weaker than the other components.

$$\omega_{rpm} = \frac{30v}{\pi r} \cos \vartheta \tag{11.7}$$

Figure 11.8(c) shows the cepstrogram of the data, and Figure 11.8(d) the CVD plot. From the CVD, it can be seen that the strongest component is localised at about 875-Hz Doppler frequency, which was already noted when discussing the spectrogram in Figure 11.8(b). The signature spans across several cadence frequencies, and it is not immediately obvious to extract a pattern of periodicity by empirical visual inspection. This may be due to the fact that the signature of the rotor blades in the original spectrogram is not exactly periodic over time. There are oscillations and small peaks for each spectral line, but no clear periodicity can be observed in comparison to simulated signatures of blades [36]. However, the energy in the CVD plot appears to be localised at the same Doppler bins where the signature was present in the spectrogram.

As the information contained in these different representations of the radar data is not trivial to extract, convolutional networks have been suggested as a suitable classifier. They enable to identify and extract the relevant information for the classification problem at hand, without explicit calculation of features or fine-tuning of thresholds and parameters by the human operator.

The network considered here is the well-known AlexNet convolutional network [43]. This network comprises five convolutional layers and three fully connected layers, with a Rectified Linear Unit (ReLu) activation function every layer. In this case, we adopted a transfer learning approach whereby AlexNet has been pretrained with optical images belonging to a data set of about 1.2 million samples and containing 1,000 classes to identify. This approach has been shown to yield better classification performances compared to training the network from scratch with radar data, which is generally much more limited in availability and diversity of the classes compared to optical images [32,44]. To adapt the network to our classification problem of drones and payloads, the last three layers were removed and replaced with a fully connected layer, followed by a Softmax classifier and output layer [41].

Regarding the classes of interest, data were considered where the drone was hovering and flying with no payload, and with payloads of weight 200, 300, 400, 500, and 600 g, respectively. For all these cases, the three data representations shown in Figure 11.8 were calculated: micro-Doppler spectrograms, cepstrograms, and CVD plots. These were also down-sampled to reach an overall dimension of 227×227 pixels in RGB images to fit the predefined input layer and format of AlexNet. Then, the classification problem was cast as a pseudo-six-class problem, where, for each of the three data representations considered, binary classification of no payload vs 200–600-g payload was performed. As commented in Section 11.2, for a 600-g payload, the drone was not able to fly properly, hence only hovering data were considered for this payload.

To (partially) address the problem of data scarcity, spectrograms and cepstrograms were split into two halves about the 0-Hz Doppler frequency, with the assumption of symmetry between the positive and negative part, or at the very least same underlying information content. They were also divided into segments of 5-s duration. CVD plots were recomputed for each 5-s long positive and negative halves of the spectrograms.

There was a noticeable class imbalance in casting the problem as binary different payload weights vs no payload, with a ratio of 5:1. As a consequence of this imbalance, initial tests showed that the CNN had significant overfit to the payload class and achieved about 80% accuracy, but yet ignoring a vast majority of the no payload samples. While there is no ready-made method to easily augment radar data such as spectrograms or the other data representations used here [32,44], the imbalance was corrected by generating synthetic no payload data by varying the intensity values between 0.02 and 0.05. This was performed to reach the same number of samples for the two classes of interest for each data representation.

On all the available data, further augmentation processing was applied for the training samples by exploiting symmetries to, e.g., mimic different direction of motions, effectively doubling the number of input samples for training the network. Other methods that are common for optical images, such as rotations and translations, were not attempted as they would produce unsatisfactory classification results in this case. Indeed, this would result in the loss of positional information in the resulting patterns, whereby pattern position in radar data is directly related to kinematic information on the movement of the targets and their parts (i.e. Doppler frequency value and cadence frequency value).

The modified AlexNet was trained with 60% of the available data, validated with another 10%, and tested with the remaining 30% (data which were completely unseen and unused at the training stage). The computations were performed on a high-end workstation with Nvidia GTX 1080TI GPU card, the training time was ~1 h per network.

In the case of the hovering data, the total amount of data was 576 samples, split in 432 for training, 36 for validation, and 108 for testing. These amounts are the numbers of samples per data representation; hence, the total amount of data is three times larger considering spectrograms, cepstrograms, and CVD plots. Figure 11.9 shows the confusion matrix with the average result of five repetitions of the process of training, validation, and testing of the network. The percentage and number of samples for each class is shown in each cell of the matrix, with an expected diagonal structure of the matrix as the problem is basically cast as three simultaneous payload classification cases using the three radar data domains. The total accuracy, averaged across the three data representations, is 95.1%, with specifically 97.2% for the spectrograms, 93.5% for the cepstrograms, and 94.4% for the CVD plots.

In the case of the flying data, the total amount of samples was 460 samples, split in 344 for training, 28 for validation, and 88 for testing. As commented earlier,

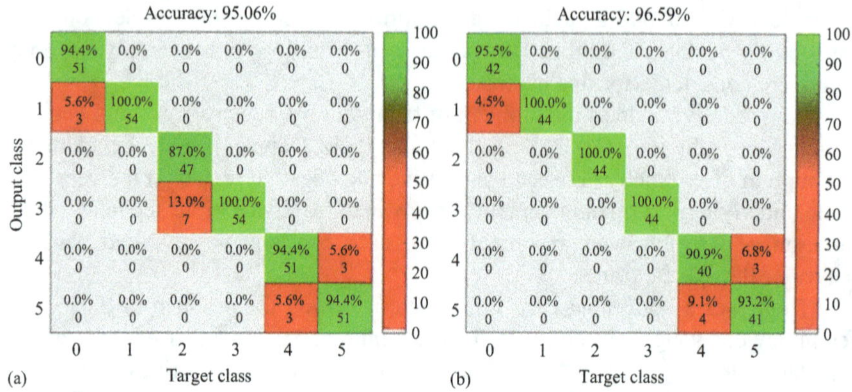

Figure 11.9 Confusion matrices for the hovering drone case (a) and flying drone case (b). Class 0 (spectrogram no payload), 1 (spectrogram payload), 2 (cepstrogram no payload), 3 (cepstrogram payload), 4 (CVD no payload), and 5 (CVD payload)

there are slightly fewer samples because the DJI drone could just about hover when fitted with the 600-g payload but not fly. The same testing procedure as for the hovering data was performed, and the resulting confusion matrix is shown in Figure 11.9 on the right-hand side. The averaged accuracy across the three data representations was 96.6%, with 97.7% for the spectrograms, 100% for the cepstrograms, and 92.0% for the CVD plots. The somewhat higher accuracy for the flying case with respect to the hovering case might be caused by the lower variability in the classes with payloads, i.e. the heavy-payload class is missing. Another test was also performed by using the testing flying data with the network trained with hovering data and vice versa. In this case, 78% average accuracy was achieved in both situations.

The approach and results presented in this section only scratched the surface of possible techniques inspired by deep learning and neural networks in the classification of drones and in this specific case a drone carrying payloads [20,45]. The results showed rather similar performances achieved using the three data representations separately for classification but jointly in the same network. An obvious extension to this investigation could consider forms of information fusion between the different representations. This could be done in different ways, for example by creating a new data representation by concatenation or 3D shaping of the three separate representations; by applying decision fusion with additional final layers within a single CNN; by using three separate, simpler CNNs for each representation, followed by a decision fusion stage. Another interesting study can investigate the effect of different dwell times, i.e. different temporal durations of the images used as inputs to the network. Dwell time is expected to be an important operational parameter in practical scenarios, where the latency in generating a label for situational awareness and the possibility of tracking the drones in the radar beam consistently over time are important aspects to consider.

11.3.3 Spectral kurtosis analysis and features

On the same data set for drone payload classification, an approach based on the spectral kurtosis of the micro-Doppler signatures as a feature was also proposed in [46].

Spectral kurtosis has the property of capturing and emphasising non-stationary and/or non-Gaussian components in the spectral domain of a given signal, which might not be well characterised when using more conventional power spectral density calculations [47,48]. As this tool has been proposed for the analysis of vibration and rotation dynamics of machines, an interesting investigation pertains to whether this can also work well to capture the rotations of the blades of small drones in the radar signatures.

By definition, the spectral kurtosis can be defined as the kurtosis calculated for each frequency component of the signal under test. If applied to micro-Doppler signatures in spectrograms originated from STFT, this can be computed as the normalised fourth-order moment of the STFT magnitude, as shown in (11.8). Essentially, the kurtosis $K(f)$ is proportional to the ratio of the fourth-order moment of the STFT and the square modulus of the second-order moment of the STFT.

$$K(f) = \frac{\frac{1}{M}\sum_{m=0}^{M-1} |STFT(f,m)|^4}{\left(\frac{1}{M}\sum_{m=0}^{M-1} |STFT(f,m)|^2\right)^2} - 2 \tag{11.8}$$

As shown, the spectral kurtosis depends on the spectrogram and its frequency resolution, which, in turn, depends on the duration of the sliding window used in the STFT calculation. The trade-off in time vs frequency resolution with the fixed-duration window of STFT is a well-known issue. A simple yet-effective way to partially address this problem is to use a multiwindow approach. The spectrogram is calculated with two different windows, and so also the spectral kurtosis computed with two different frequency resolutions, one finer and one coarser, enabling to extract different non-stationary components. A feature fusion approach can be applied, as proposed in [46], to form a single vector of feature samples by simply concatenating the values of spectral kurtosis extracted with both windows. This vector can then be used as the input to a classification algorithm.

It should be noted that additional pre-processing steps can also be performed to boost classification performances, for example different approaches to normalise the feature vector, or techniques for feature selection and dimensionality reduction, such as classic Principal Component Analysis (PCA).

Interesting results for this approach based on spectral kurtosis are reported in [46]. Each data recording was cut to segments of 2-s duration prior to computing spectrograms and extracting kurtosis. In that work, the values of the two STFT windows used for calculation of the spectrograms were 100 and 25 ms, respectively, and the size of the combined feature vector was 1,024 samples, compressed down to 64 after the application of the PCA. The applied classifier was a simple nearest neighbour classifier with the parameter number of neighbours set to 5. The split between training and testing data was done with a percentage of 70% and 30%, respectively, and the process repeated 100 times in a cross-validation approach with a random selection of data for training and testing.

The reported results yielded classification accuracy of 92.6% for the hovering drone case, 70.8% for the flying drone case, and 78.4% for the mixed classification of both hovering and flying cases, as shown in Table 11.1. In this case, the problem was cast as a six-class problem, with the no payload case and five payload cases (from 200 to 600-g weights). An analysis of the confusion matrix revealed that the majority of misclassification events happened between the two heaviest classes of payloads, 500 and 600 g, for which the normal dynamic of the drone was disrupted because of the effort to take off for hovering (difficulty to fly for the 500-g payload and impossibility to fly for the 600-g payload). The confusion matrix is reported in Table 11.2 for completeness.

While still an initial proof of concept, these results show an interesting application of a new feature, the spectral kurtosis, which was not applied to drones and payload signatures before. Further work can explore the variables involved in the process, for example the duration of the STFT windows used, the reduction factor within the PCA, and different modalities to combine and select information from multiwindow spectrograms.

Table 11.1 Probability of correct classification (Pcc) for six-class payload classification problem using spectral kurtosis features [46]

Pcc (%)	Hovering	Flying	Hovering and flying
	92.61	70.83	78.36

Table 11.2 Averaged confusion matrix for the results of classification of six classes of payloads with spectral kurtosis features in the case of drone hovering [46]

		Predicted					
		0g	**200g**	**300g**	**400g**	**500g**	**600g**
Actual	**0g**	**97.31**	0.15	0.23	2.15	0.16	0
	200g	1.23	**95.08**	1.23	2.46	0	0
	300g	4.46	6.77	**88.16**	0.23	0	0.38
	400g	0.08	0	3.15	**95.85**	0.92	0
	500g	0.08	1.15	0.15	0.77	**97.31**	0.54
	600g	0.54	0	0	0.08	12.61	**82.00**

Note: The elements in bold fonts indicate correct classification as in the diagonal elements of the confusion matrix.

11.4 Conclusions and outlook

This chapter presented techniques and results published in the recent open literature on the problem of radar-based classification of drones carrying payloads. While limited to a simple data set of experimental data and proof of concept algorithms, the presented techniques are of interest and can inform future work emerging in the very active area of research of radar signatures of drones.

Classifying payloads on drones using radar signatures is a multifaceted research problem, which can be cast as a simple binary detection of the payload (is a payload there or not?), or as a more complex payload identification (what is the type, size, material composition of the payload?). Furthermore, in a way, this is also a problem within the problem of first of all recognising the presence of drones against other confusing targets, such as birds as the main example of false alarms. This problem is expected to be even more challenging as drones can increasingly operate as a coordinated swarm with many platforms.

Drawing conclusive results as to what features are significant for the detection and then classification of payloads is premature, as at the time of writing the open literature shows only examples of a few limited experimental data sets and related results, which this chapter aimed to summarise.

From these results one can conclude that the reflectivity/RCS of the drone plus payload is not significantly different from the values measured with the unloaded drone, without payload. On the contrary, changes in the Doppler and micro-Doppler spectrum are noticeable and probably due to the increase in weight causing an increase in rotation velocity of the rotor blades. Radars set up with the correct parameters can capture these differences, and these can be represented by suitable features, although with so little data it is yet to be concluded which features are the most effective. Furthermore, it is unclear how repeatable the promising results presented here are, if, for example, changes in the Doppler/micro-Doppler patterns of the blades are induced by wind gusts or special dynamic behaviour of the drone, rather than by the presence of payloads.

Nevertheless, the potential of micro-Doppler as a source of information for the drone payload problem has also been confirmed by the very recent studies in [49,50], where the signatures of drones carrying and then dropping a payload were recorded. In this case, the reduction in rotation rate of the blade signatures in the micro-Doppler was clearly noticeable just after the payload was dropped, showing that the drone would come back to a 'normal' dynamic regime after losing the payload. These two studies add up to the body of results presented in this chapter, although the questions remain on how repeatable these results are, as well as generalisable to different scenarios with other models of drones and types of payloads.

For future work, further research questions and material for investigation can include the following:

- The characterisation of payloads that are not solid objects attached to the drone body but can be, for example, liquids to be sprayed or powder to be diffused in the air. A recent work [51] by the University of St Andrews in the United

Kingdom has, for example, presented an initial analysis of liquid sprayed by a commercial drone used for agriculture in the K and W radar bands (24 and 94 GHz, respectively). The analysis provided some values of signal-to-noise ratio and examples of micro-Doppler signatures, showing the potential of mm-wave radar to provide fine resolution and leverage stronger reflectivity of the liquid spray while being diffused in the air.

- The design of multi-frequency radar systems, whereby lower frequencies can provide detection and tracking capabilities for drones at longer distances, and higher frequencies can provide the resolution needed for better situational awareness and recognition capabilities (e.g. payloads), even if limited to a closer range. The definition of long-/short-range depends on the expected RCS of the drones and the hardware characteristics of the radar systems involved. However, the technological trend, driven mostly by automotive and autonomous vehicles, has seen relatively inexpensive chipsets and radar transceivers at mm-wave being brought to market, providing increasing performances in transmitted power, range, and angular resolutions. Furthermore, this approach can be extended to multimodal sensor frameworks, where other types of sensors, such as cameras, acoustic, and communication sensors, can be integrated.
- The investigation of classification techniques that do not extensively rely on the micro-Doppler signatures of the rotor blades. These cannot be detected well by radar systems at longer distances, as the RCS of the blades is lower than the RCS of the drone body. The development of classification algorithms that can operate on radar data representations and features other than micro-Doppler is therefore of interest, for example looking at the dynamic of the drone with metrics extracted from tracks (i.e. updates of location and velocity over time) [52]. Very preliminary results in [53] explored simple features extracted from polynomial fittings of the drone signature in range–time plots, essentially tracking velocity as the variation of location in time. This was applied on a proof of concept, short-range scenario with a quadcopter carrying a small payload consisting of 4-AA batteries attached to the rotor blade arms.
- The development of classification techniques that are explainable and trustable, therefore using features and decision processes that can easily be understood by human operators and decision makers in an operational scenario. This requirement is increasingly being considered, especially in security and safety applications, whereby the rationale behind decisions made by automatic target classification algorithms needs to be well understood. Examples of research in this direction are presented in [54,55], where simply explainable yet-effective features based on the symmetry, number, and intensity of the micro-Doppler signatures of drones were presented and validated on industrial radar systems.

Acknowledgements

The authors are grateful to M. Ritchie and H. Griffiths at UCL for assistance in obtaining the experimental data, and to the IET which funded the original experimental work through the A. F. Harvey Prize 2013 awarded to H. Griffiths.

The authors also thank J. Patel, C. Horne, F. Hoffmann, L. Pallotta and colleagues at Roma Tre University, for inputs in the original data collection and work on their analysis.

References

[1] "Unlocking the UK's High Tech Economy: Consultation on the safe Use of drones in the UK," Department for Transport, outcome released in July 2017.
[2] L. B. Haylen, Briefing Paper CB7734 on Civilian Drones, House of Commons Library, August 2017.
[3] https://www.bbc.com/news/topics/cnx1xjxwp51t/gatwick-drone-shutdown; BBC Topics "Gatwick drone shutdown," accessed in October 2020.
[4] J. S. Patel, F. Fioranelli and D. Anderson, "Review of Radar Classification and RCS Characterisation Techniques for Small UAVs or Drones," in IET Radar, Sonar & Navigation, vol. 12, no. 9, pp. 911–919, 2018.
[5] I. Guvenc, F. Koohifar, S. Singh, M. L. Sichitiu and D. Matolak, "Detection, Tracking, and Interdiction for Amateur Drones," in IEEE Communications Magazine, vol. 56, no. 4, pp. 75–81, 2018.
[6] C. Clemente, F. Fioranelli and G. Li, "Guest Editorial: Innovative Radar Detection, Tracking and Classification for Small UAVs as an Emerging Class of Targets," in IET Radar, Sonar & Navigation, vol. 14, no. 4, pp. 503–504, 2020.
[7] M. Ritchie, F. Fioranelli, H. Griffiths and B. Torvik, "Micro-drone RCS analysis," 2015 IEEE Radar Conference, Johannesburg, South Africa, 2015, pp. 452–456.
[8] A. Schroder, M. Renker, U. Aulenbacher, *et al.*, "Numerical and experimental radar cross section analysis of the quadrocopter DJI Phantom 2," 2015 IEEE Radar Conference, Johannesburg, South Africa, 2015, pp. 463–468.
[9] K. Singh and Y. Kim, "Automatic Measurement of Blade Length and Rotation Rate of Drone Using W-Band Micro-Doppler Radar," in IEEE Sensors Journal, vol. 18, no. 5, pp. 1895–1902, 2018.
[10] B. K. Kim, H.-S. Kang and S.-O. Park, "Experimental Analysis of Small Drone Polarimetry Based on Micro-Doppler signature," in IEEE Geoscience Remote Sensing Letters, 2017, 14, no. 10, pp. 1670–1674.
[11] A. V. Khristenko, M. O. Konovalenko, M. E. Rovkin, *et al.*, "Magnitude and Spectrum of Electromagnetic Wave Scattered by Small Quadcopter in X – Band," in IEEE Transactions on Antennas and Propagation, vol. 66, no. 4, pp. 1977–1984, 2018.
[12] J. Li and H. Ling, "An Investigation on the Radar Signatures of Small Consumer Drones," in IEEE Antennas and Wireless Propagation Letters, vol. 16, pp. 649–652, 2017.
[13] R. Guay, G. Drolet and J. R. Bray, "Measurement and Modelling of the Dynamic Radar Cross-section of an Unmanned Aerial Vehicle," in IET Radar, Sonar & Navigation, vol. 11, no. 7, pp. 1155–1160, 2017.

[14] S. Rahman and D. A. Robertson, "Radar Micro-Doppler Signatures of Drones and Birds at K-band and W-band," in Nature Scientific Reports, vol. 8, p. 17396, 2018.

[15] M. Ritchie, F. Fioranelli, H. Griffiths, *et al.*, "Monostatic and bistatic radar measurements of birds and micro-drone," 2016 IEEE Radar Conf. RadarConf 2016, Philadelphia, PA, USA, 2016.

[16] A. Huizing, M. Heiligers, B. Dekker, J. de Wit, L. Cifola and R. Harmanny, "Deep Learning for Classification of Mini-UAVs Using Micro-Doppler Spectrograms in Cognitive Radar," in IEEE Aerospace and Electronic Systems Magazine, vol. 34, no. 11, pp. 46–56, 2019.

[17] B. Torvik, K. E. Olsen and H. Griffiths, "Classification of Birds and UAVs Based on Radar Polarimetry," in IEEE Geoscience Remote Sensing Letters, vol. 13, no. 9, pp. 1305–1309, 2016.

[18] J. J. M. de Wit, R. I. A. Harmanny and P. Molchanov, "Radar micro-Doppler feature extraction using the singular value decomposition," 2014 Int. Radar Conf., Lille, 2014, pp. 1–6.

[19] N. Regev, I. Yoffe and D. Wulich, "Classification of single and multi-propelled miniature drones using multilayer perceptron artificial neural network," International Conference on Radar Systems, Belfast, 2017, pp. 1–5.

[20] I. Roldan, C. R. del-Blanco, Á. D. de Quevedo, *et al.*, "DopplerNet: A Convolutional Neural Network for Recognising Targets in Real Scenarios Using a Persistent Range–Doppler Radar," in IET Radar, Sonar & Navigation, vol. 14, no. 4, pp. 593–600, 2020.

[21] P. Zhang, G. Li, C. Huo and H. Yin, "Exploitation of Multipath Micro-Doppler Signatures for Drone Classification," in IET Radar, Sonar & Navigation, vol. 14, no. 4, pp. 586–592, 2020.

[22] N. Mohajerin, J. Histon, R. Dizaji and S. L. Waslander, "Feature extraction and radar track classification for detecting UAVs in civilian airspace," 2014 IEEE Radar Conference, Cincinnati, OH, USA, 2014, pp. 0674–0679.

[23] K. Kim, M. Üney and B. Mulgrew, "Coherent Track-Before-Detect with Micro-Doppler Signature Estimation in Array Radars," in IET Radar, Sonar & Navigation, vol. 14, no. 4, pp. 572–585, 2020.

[24] C. Ilioudis, C. Clemente and J. Soraghan, "GNSS-Based Passive UAV Monitoring: A Feasibility Study," in IET Radar, Sonar & Navigation, vol. 14, no. 4, pp. 516–524, 2020.

[25] T. Martelli, F. Colone and R. Cardinali, "DVB-T Based Passive Radar for Simultaneous Counter-Drone Operations and Civil Air Traffic Surveillance," in IET Radar, Sonar & Navigation, vol. 14, no. 4, pp. 505–515, 2020.

[26] A. D. Chadwick, "Micro-drone detection using software-defined 3G passive radar," International Conference on Radar Systems (Radar 2017), Belfast, 2017, pp. 1–6.

[27] N. Ghazalli, A. Balleri, M. Jahangir, F. Colone and C. J. Baker, "Passive Detection Using a Staring Radar Illuminator of Opportunity," 2019 International Radar Conference (RADAR), TOULON, France, 2019, pp. 1–5.

[28] T. Martelli, F. Murgia, F. Colone, C. Bongioanni and P. Lombardo, "Detection and 3D localization of ultralight aircrafts and drones with a WiFi-based

passive radar," International Conference on Radar Systems (Radar 2017), Belfast, 2017, pp. 1–6.

[29] T. E. Derham, S. Doughty, K. Woodbridge and C. J. Baker, "Design and Evaluation of a Low-Cost Multistatic Netted Radar System," in IET Radar, Sonar & Navigation, vol. 1, no. 5, pp. 362–368, 2007.

[30] M. Ritchie, F. Fioranelli, H. Borrion and H. Griffiths, "Multistatic Micro-Doppler Radar Feature Extraction for Classification of Unloaded/Loaded Micro-Drones," in IET Radar, Sonar & Navigation, vol. 11, no. 1, pp. 116–124, 2017.

[31] F. Fioranelli, M. Ritchie, H. Griffiths and H. Borrion, "Classification of Loaded/Unloaded Micro-Drones Using Multistatic Radar," in Electronics Letters, vol. 51, no. 22, pp. 1813–1815, 2015.

[32] S. Z. Gurbuz and M. G. Amin, "Radar-Based Human-Motion Recognition With Deep Learning: Promising Applications for Indoor Monitoring," in IEEE Signal Processing Magazine, vol. 36, no. 4, pp. 16–28, 2019.

[33] D. Tahmoush, "Review of Micro-Doppler Signatures," in IET Radar, Sonar & Navigation, vol. 9, no. 9, pp. 1140–1146, 2015.

[34] F. Fioranelli, H. Griffiths, M. Ritchie and A. Balleri (Editors), Micro-Doppler Radar and its Applications, The Institution of Engineering and Technology, In Press (2020).

[35] C. Clemente, L. Pallotta, A. De Maio, J. J. Soraghan and A. Farina, "A Novel Algorithm for Radar Classification Based on Doppler Characteristics Exploiting Orthogonal Pseudo-Zernike Polynomials," in IEEE Transactions on Aerospace and Electronic Systems, vol. 51, no. 1, pp. 417–430, 2015.

[36] V. C. Chen, D. Tahmoush and W. J. Miceli (Editors), Radar Micro-Doppler Signatures: Processing and Applications, The Institution of Engineering and Technology, In Press (2014).

[37] J. Markow and A. Balleri, "Examination of drone micro-Doppler and JEM/HERM signatures," Presented at IEEE Radar Conference 2020, Florence, Italy, 2020.

[38] F. Fioranelli, M. Ritchie and H. Griffiths, "Classification of Unarmed/Armed Personnel Using the NetRAD Multistatic Radar for Micro-Doppler and Singular Value Decomposition Features," in IEEE Geoscience and Remote Sensing Letters, vol. 12, no. 9, pp. 1933–1937, 2015.

[39] F. Fioranelli, M. Ritchie and H. Griffiths, "Centroid Features for Classification of Armed/Unarmed Multiple Personnel Using Multistatic Human Micro-Doppler," in IET Radar, Sonar & Navigation, vol. 10, no. 9, pp. 1702–1710, 2016.

[40] F. Fioranelli, M. Ritchie, S. Z. Gürbüz and H. Griffiths, "Feature Diversity for Optimized Human Micro-Doppler Classification Using Multistatic Radar," in IEEE Transactions on Aerospace and Electronic Systems, vol. 53, no. 2, pp. 640–654, 2017.

[41] J. S. Patel, C. Al-Ameri, F. Fioranelli and D. Anderson, "Multi-Time Frequency Analysis and Classification of A Micro-Drone Carrying Payloads Using Multistatic Radar," in The Journal of Engineering, vol. 2019, no. 20, pp. 7047–7051, 2019.

[42] S. Harman, "Analysis of the radar return of micro-UAVs in flight," 2017 IEEE Radar Conference (RadarConf), Seattle, WA, USA, 2017, pp. 1159–1164.
[43] A. Krizhevsky, I. Sutskever and G. E. Hinton, "Imagenet classification with deep convolutional neural networks," NIPS'12 Proceedings of the 25th International Conference on Neural Information Processing Systems, 1, Lake Tahoe, NV, USA, December 2012, pp. 1097–1105.
[44] M. S. Seyfioglu, B. Erol, S. Z. Gurbuz and M. G. Amin, "DNN Transfer Learning From Diversified Micro-Doppler for Motion Classification," in IEEE Transactions on Aerospace and Electronic Systems, vol. 55, no. 5, pp. 2164–2180, 2019.
[45] S. Rahman and D. A. Robertson, "Classification of Drones and Birds Using Convolutional Neural Networks Applied to Radar Micro-Doppler Spectrogram Images," in IET Radar, Sonar & Navigation, vol. 14, no. 5, pp. 653–661, 2020.
[46] L. Pallotta, C. Clemente, A. Raddi and G. Giunta, "A feature-based approach for loaded/unloaded drones classification exploiting micro-Doppler signatures," Presented at IEEE Radar Conference 2020, Florence, Italy, 2020.
[47] R. Dwyer, "Detection of non-Gaussian signals by frequency domain kurtosis estimation," ICASSP'83. IEEE International Conference on Acoustics, Speech, and Signal Processing. IEEE, 1983, vol. 8, pp. 607–610.
[48] J. Antoni, "The Spectral Kurtosis: A Useful Tool for Characterizing Non-Stationary Signals," in Mechanical Systems and Signal Processing, vol. 20, no. 2, pp. 282–307, 2006.
[49] J. de Wit, D. Gusland and R. Trommel, "Radar measurements for the assessment of features for drone characterization," Presented at EURAD 2020, Utrecht, the Netherlands, 2020.
[50] S. Rahman, D. A. Robertson and M. A. Govoni, "Radar Signatures of Drones Equipped With Heavy Payloads and Dynamic Payloads Generating Inertial Forces," in IEEE Access, vol. 8, pp. 220542–220556, 2020.
[51] S. Rahman, D. A. Robertson and M. A. Govoni, "Radar signatures of drones equipped with liquid spray payloads," Presented at IEEE Radar Conference 2020, Florence, Italy, 2020.
[52] F. Barbaresco, D. Brooks and C. Adnet, "Machine and deep learning for drone radar recognition by micro-Doppler and kinematic criteria," Presented at IEEE Radar Conference 2020, Florence, Italy, 2020.
[53] Y. Zhao, X. Zhang and F. Fioranelli, "Initial results of radar-based classification of commercial drone carrying small payloads," 2019 International Radar Conference (RADAR), Toulon, France, 2019, pp. 1–4.
[54] C. Bennett, M. Jahangir, F. Fioranelli, B. I. Ahmad and J. Le Kernec, "Use of symmetrical peak extraction in drone micro-Doppler classification for staring radar," Presented at IEEE Radar Conference 2020, Florence, Italy, 2020.
[55] F. Hofele, "Automatic radar target classification a new method having special regard to drones," 2019 International Radar Conference (RADAR), Toulon, France, 2019, pp. 1–6.

Chapter 12

Good practices and approaches for counter UAV system developments – an industrial perspective

Mohammed Jahangir[1] and Daniel White[1]

12.1 Introduction

Radars that can detect and track small targets such as drones are steadily becoming more common with several companies, including Thales–Aveillant, producing radar systems specifically designed for counter-UAS applications [1,2]. The need for these systems has seen a rapid increase due to recent events, including airport and military facility trespassing and attempted assassinations using weaponised drones. The number of drone incidents involving near-misses with aircraft has reached an all-time high with around 120 reports in 2018 alone in the United Kingdom [3]. Radar still seems to be the most tried and tested method of detecting, tracking and classifying this relatively new class of targets, but other methods exist that can aid the radar in detecting and classifying. To detect these small targets, a radar system requires a greater sensitivity than conventional radars as drones typically have a Radar Cross Section (RCS) that can be much less than 1 m^2. By increasing the sensitivity, there are a greater number of 'confuser targets' detected compared with traditional Air Traffic Control radar systems. The pertinent example of a confuser target is birds that have a similar airspace occupancy and RCS to drones, and hence there is a need to discriminate between these two target classes; otherwise, the radar output would become completely congested with airborne targets.

Staring radar is an important design choice for non-cooperative drone surveillance as it is able to provide the high sensitivities and high Doppler resolutions that are necessary to detect and track drones. This is due to drones having a low RCS and flying at relatively low velocities and altitudes [4] making them challenging air targets. The high-sensitivity requirement means that other slow-moving targets with a low RCS, such as flying birds and ground targets, are also detected. These (non-drone) targets are dubbed 'confuser targets' from a drone surveillance system perspective. Consequently, the operator and a supporting secondary system (e.g., optical or RF-based) can be easily overwhelmed if they must analyse or

[1]School of Engineering, University of Birmingham, Birmingham, United Kingdom

interrogate all objects. Therefore, to be useful and reliable, drone surveillance radars must possess accurate classification techniques to ensure that birds and other confuser targets do not provide false drone detections and the system does not misreport the actual drones. Accuracy is the crucial factor in assessing the performance of these counter-drone systems; a false positive means that costly countermeasures to find and neutralise the intruding drone will occur in vain, and a missed drone report could potentially lead to catastrophic damage, especially in high-risk areas such as airports or stadiums.

Several strategies have evolved for classifying drones using radar systems with supervised learning classifiers proving to be particularly attractive [5]. Supervised learning classifiers are reliant on accurately labelled training data. For drone classification, this is particularly challenging given the large volume of opportune non-drone targets (primarily birds) that are likely to be encountered in the data for wide-area surveillance sensor systems. Data from control drone trials is often used as the means to generate labelled training data for the drone class but for confuser targets, in particular birds, reliance has to be made on measurements for opportune targets which are less reliably labelled. Some approaches have been devised for the labelling of bird opportune targets [6]. There has also been a proposed method that can assess classification performance without the need to label non-drone targets explicitly [7]. However, fundamentally, supervised learning methods will benefit from more accurately labelled data and will improve classifier accuracy. Drone versus non-drone classification requires the ability to obtain more accurately labelled data for bird targets. Looking to the future of radar target classification with machine learning techniques, having access to plentiful high-quality training data will help push this frontier further into the deep learning regime, where very large datasets are required to optimise the huge dimensionality of hidden layers in a multilayer perceptron network, for example.

Truth data can come in many forms with one of the most accurate and convenient being GPS. For all target classes, GPS can be used to provide positional data with an uncertainty that is typically much less than the radar's resolution, especially if the radar is operational at ranges of the order of several kilometres. Specific trials can be conducted to collect data from aircraft and drones just by attaching almost any GPS device to them [8] or by using the built-in GPS logger that often comes embedded in commercial drones. Using specialist GPS equipment, trained birds can also be used to collect data but there can be a limit to the breed of birds and the behaviour exhibited due to their captivity. Ideally, there is a requirement for a GPS-free system that can collect data that can be compared with the radar echo history to provide class labels of the radar tracks. This system would not necessarily require a definitive 3-D position but could be an observer on the ground providing their position and orientation to a target along with the target class label which would be enough to relate it to a recorded track. This system could be used for any opportune target that presented itself during a trial such as birds, aircraft and ground targets. This system would also be easy to use and have little opportunity for human error which could corrupt the training data and lead to inaccuracies in the classifier.

In this chapter, we briefly describe the staring radar concept and show how it can be used to detect and discriminate drones against a severe background of other targets such as birds and ground clutter. The methods developed for collecting ground-truth data from both control and opportune targets are described. Throughout, examples from real-world measurements will be used to illustrate how to generate labelled training data as well as demonstrating the target recognition performance with a supervised learning-based approach.

12.2 Robust drone classification with a staring radar

Aveillant's Gamekeeper 16U staring radar system is designed for the surveillance of Unmanned Aerial Vehicles (UAVs) and is the sensor used to collect and test the methods presented in this chapter. The transmitter of the Gamekeeper radar produces a broad beam that illuminates the entire search volume continuously over time, while the receiver is comprised of a 2-D planar array that is digitised at the element level to allow the radar to generate multiple simultaneous receiver beams that cover the entire search volume. With this fixed arrangement of beams and a long integration period, the sensor is able to achieve a fine Doppler resolution enabling the capture of micro-motion signatures, such as those produced by the small spinning rotors of a micro-UAV. The radar provides raw IQ data as well as tracked outputs from the real-time processor. Figure 12.1 shows an image of the Gamekeeper 16U radar installed at the University of Birmingham, the operating parameters of which are listed in Table 12.1.

Raw radar data is processed to detect and track targets. The classifier follows and uses a multitude of features for target recognition, including the tracker calculated trajectories and potentially the Doppler spectrum of a target. Some target classes are easier to discriminate, for example, aircraft and drones. The differences between these two are seen in almost every feature evaluated as aircraft generally fly much faster, higher and have a considerably larger RCS compared with drones. Propeller-powered aircraft also produces a greater range of micro-Doppler components as they typically travel at greater radial velocities compared to the velocity of a drone's propellers. The real classification challenge comes in the form of discriminating between birds and drones as there is little separability between individual feature values. They both fly slowly, have small RCS values and fly at low altitudes. Micro-Doppler components, arising from the motion of propeller blades, have shown to aid the classification of small targets [2,9]. However, classification performance using this metric alone is limited as the stand-off range at which propeller return can be detected is dependent on echo strength, which can be 10–30 dBs less than the drone body echo strength. Due to the similarities between echoes from drones and birds with no clear singular feature to tell them apart, there is a need to search for combinations of features that together create a high performing and robust classifier. Further, there can be more than one set of features used to classify each of these targets at any one time. One approach is to use a Decision Tree (DT) classifier which uses a priori known characteristics of targets;

Figure 12.1 Aveillant's Gamekeeper 16U installed at the University of Birmingham Campus

Table 12.1 Gamekeeper L-band multi-beam staring radar system parameters

Parameter	Value
Frequency	L band
Bandwidth	~2 MHz
Transmit power	~2 kW
Receiver channels	4×16
Azimuth coverage	90°
Elevation coverage	30°
Pulse Repetition Frequency (PRF)	~7.5 kHz
Update rate	~0.25 s
Polarisation	Vertical
Drone detection range	5 km

for example, a ground target will be classified as such if its detected location is close to ground level [2,10]. Due to the complexity of this form of classifier, it can also be beneficial to use a machine learning algorithm to build the DT. This uses a set of accurate training data collected from real radar data along with the correct class labels of each of the targets to be considered. Accurate class labelling of this training file also requires accurate truth data to be collected. This is where lies the limitation of the practical implementation of such classification algorithms. The following sections describe methods for obtaining good truth labelling from field trials for both control test targets and targets of opportunity.

12.3 Methodology for ground truthing

The first stage of creating a machine learning classifier is to collect training data with accurate labelling. For control targets, such as drones and aircraft, labelling can be achieved using GPS track association. This uses an algorithm to compare the GPS truth data with any radar track that appears within a set volume of space around the truth track and has a similar velocity and heading. Section 12.3.1 discusses the approach for ground truth with control targets and presents example results from test targets, including drones and birds. For opportune targets, a whole new method has to be created due to information from GPS not being routinely available. These methods are described in detail in Section 12.3.2.

12.3.1 Control targets

12.3.1.1 Drones

Drone surveillance radar is a growing area of research and development motivated by the increasing numbers of drones and their constant improvement and usability. Modern drones are highly configurable in their trajectories, and more advanced drones are getting cheaper all the time due to their constant development. There are many applications and programs that can provide the end-to-end functionality of planning a flight route/path and recording the GPS data. A popular application for controlling and recording the flight is 'Litchi' [11] which is compatible with DJI-branded drones. This was the choice for the trials described in this section, although alternatives certainly exist. The drone used in the trials reported here is shown in Figure 12.2. It is possible to extract the on-drone GPS output as a comma-separated value file or a formatted table. Such a table can also contain many potentially useful fields, including, for example, latitude, longitude and altitude information with timestamps over the duration of its flight.

Possessing the truth data of the drone flight is an enabler for generating labelled data for machine learning methods. There are a couple of methods to use this data effectively to provide labels to a tracker output, the choice of which will depend on the nature of data that is required for a given machine learning architecture. If DT architecture is to be trained, then a comparison of the tracker output can be made with the truth trajectory, and the track ID produced by the radar will now have a correct label at all points of its trajectory. See Figure 12.3 for a comparison of the truth data and a tracker output. Note the seemingly erratic nature of the radar-derived locations information. This is quite typical and is simply a function of the parameters of the radar measurement system and an additional characteristic that a classifier has to cope with. Direct comparison between truth and tracks can be used to test the integrity of the tracker, which is useful to check,

Figure 12.2 The quadcopter DJI Inspire 2 drone used in the field trials presented in this chapter

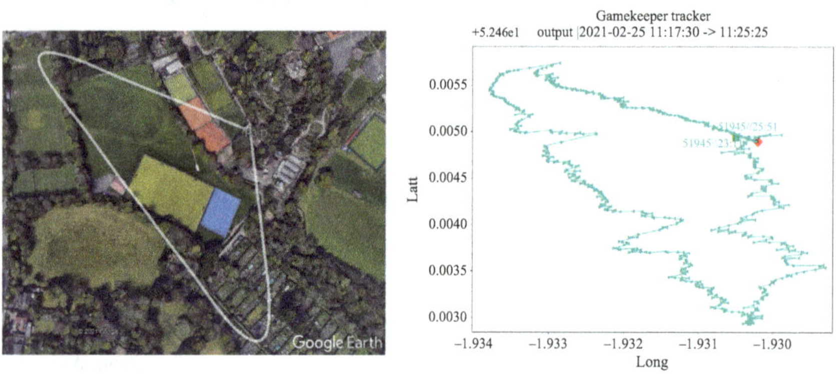

Figure 12.3 (Left) DJI Inspire 2 rotary drone GPS truth displayed in Google Earth, (right) the extracted tracker output of the radar, including ID number and timestamps

especially near the far edge of the radars instrumented range. With radar tracks now corresponding to the drone class, the information is there to package this data into a machine-learning algorithm. Figure 12.4 shows different views of a drone truth trajectory. This information may be used to produce spectrograms of the drone for the duration of the track. This is done by plotting just the Doppler axis of the radars 4-D data cube (aperture row × aperture column × range gate × Doppler bin) where the target is located. Stacking the Doppler axis along time shows the dynamics of the drone's radial motion and is the best way, pictorially, to show the effects of the rotary wing drones' blades micro-Doppler returns. An example of this is shown in Figure 12.5. The classification of radar targets using these spectrograms via a convolutional neural network classifier shows promising results [12].

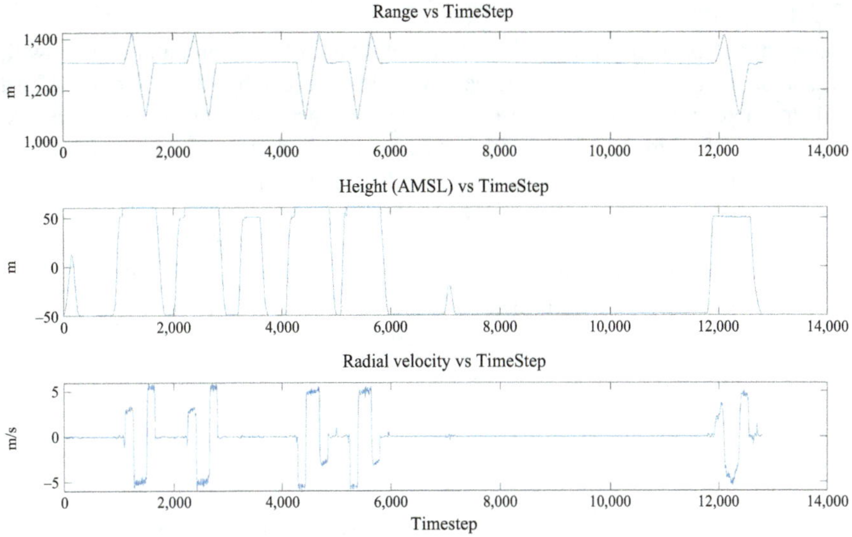

Figure 12.4 Target trajectory features obtained from GPS truth – DJI Inspire 2 rotary drones. AMSL stands for average height above mean sea level

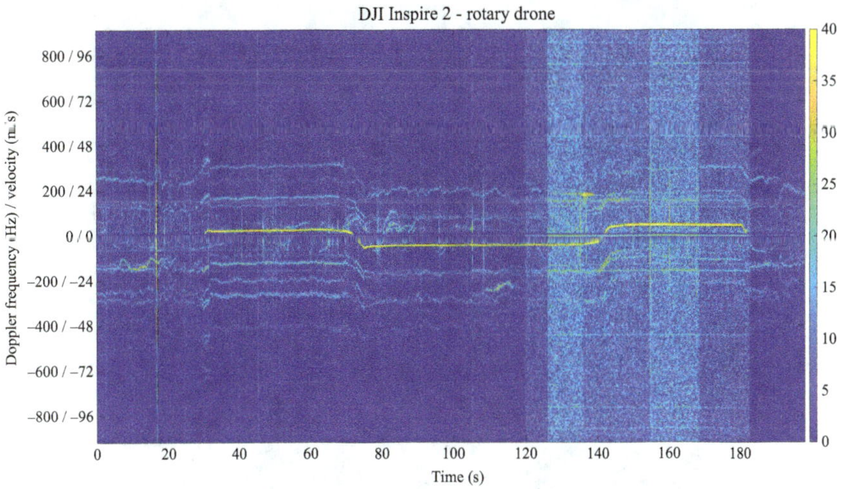

Figure 12.5 DJI Inspire 2 rotary drone spectrograms. At times after 120 s, the background noise detected by the radar increases drastically. The most likely cause of this is from the tracked target entering a new resolution cell that experiences a higher level of clutter

12.3.1.2 Birds

Whilst there has been some work on radar measurements using control trials with birds at short range [13], there has been very little research involving control bird trials at longer ranges for high-resolution Doppler measurements and/or measured trajectory data. Here we describe an experimental technique for obtaining radar data for trained birds that are equipped with GPS tags to provide accurate tracking information that will facilitate precise data labelling.

Control flights with birds require a lot of specialist knowledge and specific equipment to ensure that the welfare of the birds is not compromised whilst also ensuring that accurate tracking data is obtained to support the analysis and interpretation of the sensor data. In the experiments, each bird was deployed with a Microsensory Radiotracking GPS tag which was affixed to the tail feathers of each bird prior to a test flight and removed after the return of the bird to its handler. This tag provided high-accuracy GPS tracks of the control bird targets. Figure 12.6 shows an image of the GPS tag attached to the back of a Common Buzzard (*Buteo buteo*).

Field trials were conducted at an ex-RAF Airfield with a number of birds of prey. This location provided a rural environment with flat terrain, low-lying vegetation and wide-open areas for flying birds in good visibility surroundings. Four species of raptors were used as control targets: an African Peregrine Falcon (*Falco peregrinus*) (body mass: 720 g), a Saker-Gyrfalcon (*Falco cherrug* × *Falco rusticolus*) Hybrid (900 g), a Common Buzzard (700 g) and a Harris's Hawk (*Parabuteo unicinctus*) (900 g). Each bird was flown multiple times by professional

Figure 12.6 Microsensory GPS tag affixed to a Common Buzzard

handlers approximately 1 km from the radar, along its boresight. The Common Buzzard and Harris's Hawk both flew from handler to handler very close to the ground. The Peregrine Falcon flew multiple figure-of-eight patterns, climbing in altitude and diving to intercept a lure operated by a handler. The Gyrfalcon Hybrid flew at the highest altitude, while chasing a small handler-operated drone which allowed it to be easily detected by the radar. Figure 12.7(a) shows the image of the Gyrfalcon Hybrid together with an example of GPS track from the tag.

The truth data from the GPS tag can then be analysed to evaluate the trajectory profile. Figure 12.8 plots a number of trajectory features obtained from the GPS truth data collected for the Gyrfalcon Hybrid. The GPS data was recorded at 1 Hz and it was interpolated to match the timestamp of the radar tracker output which

(a) (b)

Figure 12.7 (a) Gyrfalcon Hybrid, (b) GPS track from control flight

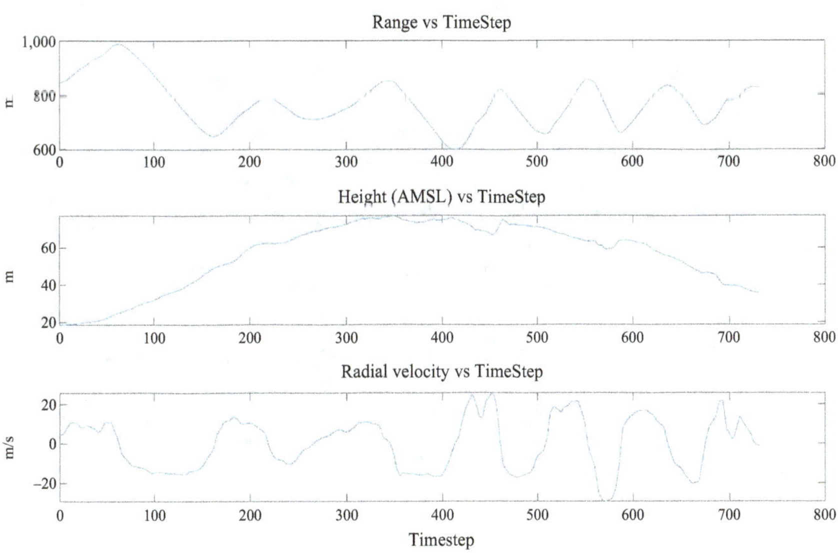

Figure 12.8 Target trajectory features – Gyrfalcon Hybrid bird

had a time-step duration of 279 ms. From these plots, it can be seen that in this instance the bird climbed up to a height of about 70 m and then descended back to ground. The bird was circling while it was ascending and descending, and this can be seen from the plot of its slant range and the radial velocity along the line-of-sight of the radar. Similar to the GPS truth data from control drone trials, the GPS truth data from the bird target can be used to label the radar tracker ID and this can be used to generate a Doppler spectrogram for the bird from the raw radar data.

Figure 12.9 compares the GPS truth with the radar tracker output (top image) along with the spectrogram obtained from the raw radar data using the information

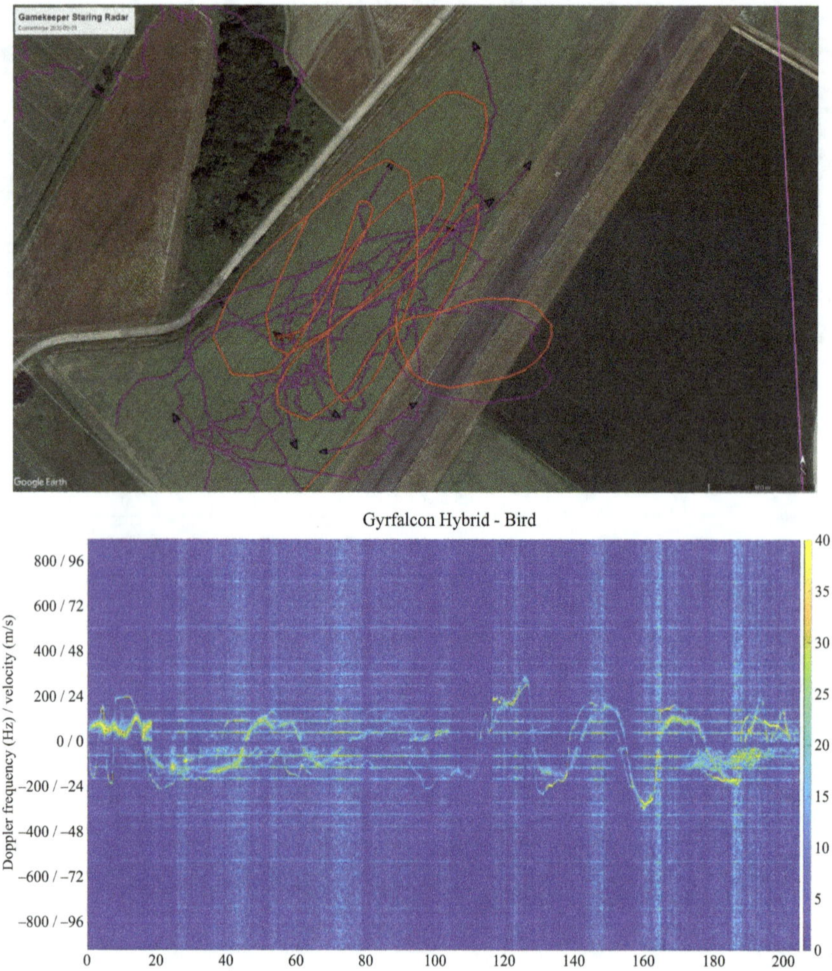

Figure 12.9 (Top) Gyrfalcon Hybrid bird GPS in red overlaid onto the radar tracker output shown as purple lines (bottom) Doppler spectrogram

from the GPS truth. There are a couple of observations to be made here. First, the radar track is nosier than the GPS truth which is to be expected given the coarse angular resolution of the radar. Second, in the spectrogram the bird body Doppler is exhibiting rapid changes, which is understandable as the bird was circling during this test flight. However, there are also some Doppler sidebands that repeat throughout in frequency. These are due to an artefact of the radar that was present during this trial and not related to the target signal in this instance.

Figure 12.10 shows a plot of the measured Signal-To-Noise Ratio (SNR) of the Gyrfalcon Hybrid obtained from the test flight. The radar measurements show that the average SNR of the Gyrfalcon Hybrid was approximately 40 dB but varied significantly over the course of the flight due to the bird changing direction and height above ground. The measurement campaign was a useful demonstration of undertaking control bird flights and successfully collecting radar data in an open rural setting, thus providing a limited means of extending the all-important air truth information.

However, with all control targets there remain significant limitations in terms of the volume and variety of data that can be collected. For example, the birds were only able to perform a maximum of three flights in the whole day with rest periods being allowed between flights. Furthermore, the birds were specially trained to conduct precise flight profiles which are not truly representative of their natural flight. Therefore, for more representative data of bird targets, it is important that training data is collected for birds in their natural habitat which necessitate being able to use data from targets of opportunity and this is addressed with the technique described in the following section.

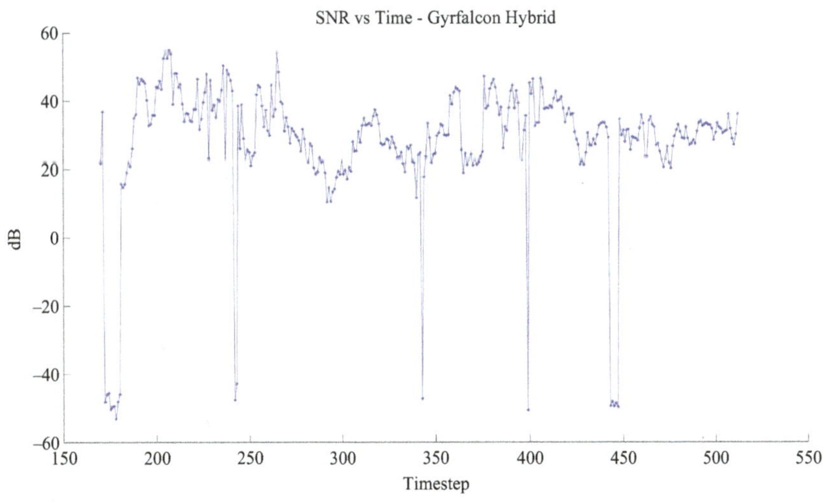

Figure 12.10 Gyrfalcon Hybrid bird SNR

12.3.2 Targets of opportunity

The dichotomy of birds and drones in regard to truth data collection is stark. On the one hand, drones are ideal test subjects for radars, with their controllability and often in-built GPS trajectory recording abilities, but are somewhat effort intensive to fly and set up. Birds on the other hand are tracked in great numbers and at nearly all times for a radar system with a long-range and a maximum altitude of detection. The radar installed at the University of Birmingham captures ~2,500 tracks every 10 min during the day. Some tracks are 'phantom' but can be mostly filtered out by removing extremely short or fast tracks, or ones that are at zero-radial velocity throughout their existence. However, providing truthing and/or labels for these related radar tracks is not easy. Getting accurately labelled truth data requires either sourcing bird handlers to join a trial and provide the birds and specialist GPS trackers, or alternately to somehow label the free bird tracks reliably. The first of these options can be undesirable as they rely on external experts, cost, transportation, organisation, etc. Additionally, as staring radar looks to engage more challenging environments such as deployment in urban/city areas, having access to large areas of open space to conduct bird of prey trials cannot be guaranteed (admittedly, drones truthing shares this same detriment as restrictions on drones in residential areas ever increase). There are also valid concerns that the trajectories flown by birds of prey will not necessarily train a classifier network well to deal with smaller urban birds such as pigeons or magpies. With these realities in mind, this section will demonstrate a method for converting free, opportune bird tracks into accurately labelled radar data to support classifier performance and progress.

The method used in this case records an observer position and orientation towards the specific target by utilising an *iPad* tablet and the *Theodolite* application [14]. Other software is capable of carrying out this task, including *Dioptra* or *Solocator* [15,16]. This combination allows an observer to use the iPad camera to capture an image of the opportune target and input, via a custom note, the targets class label, e.g., bird or, ideally, a type of bird. The image was then saved to the tablet by the application along with Exif/meta-data stored within the image. This meta-data contained all the positional and orientation information of the tablet at the time the image was taken using the device's accelerometers and wireless systems. This information includes the time, GPS position, altitude, bearing angle, elevation angle as well as the custom note input by the observer containing the class label. Once proficient in the use of the tablet, an initial measurement of a target could be recorded in less than 5 s with consecutive measurements taken every second due to the application saving the last entered custom note class label. After the first measurement, a target and related ground-truth information could be recorded at a typical rate of 1 Hz.

After all the ground-truth data had been recorded as a compilation of images, they are easily transferred to a computer for comparison with the radar tracks. An example of a captured image used to verify a track is shown in Figure 12.11. A simple script then extracts the meta-data from each image and creates a KMZ file for viewing on the Google Earth. Each measurement taken is displayed at the exact

Figure 12.11 Opportune bird target 'bird23' captured using tablet and application overlayed with tablet sensor data

time of capture by first plotting the position of the tablet, using the GPS position and altitude and then projecting a line in the direction of the target by using the bearing and elevation angles. Each line is then associated with the extracted image data and given the name of the custom note input by the observer. This provides a visual check that the note input by the observer corresponds to the captured image.

The custom note also allowed for a navigable folder structure as this string of characters is also analysed in the same script. For instance, if the observer had input a custom note of 'bird23' where the numbers '23' corresponded to the 123rd opportune bird target observed, then each image with the custom note 'bird23' could be put into a folder which corresponds to all the measurements taken of the same target. This folder is then also put into a folder along with any others containing the same string of 'bird'. All this information is contained within a single KMZ file and is subsequently easy to associate with the radar tracks once converted into a Google Earth format.

An example of an opportune bird target with ground truth compared with the radar track is shown in Figure 12.12. This method of providing ground-truth data of an opportune target was an improvement on traditional methods that are fraught with the possibility of human error. The new method has allowed for a greater number of radar tracks to be given a class label, significantly improving classifier

Figure 12.12 Opportune bird target 'bird23' images displayed on Google Earth showing orientation lines correlating with a radar track

training. In addition to its main function, the tablet provides a range of applications such as a view of the live display of the radar output for monitoring targets in the field. This acts as an early warning system for the observer to prepare for approaching targets before they are directly visible.

To indicate the degree of improvement made to the ground truthing, compared to predecessor methods, an opportune target measurement took over a minute to record (due to the number of separate measurement devices). This has now been hugely improved to a rate of 1 Hz, greatly improving utility. The previous method utilised handwritten logs which were transferred manually onto a computer to generate a KML file. Hence, the reliability of target measurement was lower due to possible human error in making inputs and the longer recording time resulting in fewer target track associations and their labelling.

12.4 Ground-truth results for drones and birds

The ground-truthing method described in the previous section was used to provide labelling of real radar tracks obtained from measurements carried out at the trial site based in Deenethorpe, Northamptonshire, United Kingdom. Gamekeeper radar data was collected for several control targets flights using a DJI Inspire drone. The

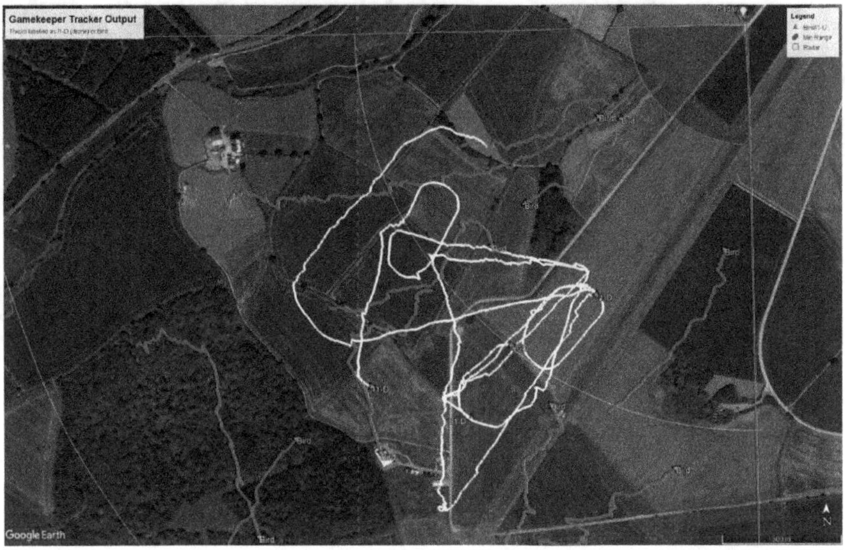

Figure 12.13 Radar tracks labelled as bird (blue) and drone (white) using the truthing method

radar data was processed, and the tracks exported to Google Earth, which were then cross referenced against the ground-truth information and track IDs were registered corresponding to drone and bird targets. Figure 12.13 shows a selection of tracks that were labelled in this manner.

The track labels are then used to extract trajectory-based features from the radar track data. The trajectory information is also used to generate spectrograms of tracks from the raw radar data. Figures 12.14 and 12.15 show the labelled trajectories and spectrogram for two targets, labelled using the procedure adopted here. Figure 12.14 shows a drone with its distinct micro-Doppler signatures clearly visible. In Figure 12.15, the target is a bird and whilst there is evidence of some very faint Doppler sidebands, they exhibit a lack of periodicity and are quite different in character from that of the drone. The labelled trajectory data is used in the machine learning classifier. Previously work used a far less reliable process of labelling unknown tracks of opportune targets which would limit the ability of the machine learning classifier to utilise labelled data from opportune targets [4].

12.5 Machine learning classification

Once radar, truth and ground-truth data had all been recorded, the next stage was to create an accurate classifier for the radar system. The results presented in this section are to be taken as an example. Extending the capabilities of the classification procedures used in the following is an active area of research [5]. Several approaches have been developed that utilised either trajectory-based features [17]

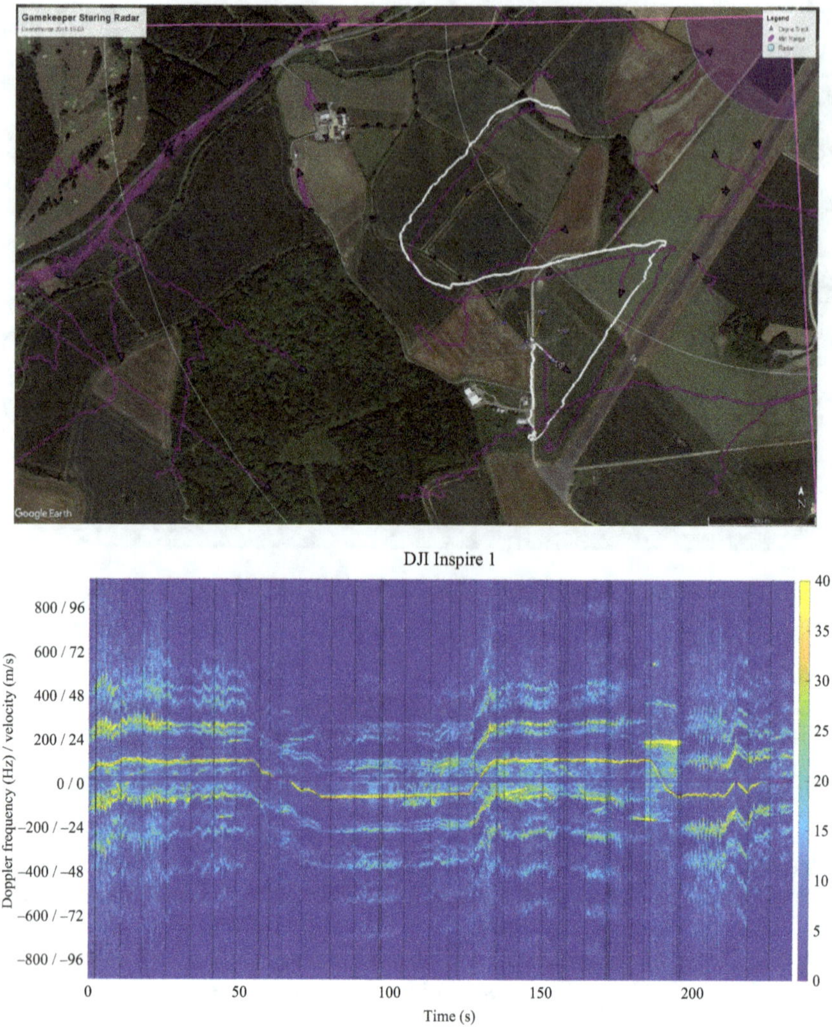

Figure 12.14 (Top) Radar tracks with a selected tracked (ID 126795) labelled as drone; (bottom) spectrogram for track ID 126795

or micro-Doppler features [18] or a combination of both [10] for discriminating drones from non-drones. Furthermore, these techniques are based both on standard feature-based classifiers as in [9,19] and those that use deep learning for discrimination as reported in [18]. Here we restrict the choice of a classifier to a DT classifier as the main motivation is not the absolute performance but the ability to understand the utility of the truth labelling performed with the method outlined in the previous section. A key motivation of employing a DT classifier is that the learned features are transparent and physically meaningful.

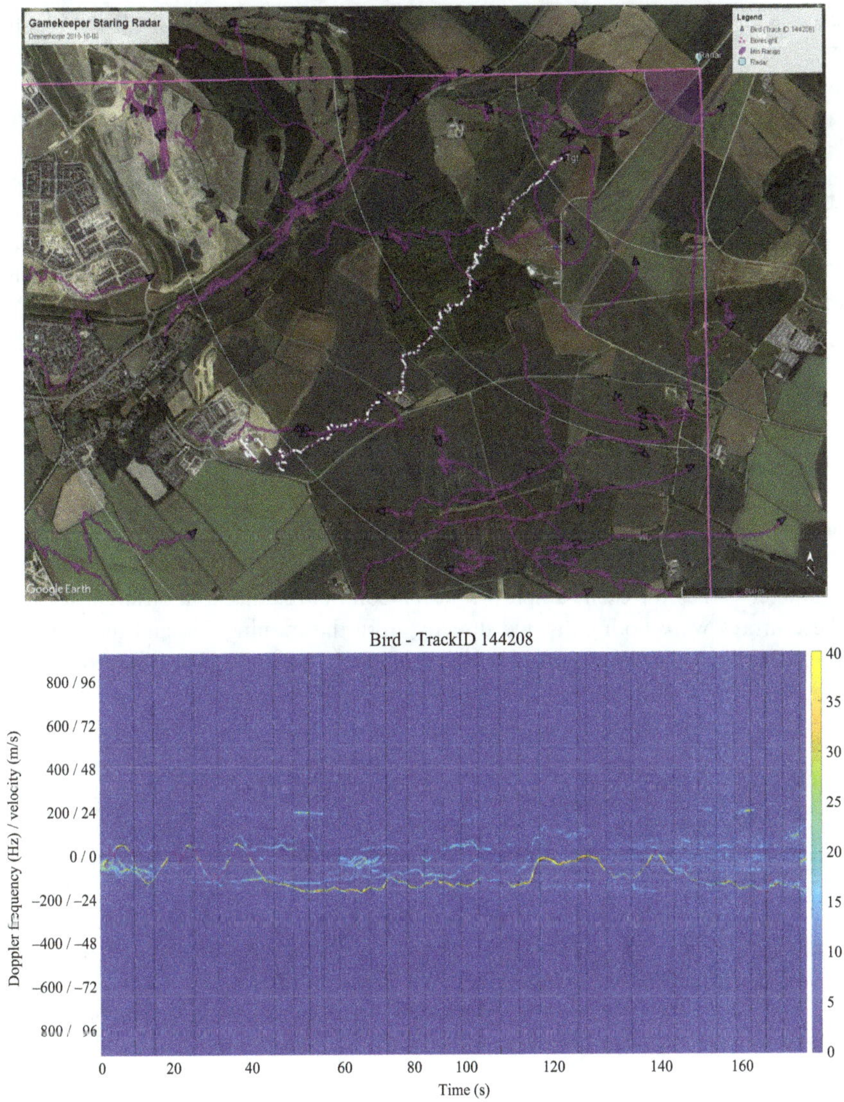

Figure 12.15 (Top) Radar tracks with a selected tracked (ID 144209) labelled as bird; (bottom) spectrogram for track ID 144208

The DT classifier is based on supervised machine learning. To use machine learning, a training file was created using truth data collected from drones consisting of samples of tracked targets collected from the radar and containing the parameters recorded of each feature. The classifier simply has two classes: (i) drone and (ii) non-drone. Each sample requires its class label to be recorded. This allows the machine learning algorithm to find any differences between the collective class features and hence enables discrimination. This is done either using a feature on its own (unlikely for drones and birds) or by combining features together to create a case-by-case classifier.

Features that the classifier selects need to avoid over-tuning to the training data, commonly known as overfitting. For instance, if trials were conducted in a particular region and the *x*–*y* position was used, then the classifier could determine that only a target of a particular class would appear in that region. In avoidance of this, the classifier is only allowed to use the positional feature of height. Derivatives of position are used such as velocity, acceleration and jerk as well as the number of micro-Doppler components associated with a target. In addition to the instantaneous features, statistical analysis over a rolling window is used to analyse the feature history of a target. The mean, minimum and maximum of features are used. In total, 13 features were used which included features such as the age of the track and the target RCS with the number of Doppler components being the most significant feature.

The DT classifier was created in MATLAB® using the *fitctree* function from the 'Machine Learning and Statistics' toolbox. This function called for a 1-D array of class labels relating to each sample and a 2-D array containing the feature parameter values for each sample with each column relating to a different feature. These arrays were both extracted directly from the training file and the function would then use these arrays to produce a DT classifier. It also allowed for the option of *k*-fold optimisation and pruning to prevent over-fitting to the provided training data. The output consisted of a 2-D array of the DT parameters with each row relating to a node and columns for node number, positive child node number, negative child node number, feature used, split value, data size and majority class.

A typical result from a machine-learned two-class drone DT is shown in Figure 12.16 along with the associated confusion matrix. Class 1 (blue) is the non-drone class and Class 2 (orange) is the drone class. The classifier was trained on 81k data points. The test was performed on nearly 10 min of radar output tracks from data not included for training. This shows a correct classification for drones to be 94.3% and a false-positive rate of 0.5%. This result shows what is potentially possible; even with a limited set of training data, the classification model is still flexible enough to provide similar results for multiple drone models and separate radar sites.

A DT classifier can also be produced to include more than two classes and is only dependent on the training data labelling supplied. If DT only discriminated between two classes such as drones and non-drones as in this case, then the radar classification algorithm could also use another two-class DT classifier of bird and non-bird in parallel to add in this class to enable it to be a multi-class classifier. The process of creating a classifier using this method can easily be automated so a new classifier can be produced whenever needed. This could be when new specific class data needs to be added due to it falling outside previous class feature parameters or if changes were made to either previous stages of the radar system's algorithm chain or to the hardware. This could result in even a small parameter change to propagate through to the classification algorithm and result in classification error.

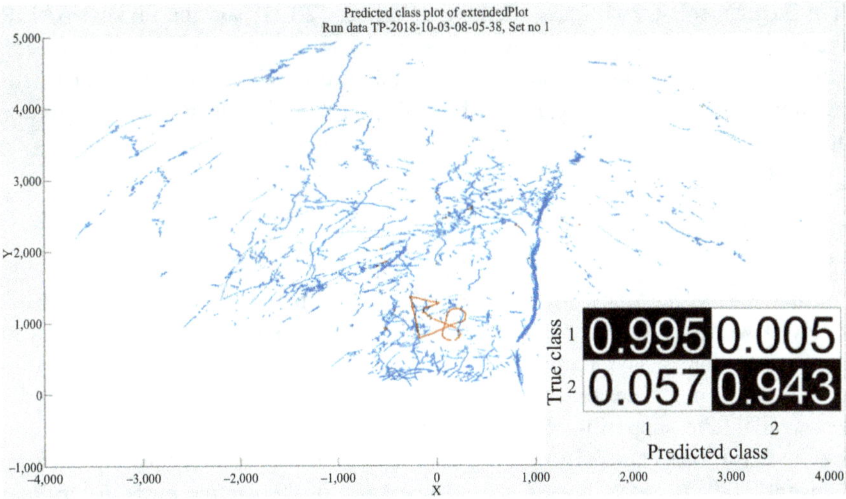

Figure 12.16 Typical results for a machine-learned decision tree classifier showing the non-drone (Class 1 blue) and drone (Class 2 orange) class with confusion matrix

12.6 Conclusions

In the modern day with an acceptance of machine learning techniques as a method to getting superhuman performance in areas of classification and regression, 'big data' or 'data science' is an area rapidly expanding all over the world. A radar, and in particular *staring radar*, produces massive amounts of data, more than could ever be interpreted by a human even if assuming a perfect tracker display and user interface. The challenge is clear to radar engineers to capitalise upon the explosion of machine learning techniques of today and tomorrow. However, to be effective in this approach, bigger and more complex machine learning algorithms that are suspected to achieve an improvement in classification performance will require ideally excess amounts of accurately labelled training data to be fully explored.

In this chapter, we presented an approach to quickly and accurately collect ground-truth experimental data of targets of opportunity, mostly birds, recorded at radar drone trials. This method enables the recording of accurate information on these targets of opportunity and compares them with the radar output to provide a greater number of class labels to be associated with the radar tracks. The combination of the tablet and applications detailed in Section 12.3.2 allows for measurements to be recorded quickly and accurately with a low opportunity of human error input. This then allows for better and more complex classifiers to be created using machine learning by providing more accurate and significant training data. The ground-truth techniques have been deployed in a number of studies of drone surveillance to provide more effective labelled data for the training and validation of machine learning classifiers, examples of which include the SESAR CLASS

Live Experiments for U-Space tracking of drones [20,21] and the SESAR SAFIR live demonstration of drone traffic management capabilities in urban areas [10,22]. Future work will compare results showing the effects of limited truth data versus the more complete set that can now be collected with the method detailed in this chapter.

Acknowledgements

The authors wish to acknowledge and thank the many people who have been involved with the development of staring radar and especially its application to drone surveillance. The authors are very grateful to Thales–Aveillant who manufacture the Gamekeeper radar for their assistance with field trials and making the data available to support the analysis. Furthermore, the authors would also like to acknowledge the contribution by the International Centre for Birds of Prey (Newent, UK) for the arrangement and operation of the captive birds in the field trials with birds of prey.

References

[1] Oswald, G.K.A., "Holographic surveillance radar", In the Proceedings of SPIE 7308, Radar Sensor Technology XIII, 73080D, 29 Apr. 2009.

[2] Jahangir, M. and Baker, C.J., "Robust detection of micro-UAS drones with L-band 3-D holographic radar", In the Proceedings of the Sensor Signal Processing for Defense, Edinburgh, UK, Sep. 2016.

[3] "Airprox involving UAS Drones". In: (2018). https://www. Airprox board. org.uk/Reports-and-analysis/Statistics/Airprox-involving-UAS-Drones/.

[4] Jahangir, M., Baker, C.J. and Oswald, G.K.A., "Doppler characteristics of micro-drones with L-band multibeam staring radar", In the Proceedings of the IEEE Radar Conference, Seattle, WA, US, May 2017.

[5] Taha, B. and Shoufan, A., "Machine learning-based drone detection and classification: state-of-the-art in research". IEEE Access, 2019. DOI: 10.1109/ACCESS.2019.2942944.

[6] Sim, J., Jahangir, M., Fioranelli, F., Baker, C.J and Dale, H., "Effective ground-truthing of supervised machine learning for drone classification", In Inter. Radar Conf. 2019, Toulon, France, Sep. 2019.

[7] Jahangir, M., Ahmad, B.I. and Baker, C.J., "The application of performance metrics to staring radar for drone surveillance", In EuRAD 2020, Utrecht, The Netherlands, Jan. 2021.

[8] Jahangir, M. and Baker, C.J., "Characterisation of low observable targets with a multi-beam staring radar", In IET Radar Conference, Belfast, UK, Oct. 2017.

[9] Molchanov, P., Egiazarian, K., Astola, J., Harmanny, R. I. A. and de Wit, J. J. M., "Classification of small UAVs and birds by micro-Doppler signatures", In EuRaD2013, Germany, Oct. 2013.

[10] Jahangir, M., Ahmad, B.I. and Baker, C.J., "Robust drone classification using two-stage decision trees and results from SESAR SAFIR trials," In Proc. of the IEEE International Radar Conference (RADAR), 2020, pp. 636–641.

[11] https://flylitchi.com/Autonomous flight app for DJI drones for Android and iOS. Accessed March 2021.

[12] Dale, H., Baker, C.J., Antoniou, M. and Jahangir, M., "An initial investigation into using convolutional neural networks for classification of drones", In 2020 IEEE International Radar Conference (RADAR), Washington, DC, USA, 2020, pp. 618–623, doi: 10.1109/RADAR42522.2020.9114745.

[13] Rahman, S. and Robertson, D.A., "Radar micro-Doppler signatures of drones and birds at K-band and W-band". Sci Rep 8, 17396, 2018.

[14] https://apps.apple.com/gb/app/theodolite/id339393884 *Theodolite* application for iOS. Accessed March 2021.

[15] https://play.google.com/store/apps/details?id=com.glidelinesystems.dioptra&hl=en_GB&gl=US *Dioptra* application for Android. Accessed March 2021.

[16] https://play.google.com/store/apps/details?id=com.solocator&hl=en_GB&gl=US *Solocator* application for Android. Accessed March 2021.

[17] Bækkegaard, S., Blixenkrone-Møller, J., Larsen, J. J. and Jochumsen, L., "Target classification using kinematic data and a recurrent neural network", In the Proceedings of the International Radar Symposium, Bonn, Germany, Jun. 2018.

[18] Brooks, D.A., Schwander, O., Barbaresco, F., Schneider, J.Y. and Cord, M., "Complex-valued neural networks for fully-temporal micro-Doppler classification", In the Proceedings of the International Radar Symposium, Ulm, Germany, Jun. 2019.

[19] Bennett, C., Jahangir, M., Fioranelli, F., Ahmad, B.I. and Kernec, J.L., "Use of symmetrical peak extraction in drone micro-Doppler classification for staring radar", In 2020 IEEE Radar Conference, Florence, Italy, Sep. 2020.

[20] CLASS – Clear Air Situation for uaS. [Online]. Available: https://www.sesarju.eu/index.php/projects/class.

[21] Jahangir, M. and Baker, C.J., "CLASS U-space drone test flight results for non-cooperative surveillance using an L-band 3-D staring radar", In International Radar Symposium IRS 2019, Ulm, Germany, Jun. 2019.

[22] SAFIR – Safe and Flexible Integration of Initial U-space Services in a Real Environment. [Online]. www.sesarju.eu/index.php/projects/safir.

Conclusion

Carmine Clemente[1], Francesco Fioranelli[2], Fabiola Colone[3] and Gang Li[4]

The aim of this book was to provide an overview of existing challenges and solutions in the field of counter unmanned aerial vehicles (UAVs) from both an industrial and an academic perspective, with a particular focus on radar techniques.

When editing this book, we have been particularly intrigued by the variety of solutions and approaches that could have been represented in the book, and we decided to include a number of contributions covering fundamental aspects in this area, such as radar counter UAV systems design, investigation of active and passive radar sensors, and algorithm development for the detection, localization, and classification of this particularly challenging class of targets. Our goal was to be able to communicate useful design strategies and solutions to a range of readers, from the radar engineer developing novel systems, to the master students starting their final year projects and looking for some inspiration.

Despite the fact that this topic has attracted a huge interest from industry and academia in the last decade and has led to the re-engineering of existing solutions as well as to the creation of novel concepts and systems, there are still a number of challenges that have not been entirely solved yet. These would need to be addressed in order to ensure higher levels of security as well as enabling a smooth integration of UAVs in the airspace to truly revolutionize daily life.

A challenge on which most of the research community is currently focusing is the discrimination of UAVs from other targets such as birds. As explained in the book, this is a difficult task to accomplish as birds have similar RCS to the one of the drones. However, both the industrial and scientific communities have demonstrated that this task can be accomplished with good reliability, even if at much shorter ranges than the nominal detection range. Extending the range at which systems can reliably discriminate UAVs from birds and move towards finer classification levels such as the UAV type and model discrimination are topics that

[1]Department of Electronic and Electrical Engineering, Centre for Signal and Image Processing (CESIP), University of Strathclyde, Glasgow, United Kingdom
[2]Department of Microelectronics, Microwave Sensing Signals & Systems (MS3) Group, TU Delft, Delft, The Netherlands
[3]Department of Information Engineering, Electronics and Telecommunications (DIET) – Sapienza, University of Rome, Rome, Italy
[4]Department of Electronic Engineering, Tsinghua University, Beijing, China

require further investigation. The use of algorithms exploiting tracking information and perhaps working at low SNR in a track-before-detect approach could help boosting the range at which birds can be discriminated, while micro-Doppler, high-resolution range profiling, and machine learning are good candidates to be investigated for the purpose of finer UAV classification.

Practical challenges include the capability to provide systems that could be deployed to monitor environments such as urban areas that are expected to see significant increase in the presence of UAVs but where the installation of dedicated sensors is typically unwelcomed. In such scenarios, the deployment of passive radar sensors has been demonstrated to represent an effective solution for the detection and accurate localization of drones by exploiting a number of sources of opportunity with complementary characteristics. Further enhanced capabilities can be expected from passive radar sensors exploiting the 5G and 6G technologies thanks to the high frequencies and bandwidth that these systems will be increasingly using, thus allowing enhanced target discrimination capabilities. Moreover, based on transmissions for mobile communication and networking, a hybrid active and passive localization of the targets can be foreseen by jointly exploiting radar measurements, passive source location approaches, and even self-reported positions. Another area for future development is the investigation of adaptive active systems, which are able to change their functionality dynamically by exploiting adaptive waveforms and receiver designs derived by cognitive frameworks, or taking the latest advances in artificial intelligence to enhance clutter rejection and target discrimination. Artificial intelligence, in particular, has demonstrated the capability to enhance radar performance in a number of domains, including micro-Doppler-based target recognition, an area with high relevance for counter UAV systems.

The capability to predict the intent of a UAV is another feature that future systems could potentially integrate, and in order to develop such a capability the solutions could consider frameworks fusing tracking and target fingerprinting information with pattern of life analysis.

Finally, as the technology progresses, it is also possible that the threats will evolve by developing countermeasures such as the capability to deploy deception jamming in order to confuse classifiers. This could pose an additional challenge for the counter UAV system that would then benefit by the fusion of the information collected by the radar with other sources such as optical and RF scanning sensors.

The previous list is non-exhaustive, and as probably many other research questions and technical challenges are still open in this area, we hope that this book has contributed to inspire the readers with new ideas in order to be able to solve them.

Index

www.ingramcontent.com/pod-product-compliance
Lightning Source LLC
LaVergne TN
LVHW011833200926
848484LV00002B/348